S0-ACB-503

HKAr

NOV - - 2020

Peterson Reference Guide to
BIRD
BEHAVIOR

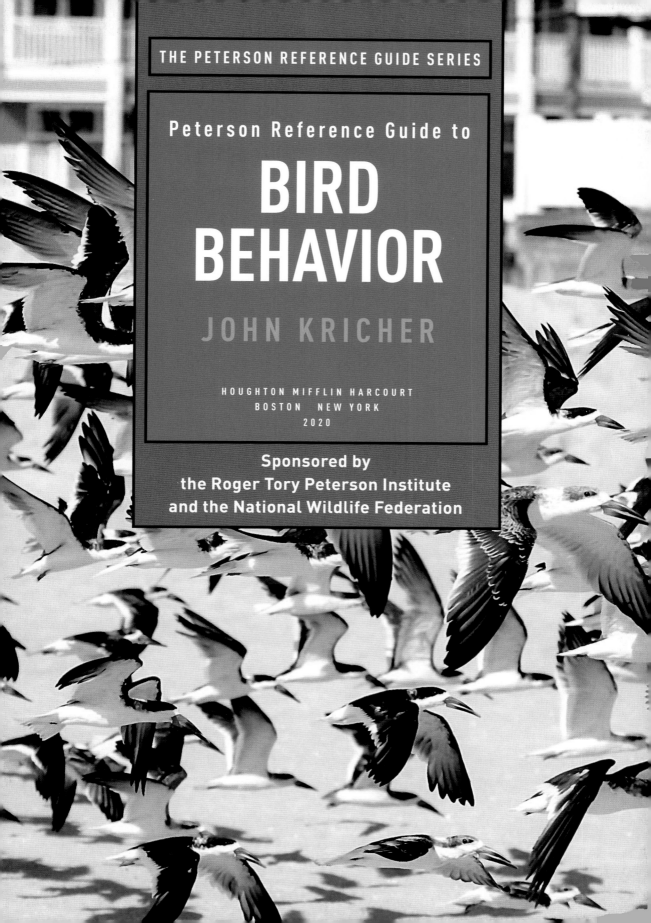

THE PETERSON REFERENCE GUIDE SERIES

Peterson Reference Guide to

BIRD BEHAVIOR

JOHN KRICHER

HOUGHTON MIFFLIN HARCOURT
BOSTON   NEW YORK
2020

Sponsored by
the Roger Tory Peterson Institute
and the National Wildlife Federation

Copyright © 2020 by John Kricher
All rights reserved

All photos by John Kricher except where indicated.

For information about permission to reproduce selections from this book,
write to trade.permissions@hmhco.com or to Permissions,
Houghton Mifflin Harcourt Publishing Company,
3 Park Avenue, 19th Floor, New York, New York 10016.

hmhbooks.com

Library of Congress Cataloging-in-Publication Data is available.
ISBN 978-1-328-78736-1

Book design by Eugenie S. Delaney

Printed in China

SCP 10 9 8 7 6 5 4 3 2 1

ROGER TORY PETERSON INSTITUTE
OF NATURAL HISTORY

Continuing the work of Roger Tory Peterson through Art, Education, and Conservation

In 1984, the Roger Tory Peterson Institute of Natural History (RTPI) was founded in Peterson's hometown of Jamestown, New York, as an educational institution charged by Peterson with preserving his lifetime body of work and making it available to the world for educational purposes.

RTPI is the only official institutional steward of Roger Tory Peterson's body of work and his enduring legacy. It is our mission to foster understanding, appreciation, and protection of the natural world. By providing people with opportunities to engage in nature-focused art, education, and conservation projects, we promote the study of natural history and its connections to human health and economic prosperity.

Art—Using Art to Inspire Appreciation of Nature

The RTPI Archives contains the largest collection of Peterson's art in the world—iconic images that continue to inspire an awareness of and appreciation for nature.

Education—Explaining the Importance of Studying Natural History

We need to study, firsthand, the workings of the natural world and its importance to human life. Local surroundings can provide an engaging context for the study of natural history and its relationship to other disciplines such as math, science, and language. Environmental literacy is everybody's responsibility—not just experts and special interests.

Conservation—Sustaining and Restoring the Natural World

RTPI works to inspire people to choose action over inaction, and engages in meaningful conservation research and actions that transcend political and other boundaries. Our goal is to increase awareness and understanding of the natural connections between species, habitats, and people—connections that are critical to effective conservation.

For more information, and to support RTPI, please visit rtpi.org.

To my wife, Martha Vaughan, who puts up with my
behavior and revels in watching bird behavior

Peterson Reference Guide to

BIRD
BEHAVIOR

CONTENTS

INTRODUCTION

It was the middle of April, the 15th to be exact, in 1967. I was with a fellow graduate student at Muckshaw Pond in north-central New Jersey. Spring had arrived, bestowing warmth and gentle breezes. Red maple buds were opening throughout a forest of maples, oaks, and hemlocks. We came upon a glade surrounded by rock-strewn hills. A piercing, strident *kuk-kuk-kuk* immediately got our attention. An impressively large black woodpecker with a bold red crest swooped over our heads and across the glade to another hillside, wide black wings flashing conspicuous white underwing patches. The bird landed in plain sight on a tree snag, and we marveled at our good fortune to have such a fine look at a male Pileated Woodpecker. His erect red crest highlighted blazing yellow eyes teeming with life.

He was not alone. Soon a female joined him and both birds noisily engaged in short flights from one tree to another, which we took to be courtship behavior. They ceased vocalizing but moved together around the glade, stopping occasionally to whack at a tree limb. A glance at the trees revealed that the woodpeckers were no strangers to this glade, as many tree snags displayed obvious and often large oval carvings, the unambiguous mark of a Pileated. We saw the birds well but it was difficult to approach them. They saw us too. Both were somewhat wary, each tending to land on a tree but then quickly moving to the side of the trunk where we could not see them. The woodpecker pair soon took their leave of us, off to another glade, off perhaps to what would be their nest cavity.

Birds are everywhere, from cities to prairies, over the oceans, in snow during the depths of winter. They are diverse, obvious, and colorful; produce remarkable sounds; and make elegantly structured nests in which to raise their young. They are out there for us to observe, over 10,000 species worldwide. You see more species of birds on a typical day than you do mammals. They challenge and they entertain. They connect with us, indeed puzzle us, with a unique form of cognition and sentience known only to them. And in doing so, they connect us with nature. We humans who watch birds develop a feeling of empathy for a creature that, yes, *thinks,* as it tries to survive for another day.

This is a book about bird behavior, how to under-

A good view of a Pileated Woodpecker, such as this male, is a sight birders remember.

stand some of the underpinning and meaning to what birds do, how they do it, and why they do it. Joseph Hickey (1907–1993), an outstanding ornithologist and close friend of Roger Tory Peterson (both were members of the famed Bronx County Bird Club), authored a book published in 1943 with the title *A Guide to Bird Watching.* Many birders in today's world of digital cameras, high-end optics,

This Blue Jay momentarily paused from foraging to look directly at me as I took its photo. We shared a moment. I have no idea what the bird thought as it stared at me, but I did wonder. Blue Jays are highly intelligent.

superb field guides, YouTube videos, global birding tours, and crowded bird festivals would likely still find Hickey's book surprisingly insightful in many regards. Birds offer the curious observer far more than identification challenges. There is so much more to knowing birds. Hickey knew that the popularity of bird listing was increasing in leaps and bounds, thanks in large part to Roger Tory Peterson's *A Field Guide to the Birds* (first published in 1934). Hickey's book raised field ID to the next level of involvement, so to speak. He did that by introducing bird behavior to his readers, and instructing them how to observe behavior in the field. After identifying the bird, be curious, watch it, and see something of how it lives its life.

It is very satisfying to identify various bird species and tick them on your daily list. But that said—and this was Hickey's reason for writing his book—a bird is an active and sentient creature whose life is pretty darned interesting, worth knowing some-

thing about. Sure, go to Paris, enjoy some moments with the *Mona Lisa,* but no matter how much you stare, she's not really staring back at you. Birds stare back. Birds react. Birds have an agenda.

Say you have a Blackburnian Warbler in full view, but nearby a Bay-breasted Warbler has been spotted, and someone just called out a Canada Warbler in the understory. So you take a quick look at the orange-throated male Blackburnian and move on to other species. But what if you watched the Blackburnian Warbler exclusively for, say, five minutes? What might you learn about it that would go beyond your obvious satisfaction at having had a fine if rather momentary look at the bird?

Yellow-rumped Warblers are perhaps the most abundant of the various North American wood-warbler species. During spring and fall migration, birders typically pick through scores of yellow-rumps to find "good birds," such as, say, a Cerulean Warbler. But hang on a minute. That particular male

Large oval carvings reveal the work of a Pileated Woodpecker. *(Garry Kessler)*

Yellow-rumped Warbler has been, as birders are prone to say, "working that outer branch" for some time. Look more closely. The bird is being methodical and, wow, what a predator. Like a kid eating popcorn, the bird is plucking insects, one after another, in this case small lake flies that are typical of the spring insect flush that migratory birds depend upon for sustenance and, later, for food for their nestlings.

Birds, one and all, are creatures with impressive, complex brains. A brain can do a lot for you, and birds have big ones. Bird brains are different from ours, having evolved along a very different evolutionary trajectory, but are still compellingly complex and capable of remarkable feats.

A Black-capped Chickadee comes to your bird feeder and carries away a sunflower seed. Where does it go? Sometimes you see it land on a branch and hack the seed husk to open it, eating the seed within. Sometimes it does this while still at the feeder. Other times it flies away with the seed in its beak. Chickadees are "scatter hoarders" and thus often hide or cache a seed rather than devour it. They hide seeds in scattered locations—under leaves, between flakes of loose bark, whatever appeals

In the open (for now), a male Blackburnian Warbler. Any idea what he was doing?

Even if the bird was silhouetted, it would be easy to identify it as a Hermit Thrush from the characteristic cocked tail and drooped wings.

Tiny lake flies cover the leaves and are in the air as this male Yellow-rumped Warbler easily captures them by the score in but a few minutes, never leaving the branch.

to the chickadee mind, which makes a great many daily decisions. And they remember. Ask yourself: Could you take, say, 50 cashew nuts and hide each one at a different place around your yard, and then a few days later, at happy hour, go back and retrieve each and every one? Chickadees are capable of finding most of the seeds they have hidden, even after days and weeks go by. That is how they provision for challenging conditions that might lie ahead. Do they plan? Do they anticipate leaner times? How do they learn? Are they mere automatons that mindlessly hide seeds but equally mindlessly retain a sufficient spatial memory to relocate them at a future date?

Many bird species have distinctive postures that are components of their behavior. A Hermit Thrush characteristically drops its wings and raises its tail when it is on full alert. Many bird species are habitual tail flickers or tail bobbers. Phoebes are an outstanding example, but so is the Palm Warbler. The repeated tail flicking of these species has been a long-standing ID tool. Spotted Sandpipers and waterthrushes continually bob their posteriors up and down as they move about. Much of bird behavior is strongly stereotyped and invariable within a species. No conclusive explanation has been offered

This museum tray of oriole study skins provides ornithologists with much information on plumage variation in these birds as well as other valuable information, but much of the "oriole" is missing because oriole behavior is missing.

for why these tail bobbers bob their tails, but one study on Black Phoebes concluded that increased tail bobbing occurs when they detect a predator. But such behavior may be due to anxiety on the part of the bird. Phoebes tail bob when predators are no-where to be seen.

Behavior is part of what is called the *extended phenotype* of birds. A phenotype is the appearance of an organism. Bird specimens, called study skins, housed and studied in museums, are prepared in a standard manner, and thus ornithologists are able to compare them with ease. But bird speci-mens are dead, so behavior is missing, and thus a large part of the bird's phenotype is missing. Because much of bird behavior is stereotyped,

instinctive in nature, characteristics ranging from tail bobbing to the shape of the holes made in tree trunks by Pileated Woodpeckers, the distinctive mud nests of Cliff Swallows, the stiff-winged flight of swifts, the S-shaped curve of heron and egret necks when in flight, and the distinctive posture of an American Bittern when it mimics marsh grass are all part of how live birds behave. Observing behavior is the essence of birding.

It is the aim of this book to introduce you to vari-ous elements of bird behavior and explain some-thing of the ecological and evolutionary bases for them. This approach will connect you more inti-mately with these remarkable and beguilingly per-ceptive animals.

HOW TO USE THIS BOOK

This book is focused primarily on North American birds, though of necessity I've included a few global examples. The book is meant to be informal and conversational, with technical jargon kept to a minimum. Certain ornithological terms, such as *kleptoparasitism, dynamic soaring,* and *geolocator,* are defined within the main body of the text. Because I have used many sources, I have included both a list of comprehensive ornithology books as well as a chapter-by-chapter reference section at the end of the book with select annotations accompanying many of the references listed. Some references, by necessity, are technical, but I have tried to include many that are written for lay readers and interested birders. The reference list for each chapter includes all of the studies I have discussed in the chapter.

Bird identification is rich with behavioral field marks. Behavioral clues help identify birds as to species, but behavior also allows you to better understand the bird. Birds think. Their actions are overwhelmingly purposeful, not random. Birds are constantly making decisions—hundreds if not thousands of decisions daily. Behavior tells us about birds as living beings, not just as "things."

Each chapter provides a concise overview of a particular category of bird behavior. Some forms of bird behavior are universal among bird species: waterfowl have much behavior in common with flycatchers, including flying, preening, courtship behavior, nest building, and aggressive behavior. Some behaviors are specific to certain bird families and even to bird species. For example, American Avocets feed in water by sweeping their long upturned bills in a unique manner. Look closely and notice that female avocets have shorter but more sharply curved bills than those of males (chapter 12). Reddish Egrets "dance" as they scurry around shading the water with their wings, exposing fish that they then capture. White-breasted Nuthatches sweep their bills on the surface of bark surrounding

Reddish Egrets become very active when foraging—"dancing" and shading the water with their wings. Their foraging behavior is distinctive, but other herons, including Tricolored, also make similar movements on occasion. *(Jeffory A. Jones)*

Rock Pigeon courtship as well as other behaviors obvious when watching this species are both insightful and entertaining, and there is much you can learn from watching them.

OPPOSITE: Courtship behavior between two Snowy Egrets. Behavior is essential in establishing a pair bond.

The Song Sparrow (*top*), the European Robin (*center*), and the Herring Gull (*bottom*—male on right, female on left) have been subjects of intense study in classical ornithological literature, and each study focused on behavior.

their nest hole, sometimes with an insect in the bill (often a noxious blister beetle), perhaps to divert animals such as squirrels from the nest cavity. Even the most common birds, such as Rock Pigeons, offer a broad array of readily observable behaviors. For many years my ornithology classes studied Rock Pigeon behavior. They loved it!

Three classic books about ornithology are Margaret Morse Nice's study of the Song Sparrow (and other species), detailed for the lay reader in *The Watcher at the Nest* (1939); David Lack's *The Life of the Robin* (1943), about the European Robin, not the American Robin; and Niko Tinbergen's *The Herring Gull's World* (1953). Each of these books, in addition to numerous others, shows how much is to be learned by carefully observing common birds. You have access to birds no matter where you are.

With practice, you'll be able to note how behaviors differ not only among species but also among individuals. For example, I watched two Pine Warblers visiting my suet feeder, a metal cage feeder with a suet chunk inside. I saw both birds at the same time on my deck rail where the suet hung, so I knew there were two of them. Soon one bird flew to the suet cage, briefly perched, and took its tiny piece. After it flew off, the other bird came up from below the suet, briefly hovered underneath the suet cage, not landing, and snatched its piece. The two birds fed differently from each other. Subsequently I observed these birds as single individuals coming to the feeder and each bird grabbing its piece of suet, either perching on the feeder or hovering below it.

Could I prove that I knew one from the other? The birds were not banded or otherwise marked, so I am hypothesizing that each individual adopted a different feeding technique unique to it and stayed with it. I noted that one bird was more intensely colored than the other, and it was the individual consistently landing on the suet feeder. It also tended to displace the other bird when it was perched on the deck rail, suggesting it was dominant to it, perhaps reflected in its brighter plumage. So I am confident about my observation.

The very observation of Pine Warblers utilizing a suet feeder poses questions summed up in one word: Why? It was early April and it was cold outside. The Pine Warblers, which may have been early migrants or overwintering birds, were deriving high levels of nutrition from suet, a substance rich in caloric fat. On a per gram basis, fat has twice the energy payoff of carbohydrate or protein. Okay, that makes sense: a little suet stokes the warblers' metabolic furnaces, helping them endure the cold. But how did the warblers know that suet is food? They normally forage for arthropods on pine bark and branches. They do not normally encounter a big

This male Pine Warbler began eating seeds on a regular basis. Pine Warblers have often been observed ingesting seeds at bird feeders. Do they "know" that the seeds are food or do they "learn" from watching other birds?

blob of white goo inside a dangling metal cage in pine woodlands.

Perhaps the warblers, who associate with chicka-dee flocks, noticed that chickadees, titmice, nut-hatches, woodpeckers, and other birds were routinely visiting the suet feeder. If so, they learned by observation of other species. As you watch birds in both single- and mixed-species flocks, you will quickly catch on to the fact that they use each other as sources of information about food availability. Of course, this begs the question: How do chickadees and other species know suet is food? Chickadees are general foragers, curious, exploring many potential food sources. Therefore they're apt to encounter and try out unique food sources such as suet. The knowledge that suet is food is likely a cultural adaptation passed on from bird to bird, species to species, by observation.

If Pine Warblers are capable of learning that suet is food, do they also know about birdseed? The suet feeder was near a deck rail as well as other feeders upon which I placed sunflower seed hearts. One might predict that watching goldfinches and chick-adees consume seeds might act to stimulate Pine Warblers to ingest seeds. In fact, I have often observed Pine Warblers taking seeds, and it is a well-documented behavior.

Pine Warblers are adapted to be versatile in their foraging behavior. They are habitat specialists, confined essentially to pine woodlands where most foraging is done along the bark and branches of pine trees. Pine Warblers often go to the ground and search among the leaf litter and are quick to spot unique feeding opportunities. I watched as several were feeding on insects atop the mud and debris of a muddy, drought-stricken pond. The only birds feeding in close proximity to the warblers were two Solitary Sandpipers.

The more you watch birds, the more they reveal. Watch flocks, watch individuals, and ask repeatedly, why is the bird doing what it is doing? What is it telling you about itself?

2

LEARNING TO WATCH BIRDS BEHAVE

BIRDS ARE ALWAYS EXHIBITING BEHAVIOR

With accelerated metabolic rates, keen sensory organs, and large brains, birds live high-octane lives. I once spent a few moments watching a juvenile European Starling (juvenile plumage is distinctive) perched atop a light pole in Washington, DC. I was in a cab, waiting for a long-sequence red light to change to green. The starling did not fly in the time that I observed it, and it kept methodically moving its head, first to the right, then left, then looking up, then a look behind. It looked down too, but less often. Why? It was perched in the open, exposed, alone (somewhat unusual for a starling). It was inexperienced. My guess is that it perceived that it was vulnerable, a possible target for a predator. Somehow the naïve starling realized predators exist and pose a threat. With millions of generations of starling instincts programmed in its brain, it "knew" to be cautious.

There are no do-overs in nature. Make one serious mistake at any age, and you are gone from the gene pool. Starlings are commonplace bird species ignored by many (most?) birders. This one did not fly and was not feeding; it was just alone, idle on a pole, but its behavior was nonetheless interesting to me.

BIRDS INHABIT A MULTISPECIES ENVIRONMENT

An extensive mudflat at low tide may contain ten or more species of plovers and sandpipers, all simultaneously foraging in various ways. Does a Semipalmated Plover pay any attention to a nearby Black-bellied Plover? How might we know, and why might it be important to know? During spring migration, a single tree may have nearly ten different wood-

OPPOSITE: Maintenance behavior looms large as birds typically spend much time preening and performing other types of self-maintenance. This Pied-billed Grebe, just awakened, is having a good stretch.

This western Red-tailed Hawk is certainly interested in the Common Raven, and the raven is very much aware of it. Birds live in a world not only of their own species, but also of different species that constantly interact.

warbler species feeding among its branches. American Robins, Common Grackles, Brown-headed Cowbirds, Chipping Sparrows, and Song Sparrows may all be seen on the same lawn at the same time, each searching for food of various sorts. How aware are they of one another?

Have you ever dined in a multispecies restaurant? Neither have I. The bird world is rather like the famous bar scene in the first *Star Wars* film, composed of highly different species sharing a common area and resource base, all with reasonably similar levels of intelligence.

For birds, normalcy is to live in a world of multiple bird species of various sizes and proclivities. That is why bird feeders are apt to be stressful places for birds (chapter 12). Imagine going into a restaurant where there are no rules. Big guys get to push little guys out of the way and take the food. A thug might suddenly come in and injure or kill you (and eat you). Everyone runs away, but once the thug leaves, they all slowly return. They still have to eat.

Displacement is common among gulls since perching areas such as dock pilings are often limited. Here an immature Herring Gull displaces an adult.

BIRDS KNOW THEIR OWN SPECIES

All Black-capped Chickadees look alike to us humans but not to each other. In chickadee society, individuals know each other and act differently depending upon the bird with whom they are interacting (chapter 11). Bird society is layered in degrees of dominance, some individuals having high rank, some low rank. The distinctions are often subtle. Species such as waxwings displace one another from particular positions on branches. A gull will land in precisely the spot occupied by another gull, displacing it.

A flock of about 40 Royal Terns and Laughing Gulls are standing on a dock in close proximity. So what? But look, see how the rest of the dock is devoid of any terns. Why are they all bunched together on one small section of an otherwise expansive dock? Are these terns from the same breeding colony? There is a mix of juveniles and adults. Are the juveniles, all now fully fledged, still associating with their parents? Do they interact with Laughing Gulls?

Four Fish Crows are in close proximity atop a cabin cruiser roof at the same dock. There is an abundance of possible perches, but these four Fish Crows associate together, almost touching one another. The Fish Crows continually vocalize, bob their heads, and seem to be having a conversation. One might speculate that the crows know each other better than the Royal Terns do. The terns' only apparent interaction comes when one flies off and captures a fish, immediately to be chased and harassed by several other terns attempting to make

the individual with the fish drop it. The crows seem rather like a group of friends or perhaps family. The terns are more like people waiting at an airport gate, clustered together but with no loyalties to one another.

FAILURE IS NOT AN OPTION

I was once showing a group of eager college students how various birds forage in a salt marsh. An adult Yellow-crowned Night-Heron was standing in a shallow channel. I told the students that it was likely hunting minnow-type fish or crabs. One of the students laughed and proclaimed the bird to be "dumb." The student had been out early in the morning crabbing in the same channel and had caught nothing.

Within a minute or so after my student's proclamation, the night-heron shot its neck forward and captured a blue crab. It then adeptly proceeded to disarticulate the crustacean and devoured it. "Got lucky," my student opined. Its crab eaten, the night-

This Great Blue Heron has its work cut out for it. It must manipulate this fish, an eel-like prickleback, to swallow it headfirst and avoid choking on the fish's spines. *(Jessie Knowlton)*

This male Black-and-white Warbler makes hundreds of daily decisions as he goes about doing what he does. He thinks.

heron went back to its hunting posture and soon had another crab. Made it look easy. The students were impressed. I told them that for the student who was out crabbing in the morning, crabbing was a simple pleasure, one in which he is unskilled, but it didn't matter because his survival was not at stake. Not so with the night-heron. First, it was foraging by day, a bit unusual for this species. It was nesting season, and perhaps it was provisioning for its young in a nearby colony. Second, unlike the student, who led a multidimensional life, this is pretty much all the night-heron does. Little surprise that over millions of night-heron generations, it has become good at it.

BIRDS LIVE HIGH-SPEED LIVES

To understand bird behavior, it is essential to be mindful of the intensity with which these high-metabolism animals live. Imagine a fictional video game called *Bird!* in which you might play the role of a Black-and-white Warbler flying through a structurally complex three-dimensional woodland. You must weave accurately among tree branches at what appears to the human eye to be an impossible speed, then deftly land on a tree trunk where you begin to hyperactively search for food items, remaining aware of the potential for predators, all the while moving rapidly from trunk to branches, often upside down. That is part of the world of a Black-and-white Warbler, a world that moves very quickly.

Were you to wake up with the body rhythms, sense organs, and perceptive qualities of a bird, the world would not look or feel at all the same. It would be alien. Though you would see in color, the colors would be more vivid because you could see into the ultraviolet part of the color spectrum as well as the visual spectrum of humans, ranging from blue to red. Your bird eyes would be disproportionately large in your skull and able to discern a level of detail that requires very rapid processing for the brain to comprehend. Your reflexes would be disconcertingly fast as you would constantly process images and sounds as well as the feel of air on your feathers. Your body would be incredibly lightweight and thus you would easily perceive small variations in air currents as you flew. If you were a songbird, your sense of hearing would allow you to easily separate complex songs of others of your species, songs in which the jumbled notes occur so rapidly that human ears are unable to separate the notes unless the songs are recorded and played back at a significantly slower speed.

Watch a flock of Ruddy Turnstones sleeping on a beach. For most of the individuals, one eye will be open, one shut. Birds have bilaterally organized brains and, for at least some species, the brain seems capable of sleeping on one side while staying awake on the other (see chapter 9). Likely most of the turn-stones will be standing on one leg, balancing effort-lessly, while one-half of their brain is asleep.

BIRDS CHOOSE THEIR HABITATS

As Roger Tory Peterson wrote long ago in his classic book *How to Know the Birds* (1949), "Although I have seen thousands of Meadowlarks, perhaps tens of thousands, I have never seen one in a woods. And I have never seen a Wood Thrush in a meadow." But how do individual birds "know" their appropriate habitat?

Habitat selection is on a spectrum. At one end of the spectrum, some species are extreme generalists, able to occupy many habitats. At the other end, some are extreme specialists. There are many species that fit between those extremes. Some species, such as American Robins and Song Sparrows, occupy a wide diversity of habitats, adapting well to human-influenced habitats. Common Ravens are among the most generalized of bird species, residing in habitats from the frigid Arctic to hot deserts; different popu-lations can be equally successful in urban and wood-land environments. You can observe Common Ravens in downtown Oakland, California, or flying above the densely wooded Appalachian Mountains in Shenandoah National Park. Other species, such as Kirtland's Warblers and Sedge Wrens, are far more specialized in their habitat requirements.

One principle of habitat choice is that specialized species are often the most threatened, since they have little flexibility should habitats deteriorate. Generally, the degree to which a species is common or rare correlates with the rigidity of its habitat requirements.

BIRDS ADAPT, INCLUDING TO HUMANS

We share the world with birds as they do with us. Some birds' names reflect their adaptive histories with humans. Barn Owls and Barn Swallows make nests in barns. Chimney Swifts reside in chimneys. House Wrens, House Sparrows, and House Finches nest around houses. Houses, chimneys, and barns are human constructs, and various bird species have adapted to use them. For hundreds of years, Rock Pigeons have made nests on building ledges, living among humans in cities. More recently Peregrine Falcons have done the same. The story of Pale Male, a Red-tailed Hawk that nested on a building in New York City near Central Park, was told in several books (some for children) and in a documentary film shown on *Nature*.

The return of Eastern Bluebirds from threatened to thriving population is thanks to humans supply-

Ruddy Turnstones plus three Dunlin and one Sanderling idling on a winter beach in coastal Georgia. Most are standing on one leg, normal for many roosting birds.

ing an abundance of nest boxes and keeping these boxes clean and free of parasites. Numerous other species, such as Purple Martins, Eastern Screech-Owls, House Wrens, Tree Swallows, Wood Ducks, and American Kestrels, take readily to nest boxes. Such behavior demonstrates adaptive flexibility among these species.

Even birds such as Snowy Owls, which inhabit windswept tundra where human presence is generally minimal, adapt readily to large municipal airports when the birds migrate south in winter (chapter 5). The flat grassy areas that are interspersed between runways and taxiways apparently provide habitat for rodents, and the owls thrive (requiring their periodic capture and relocation to avoid collisions with aircraft).

Bird behavior is at the cutting edge of evolutionary adaptation. It should not be surprising that as humans change habitats, as climate change intensifies, birds that adapt do so through behavioral changes. Gulls adapted to garbage dumps as prime feeding areas, a reality that is in part responsible for major increases in some gull species during the twentieth century (after suffering major losses from human persecution in the nineteenth century). Least Terns have located colonies on rooftops in some towns and cities. Common Nighthawks and Killdeer are also gravel rooftop nesters. For many years Lake Merritt in Oakland, California, has attracted hundreds of wintering ducks such as Lesser Scaup, many of which swim near shore and show no aversion to nearby human walkers along the lakeshore.

THE BIRDS YOU WATCH ARE WATCHING YOU

They see you. You may think they don't, but they do. The sensory abilities of birds by and large exceed yours. If you are looking at a Loggerhead Shrike, a Brown Thrasher, or a Hermit Thrush perched on a tree, it is seeing you, too, unless you are looking from a considerable distance using a spotting scope. The swallows skimming near you as you stand near a marsh see you. As you walk a sandy beach, each and every Sanderling running along the tideline searching for food sees you.

Often the birds will be unconcerned about seeing you. Sanderlings will ignore you if you don't approach them too closely, but if you do, they'll take wing and relocate at what they perceive to be a safe distance from you. A soaring Wood Stork will pay no attention to people on the ground, but it sees them. Birds learn quickly about the degree of threat posed by our species. In parts of Europe where small birds are routinely hunted, they avoid humans. European Robins in much of Europe do not allow close approach. But robins in England and

These two male Wild Turkeys (*top*) show no aversion to strolling down a suburban street, exhibiting little fear or concern for the humans they may encounter. Boat-tailed Grackles (*bottom*), along with other bird species, now frequent fast-food restaurants, finding food around the parking areas.

Ireland, where they are never persecuted, are familiar and confiding dooryard birds. Wild Turkeys were once highly wary and avoided humans. Now in many places in North America Wild Turkeys stroll nonchalantly down neighborhood streets. We don't shoot them in neighborhoods, and they have learned that they have little to fear.

Laughing Gulls soon learn that beaches frequented by *Homo sapiens* are often rich potential food sources as people discard bits of sandwiches and crackers. Some are known to have become aggressive, snatching food directly from people. Ring-billed Gulls, Fish Crows, and Boat-tailed Grackles learn that fast-food restaurants are lucrative hunting grounds. Seabirds and gulls routinely follow fishing boats. Gulls in particular learn

This Canada Goose has become aggressive because my group, walking on a road, had to pass closely by where the goose was watching over its goslings.

quickly how to find food. My friend Dan Tankersley wrote me: "At Oracle (formerly AT&T) Park in San Francisco, toward the end of every baseball game, right around the seventh inning, mostly Western Gulls begin gathering outside the park just beyond the outfield walls. They know there will be lots of food in the stands and somehow know when the game is about to end, be it a day or night game. By the ninth inning, some are already in the bleachers in areas vacated early by fans. As soon as the final out is recorded, the gulls fill in as the fans leave. Some folks regret that they didn't leave sooner as they will have a smelly souvenir for their ride home. It is an amazing event to watch."

SOME POINTS ON ETIQUETTE

If you are watching a Red-headed Woodpecker foraging in a tall pine tree, you may conclude that the bird typically forages at least halfway up the trunk of the tree, never lower. The fact that when you made the observation you were standing close to the base of the tree may, however, have had something to do with the woodpecker's location being no lower than midway up the tree. It saw you and was keeping its distance. We often unknowingly affect the behavior of the birds we are watching.

Some bird species may become aggressive toward humans. Colonial nesters such as terns dive and strike humans on the head if they approach too closely. Northern Goshawks are legendary for their aggressiveness in defending their forest nests. Canada Geese and Mute Swans exhibit strong aggressive behavior if a human approaches their nest or young too closely (chapter 15).

Observing bird behavior allows you to develop empathy for birds—*all* birds. I have been asked numerous times if I have "seen any good birds." It's an innocent question, but I detect a surprised look when I reply that I was mostly watching the interactions within a grackle flock. Those interactions were interesting, at least to me. Good birds.

AUDIO PLAYBACK AND PISHING

Birders have birdsong and call note recordings on their smart phones, and using these in the field can often draw a species much closer for viewing. Numerous published studies in the ornithological literature have relied on use of playback in various forms, mostly to study birdsong and its adaptive significance. Playback studies by ornithologists have contributed much to our understanding of how birds use song, how they respond to songs of other birds, and how they communicate with call notes (chapter 13). When used as part of ornithological research, playback is a vital tool in understanding much of bird behavior.

Playback used by birders is more problematic. When a bird hears the recorded sounds of another of its species it may come to take a closer look, particularly if it is actively defending a territory. Thus, you see it better, whereas without the draw of playback a skulking bird species may be almost impossible to observe in the open. But if you use playback, you are artificially exciting the bird the way another of its species would. So playback should be used sparingly.

Never use playback where it is prohibited, such as at various wildlife refuges and sanctuaries. That is obvious. Don't use it in areas visited by numerous birders. Doing so sends the wrong message. With many birders around, which in itself could stress a bird, adding frequent playback almost certainly contributes to raising stress levels on a continual basis. It is the continual basis that is the problem.

Pishing (or, as it is sometimes called, spishing) and the use of screech-owl or pygmy-owl calls also act to draw groups of birds closer to the viewer. Pish-ing approximates distress and/or agitation calls of some bird species, causing the birds to gather in an attempt to ascertain the source of the distress, while owl vocalizations attract birds that mob owls. For a few minutes, you may be surrounded by chickadees and associated species while bunches of skulking sparrows momentarily hop to the top of the shrubs to have a look. This assemblage of active, agitated, and vocal birds is an example of mobbing behavior (chapter 12). Pishing and the use of owl calls convey temporary and false danger signals to birds. As with playback, birds are adapted to respond to pishing and owl calls, so no real harm is done. But again, it is a matter of frequency. In heavily birded areas, frequent pishing and owl calls should not be used. Many birds will soon choose to ignore them, a form of learning called habituation (chapter 4). The birds have learned that these sounds are meaningless and other birders are likely annoyed.

I recommend all birders read and adhere to the American Birding Association Code of Birding Ethics, available online from the ABA (listing.aba.org/ethics/).

This female Mute Swan on her nest near a road and walkway has habituated to humans being in close proximity and does not act aggressively. Note the crackers that someone tossed on the nest, offering "food" to the incubating swan.

3

ON BEING A BIRD:
ANATOMY AND PHYSIOLOGY

WHY ANATOMY MATTERS

Flight, supported by a unique anatomy and physiology, affects everything about birds, including all aspects of their behavior. This chapter will enhance your understanding of bird behavior by providing an explanation of what birds really are, both structurally and functionally. As a friendly suggestion, you may want to have a tasty rotisserie chicken handy. You can enjoy dining on it and see many of the skeletal and muscular characteristics discussed below.

A bird's anatomy, physiology, and behavior are the products of a temporal trajectory that began more than 150 million years ago at a time called the Jurassic period of the Mesozoic era, the Age of Dinosaurs. It was then that birds first took to the air and in essence became birds, heavier-than-air feathered vertebrates. Not all birds fly. Ostriches do not, and it is no coincidence that ostriches are the most massive of modern birds, males weighing up to 250 pounds (113.4 kg). But look at an ostrich skeleton from beak to tail, compare it to that of a Herring Gull or any other species of flying bird, and the unmistakable remnants of an animal whose ancestral lineage was adapted to fly become obvious. Other than in size and proportion, ostrich skeletons look barely different from those of gulls and crows. Ostriches and their relatives are *secondarily flightless*. Their ancestors flew. The same is true of all extant flightless birds.

The term *modern bird* refers to a group called the Neornithes, which is divided into two major groups, or superorders. The Paleognathae are so named because the prefix *paleo* means "ancient" and the suffix *gnathae* refers to characteristics of jaw structure, so the name of this superorder means

The Common Ostrich is the world's largest extant bird, but it is flightless. Nonetheless its skeleton abounds with evidence of its descent from ancestors that flew.

"ancient jaw." This group consists of the Neotropical tinamous (all of which fly poorly) as well as the large flightless birds called ratites: the ostriches, emus, cassowaries, rheas, and kiwis, plus some extinct forms. The Neognathae ("recent jaw") comprise all of the remaining groups of living birds. The Neornithes survived the mass extinction that ended the Cretaceous period and Mesozoic era some 65.5 million years ago. Today's birds, while many are

OPPOSITE: "Free as a bird." This Black-legged Kittiwake is but one example of the elegance that flight brings to birds and why we humans admire them so much.

recently evolved, nonetheless trace their ancestries back into deep time, to the Age of Dinosaurs.

FEATHERS MAKE THE BIRD: AT LEAST THEY DO NOW

Feathers are *the* defining characteristic of modern birds, but feathers were around before modern birds existed. Many non-avian, nonflying dinosaurs, now long extinct, evolved feathers, and thus feathers came first, birds second. (And eggs came before chickens, but we'll deal with that later!) Feathers are presented in more detail in chapter 8, but here is a brief introduction.

Feathers are unique structures derived from the epidermis layer of the skin, as are the scales of reptiles and the hair of mammals. But feathers are different from both scales and hair. Feathers are strong but lightweight, made of a form of keratin, similar to the material composing finger- and toenails. Feathers possess remarkable properties of insulation (think of down jackets, for example). They are diverse in structure, color, and function. An American Robin has about 3,000 feathers, an amount typical of most passerine birds. A Ruby-throated Hummingbird, being small, has only about 940 feathers. Swans have many more feathers, up to 25,000, but the majority are on the head and neck. Feathers wear out and require molting and replacement, and molt is both a critical component of the annual cycle (chapter 7) and part of basic avian maintenance behavior (chapter 9).

The original function of feathers was likely for display, such as colorful crests that could serve for spe-

The Greater Roadrunner, shown here with its reptilian prey, shares numerous characteristics with its early dinosaur ancestors, but it is very much a modern bird. *(Dan Tankersley)*

cies, sex recognition, or intimidation. Feathers also became essential for insulation, to maintain a warm body. Birds and mammals are the only vertebrates that are fully endothermic, or warm-blooded (see "Blood Circulation and Gas Exchange," page 30). Feathers greatly reduce heat loss.

Birds' dinosaur ancestors were likely colorful, small, warm . . . and flightless. But that would soon change. Going back about 150 million years to the Upper Jurassic period, the *urvogel,* or "first bird,"

This Mute Swan shows the density of neck feathers characteristic of swans.

The Calliope Hummingbird has the distinction of being the smallest North American bird. (Bruce Hallett)

Archaeopteryx, did fly, at least to some degree. Its fossilized wing and tail feathers were discovered to be structurally the same as those typical of modern birds.

BEING SMALL AND LIGHTWEIGHT IS ADAPTIVE IF YOU WANT TO FLY

Birds are generally small. The smallest North American bird is the Calliope Hummingbird of the southwestern states. It weighs 0.1 ounce (2.7 gm). The heaviest North American birds are the Wild Turkey, which is barely able to fly, and the endangered California Condor. Adult male turkeys may weigh around 30 pounds (13.6 kg) or more, and an adult condor weighs 23 pounds (10.5 kg). North American terrestrial mammals show a much broader weight range. The smallest of the various shrew species weighs in at about 2 to 4 grams, similar to small birds. But the largest terrestrial mammal in North America, the moose, weighs up to 1,323 pounds (600 kg), about 57 times more than the largest flying bird. But of course, moose don't fly.

Weight is a huge deal in flight because a volume of air contains very little mass. Getting a heavier-than-air object into sustained, powered flight, providing and maintaining adequate thrust and lift against the forces of drag and gravity, is best served when that object is not very massive. Smallness matters because weight (mass) matters.

Lighter body weight is facilitated by the presence of pneumatic bones—thin and hollow—supported internally by struts that add strength but not much weight. Bones contain extensions of air sacs, a system of thin membranous structures that permeate the bird's body. The skeleton of a bird weighs but a fraction of that of a comparably sized mammal.

Small bodies not only are efficient for flight, but also allow birds to avail themselves of a wide spectrum of food sources, which promotes the evolutionary diversity of birds (chapter 12). Agility, too, is made possible by light weight. Chickadees, crossbills, and numerous other bird species typically hang upside down on a tree branch as they search for prey. Woodpeckers and nuthatches hitch along on the underside of tree branches, held by their conspicuously large clawed feet, their lightweight bodies not succumbing to gravity. The incredibly light weight of birds gives them an advantage we humans cannot imagine in being able to go about their lives in the most agile of ways. Gravity has largely been conquered by avian evolution. If you were a bird, you might feel somewhat like astronauts felt experiencing far less gravitational effect when they ventured across the surface of the moon. The expression "light as a feather" is accurate, but it goes beyond feathers to encompass the entire bird.

THE UNIQUE AVIAN SKELETON

Birds are bipedal. Very few terrestrial vertebrates are bipedal. Even kangaroos rest on their tails, sort of tripod-like. But birds, like humans, are fully bipedal, committed to movement on land or water using paired hind appendages. There are no exceptions. Birds' legs and feet have adapted for an impressive diversity of functions, from the webbed toes of many waterfowl to the formidable talons of

Foraging wood-warblers, such as this Northern Parula (*left*), have such little mass (and weight) that they easily hang upside down when foraging. Crossbills (*right*), such as these White-wingeds, with little problem from gravity, often hang upside down as they extract seeds from cones.

raptors, to the unusually long toes of gallinules and jacanas.

Wings are modified pectoral appendages that include the equivalent of a hand, wrist, forearm, upper arm, and shoulder girdle. Wings function primarily for flight, though birds use their wings for other behaviors. Swans and geese use their wings to strike at aggressors. Ospreys and eagles use their wings to swim short distances, and numerous bird species, ranging from owls to gulls to nuthatches, spread their wings to intimidate would-be predators or competitors. Herons and egrets use their wings while hunting fish. Wings are frequently part of courtship displays. Alcids and penguins as well as some waterfowl use their wings underwater when searching for food. Finally, wings play a major role in balancing birds as they perform the act of copulation. Wings are discussed in more detail in chapter 8.

Birds' necks are highly flexible. Birds can extend and swivel their necks with ease. Almost all mammals, including humans, have necks composed of seven cervical (neck) vertebrae. Having but seven

In the *top right* photo, the pectoralis major is contracting on this Canada Goose, providing the downstroke, the power stroke of the wing; the supracoracoideus is contracting in the *bottom right* photo, raising the wing, the recovery stroke. See chapter 8 for details.

Birds use wings to balance when they copulate, as shown by this pair of Roseate Spoonbills.

neck vertebrae limits flexion of the neck. Bird necks may contain as many as 25 neck vertebrae (swans, for example), though most bird species have about 14. The S-shaped neck of a flamingo contains 19 vertebrae. Neck vertebrae, saddle-shaped at the ends, articulate in a unique manner (termed *hetero-coelous*) that provides for a greater range of flexion. Thus birds, lacking hands as we know them, use their heads and beaks to explore their world, capture food, craft their nests, look out for danger, and engage in combat, all aided by a neck that allows a bird to swivel its head 180 degrees or more. A bird facing away, its back to you, can turn its head and pretty much look you directly in the eye.

Bird skulls have a large cranium, massive eye sockets, and no teeth. Though dinosaur ancestors and the first birds possessed teeth, birds eventually evolved jaws lacking teeth (but the genes to make teeth remain in the genomes of modern birds). Teeth are dense and heavy, composed of dentine and

OPPOSITE: This feeding Mute Swan (*top*) amply demonstrates the flexibility of birds' necks. The Red-headed Woodpecker (*bottom*) has turned its neck almost 180 degrees as it looks behind itself.

enamel, and thus would add considerable weight to the extreme anterior end of the bird, not good for flying. Mergansers, ducks that feed primarily on fish, appear to have beaks lined with tiny teeth, but these serrations are not teeth; they are keratinous components of the bird's bill. Other waterfowl have smaller and less noticeable serrations.

Beaks perform the functions of food acquisition, crafting nests, feeding young, preening, and attack and defense. The inside of a bird's beak and mouth may, depending on the species, be lined with touch detectors to help sift for potential food items. The inside of some beaks may have sharp ridges that aid in food processing. Some birds, including various shorebirds and hummingbirds, exhibit rhynchokinesis, a flexibility of the upper bill such that the tip of it can bend upward. Birds such as starlings and meadowlarks have sharply pointed bills that allow the bird to see down the length of its bill. This is helpful in a form of foraging behavior called *gaping,* in which the bird stabs the bill into the ground and then quickly opens it to capture food items. Birds also have tongues that are variously specialized for different feeding techniques. For example, the tongue of a sapsucker is tipped with small bristles that aid in collecting sap. Other woodpeckers have small barbs on their long, extensible tongues that aid in securing prey.

The upper mandible of a bird is unique compared with that of reptiles and mammals in that it is slightly movable. Mammalian upper jaws are firmly attached to the cranium of the skull and are thus fused as part of the face. But birds have a kinetic upper jaw, able to be raised somewhat relative to the cranium as the lower mandible opens. Having such a feature aids in food processing, particularly for seed-eating birds.

A bird's cranium, the part of the skull in which the brain is housed, is thin, lightweight, and bulbous, adapted to shelter a large brain. Eye sockets are massive, making the skull in some birds, such as owls, appear distorted by the need to accommodate such huge eyes. A bird's eyes represent about 15 percent of the weight of its head, compared to less than 1 percent in humans (chapter 5).

The avian vertebral column, aside from the neck, is largely composed of fused and rigid vertebrae. The thoracic vertebrae, those that protect the lungs and heart, are largely fused, the rib cage reinforced by small knobs of bone extending laterally from the ribs, called *uncinate processes*. Posterior thoracic vertebrae are fused with the lumbar, sacral, and

Owls such as this Burrowing Owl are able to easily swivel their necks to look behind them. The *top* photo shows the back of the bird. So does the *bottom* photo, except that the bird is looking at you.

These two juvenile Green Herons demonstrate how different a bird's posture looks when the neck is compressed and when it is stretched.

some of the caudal vertebrae as well as the three bones of the pelvis (the ilium, ischium, and pubis) to form a massive, lightweight but strong bone, the synsacrum, a "hip bone," unique to birds. The tail vertebrae are fused into a compact pygostyle, which anchors the tail feathers.

The thigh bone, or femur, is relatively thick and short in comparison with the lower leg bone. In the parlance of bird cuisine, it is called the short joint. It is oriented so as to sharply angle forward as well as slightly down. To simulate this orientation you would squat, your knees bent. That is the normal position of an avian femur, though it is not comfortable for humans to assume such a posture. The

Vision is highly developed in birds and, as a group, birds have large eyes, as shown by this Prothonotary Warbler. As big as the eyes appear, most of the eye mass is within the skull, not visible.

The beak or bill of a bird consists of a bony upper and lower mandible covered by a rhamphotheca, the horny outside of the beak, shown in this immature Herring Gull (*above left*). Some bird beaks exhibit rhynchokinesis, the ability to bend the upper mandible, as shown in this Wilson's Snipe (*above right*). Woodpeckers such as this Red-bellied Woodpecker (*right*) have very long tongues with barbs at the tip. (*Above left: John Kricher; above right: Garry Kessler; right: Jeffory A. Jones*)

femur and knee of a bird are covered by contour feathers of the body so what you normally see as leg is actually only the lower part of the bird's leg with its (often) big feet.

The long lower leg is the tibiotarsus, a fusion of the tibia and fibula bones that comprise your lower leg, what we call our calf. The tibiotarsus articulates with the ankle and foot, whose major bone is the tarsometatarsus, the long ankle bone that many mistake for the lower leg bone. Because of the way bird legs are structured, some people think the knee joint of birds is somehow reversed, the knee facing backward, but that assumption confuses the knee (not normally visible) with the ankle (usually visible).

Birds walk on their phalanges, or toes. Usually three digits (numbered 2, 3, 4) face forward and digit number 1 (equivalent to your big toe) faces backward (and has its own name, the hallux). Perching birds are so named because in this anisodactyl toe arrangement—three toes facing forward and one facing backward—the toes have tendons that allow the bird to firmly perch when its body weight is placed on its lower legs, the toes tightly closing around a branch or other object. Some birds, such as woodpeckers and kingfishers, have two toes facing forward and two backward (called a zygodactyl toe arrangement). Other, less common toe orientations exist as well. Look carefully at bird feet and you'll soon see how large they are, toes included. The smallest feet are found in swifts, whose family name, Apodidae, means "without feet."

The avian femur, positioned such that it angles forward with respect to the vertebral column, allows the bird to maintain a center of gravity essential for efficient movement in flight. A flying bird lifts its femurs closer to the vertebral column and sharply bends its knees so that the tibiotarsi face backward (forming the so-called fetal position that humans sometimes assume, with knees tightly drawn up). In flight, the tarsometatarsi typically face backward. Feet trail behind or are held close to the body, positioning the bird's landing gear so as to reduce drag in flight. When in flight, the bird's elbows and knees are almost in perfect vertical alignment, making the center of gravity well balanced.

The femur is surrounded by dense musculature that works the bird's leg. The position of the femur remains near the center of gravity of the bird body, critical to balance and successful flight. The upper part of the tibiotarsus is also well muscled. We call it the drumstick. Various muscles of the tibiotarsus attach by long tendons to the distal part of the leg and foot. This is why the foot is thin and not fleshy.

Skeletal distinctions among birds represent adaptations of various bird groups. For example, swifts, which spend the vast majority of their lives airborne, have tiny legs and feet. Not so with flamingos and gallinules. Loons, whose dive is propelled by strong legs and feet, do not have a favorable center of gravity on land. Loons' feet are located way back on the body, so much so that loons can barely move on land. But they fly well and are outstanding divers. Grebes are similar.

WHY PHYSIOLOGY MATTERS

Birds require at least ten times more energy inputs per unit of body weight than do reptiles, such as lizards, alligators, and turtles. Mammals approach birds in daily caloric requirements, but birds nonetheless exceed most mammals in that regard. Only very small mammals, such as shrews and bats, rival birds in how much fuel per gram of weight they require. On average, birds maintain a higher body temperature and more rapid heart rate than mammals of equivalent mass.

Birds and mammals are endothermic and homeothermic, terms that are roughly equivalent to the common term *warm-blooded,* meaning that the animals feel warm to the touch. These animals typically maintain a high and constant body temperature (higher than ambient) that is generated and controlled by internal metabolic processes regulated by the hypothalamus of the brain. Metabolic rate is high, oxygen use is high, and so the animal requires lots of food calories and must have an efficient circulatory system to transport oxygen and dissolved food molecules throughout the body.

There is an inverse relationship between body mass and energy required per gram of body mass. In an absolute sense, a large bird such as an eagle requires more food mass per day than a kinglet. However, the kinglet requires more food per gram of body mass than an eagle. Kinglets must take in proportionally more food than eagles because eagles have slower metabolisms than kinglets. Birds, when compared with mammals, have higher rates of gas exchange, higher heart rates, and proportionally larger hearts. The resting heart rate of a Turkey Vulture is about 132 beats per minute. Passerine birds have heart rates that vary from 300 to 500 beats per minute. A typical hummingbird, such as a Ruby-throated Hummingbird, has a resting heart rate of 615 beats per minute and the heart rate exceeds 1,000 beats per minute when the bird is in flight.

Why be endothermic? Seems like a bad idea, in a way. Life revolves around constantly finding food and staying warm. Starvation is potentially never far away. The cost of being an endotherm is high. But there is also a huge payoff to being an endothermic animal—three payoffs, actually. One is that birds and mammals are able to maintain high levels of activity because of endothermy. Sustained activity in the form of flight or foraging is advantageous. It enables enhanced predation skills, extreme alertness,

ANKLE
JOINT

FOOT

This adult Little Blue Heron (*top*) aptly demonstrates the foot, ankle, and lower leg of a bird, making it appear to some naïve observers that the bird has a reversed knee. But the knee is not visible in this photo because it is enshrouded in feathers. The Purple Gallinule (*bottom*) navigates atop lily pads and other vegetation with the aid of its immensely long toes. (*Top: John Kricher; bottom: Garry Kessler*)

and many other useful behavioral attributes. A second payoff is access to diverse habitats, such as high mountains, cold regions, lakes, and oceans. As long as there is sufficient insulation—feathers in the case of birds, fur or blubber in the case of mammals—endotherms can endure cold and windswept environments. And third, endothermy provides for the ability to evolve highly tuned sensory organs and extremely quick reaction times.

This brings us directly to behavior. Bird behavior, in all its magnificent complexity, is a byproduct of endothermy. But why?

The brain is why. Human brains represent about 2 percent of body weight but require 20 percent of the overall oxygen taken into the body. The brain is the organ that fails first when starved of oxygen, particularly the forebrain with all the complex neural connections that function for memory, intelligence, and cognition. Endothermy provides sufficient oxygen and food to sustain complex brains.

There is one final, immense evolutionary dividend from endothermy as it affects brains, both mammalian and avian. Some animals have the potential to get smart. The adaptation known as intelligence is rooted in the high metabolic rates of millions of generations of what became birds and mammals. Not all birds are equally intelligent, just as not all mammals are. But we humans clearly owe our intelligence to our endothermic heritage. So do ravens.

BLOOD CIRCULATION AND GAS EXCHANGE

Birds have a closed circulatory system of arteries and veins interconnected by dense capillary networks that feed the various body organs. Blood is red, rich in hemoglobin, the compound that transports oxygen. Blood also transports essential food molecules to various organs and collects carbon dioxide, the principal waste product of metabolism. Like mammals, birds have a four-chambered heart consisting of two atria and two ventricles. The right atrium and right ventricle receive deoxygenated blood from the body organs and pump it to the lungs, where blood acquires oxygen and gives up carbon dioxide. From the lungs, the newly oxygenated blood moves to the left atrium and then into the left ventricle and is powerfully pumped through the bird's aorta to be distributed to the body organs.

It is essential that birds obtain sufficient oxygen for flight, and they have an astonishingly efficient respiratory (or gas exchange) system to accomplish that. Birds have a complex system of air sacs that fill their body cavities and penetrate into the bones. Nine major air sacs help facilitate air movement during respiration as well as contribute to the light body mass of birds. Birds' lungs are unique, structured somewhat like a radiator in which the air stream moves unidirectionally through the lung.

The bird's blood vessels and capillaries are situated such that oxygen efficiently moves into the bird's blood and carbon dioxide leaves it.

Here's how it works. When a bird inhales the first time, the air moves from the trachea, not to the lungs but directly to the posterior air sacs. During the first exhalation, this air moves from the posterior air sacs into the lungs. When the bird takes a second inhalation, the first volume of air moves out from the lungs into the anterior air sacs. On the second expiration the first volume of air, now high in carbon dioxide, moves via the trachea out of the bird's nostrils or mouth and back into the atmosphere. The exchange of oxygen and carbon dioxide is nearly 100 percent efficient (far better than that of the bellows-type lungs of mammals). It requires two inhalations and two exhalations for a single unit of air to pass entirely through a bird's respiratory system. Air movement is facilitated by the air sac system, by the muscles of the bird's rib cage, and, when in flight, by the springlike movement of the furcula (the wishbone).

EFFICIENT DIGESTIVE AND EXCRETORY SYSTEMS

Birds may be omnivores (eating both plant and animal material), herbivores (eating only plant material), or carnivores (eating only animal matter). The degree of dependency on plant and animal food may vary seasonally. For example, many seed-eating birds feed their nestlings primarily animal matter, rich in protein, enhancing the rate of growth. American Robins feed heavily on animal material (worms come to mind) during the spring and throughout breeding season but switch to a fruit diet during the nonbreeding season. On the other hand, raptors and seabirds are exclusively carnivores, feeding entirely on animal matter.

Birds move food into their bodies with their beaks. The beak lacks teeth but is efficient in food processing before swallowing. American Goldfinches and Northern Cardinals adeptly strip the husk off a sunflower seed before ingesting it. Sensory structures in the beaks of waterfowl and other waterbirds allow them to quickly filter out nonfood items and ingest the food they seek. Once food is swallowed it moves into the bird's esophagus. In some birds, such as fish-eating species, the esophagus is thick, an adaptation that allows the bird to ingest still-living, protesting fish.

Food then travels to a structure called a crop, an expansion of the esophagus. This is a temporary storage chamber. Jays cram multiple acorns into their crops until they either bury them or eat them. Raptors' crops fill as they devour prey, the crop visibly bulging from the neck area. Food in the crop passes into the stomach to be chemically digested.

Flight is enabled by the remarkable and unique respiratory and circulatory systems of birds. This American Oystercatcher, like all birds, processes oxygen far more efficiently than mammals do.

Pigeons and doves have a unique crop in that it sloughs off cells that become nutrient-rich "milk" they feed to their nestlings. Why a crop? Without teeth to process food, the expansible crop is useful for quickly storing food items such as seeds while the bird is still ingesting.

Once the crop is full the bird typically flies to a safer location and the stored food items leave the crop, moving to the stomach, the proventriculus, where digestion starts: the food is exposed to powerful enzymes and hydrochloric acid, and begins being broken down into less complex molecules. From the proventriculus, the food mass then moves to a thickly muscled ventriculus, or gizzard. The gizzard facilitates maceration. Some birds, such as chickens, grouse, doves, grosbeaks, and other seedeaters, regularly swallow sand grains and small pebbles that go to the gizzard and aid in grinding up food. Gizzard action serves to make food particles increasingly smaller, maximizing the surface area per particle and thus increasing the efficiency of digestion as the food moves on into the small intestine.

The small intestine breaks down the food into molecules of protein, lipid (fat), and carbohydrate. The pancreas secretes digestive enzymes into the small intestine and also releases insulin, essential for carbohydrate metabolism. Intestine length varies. Species dependent mostly on plant material have longer intestines than do meat consumers. Plants, composed largely of complex carbohydrates such as cellulose, are more difficult to digest than animal muscle, which is largely protein. Blood vessels transport digested food to the liver for final processing. The liver is a two-lobed structure that, as in all vertebrates, is the biochemical factory of the animal. The liver stores compounds, detoxifies, and distributes food molecules.

The large intestine of a bird is short, but at the juncture of the small and large intestines, many bird species have paired sacs called caeca (singular: caecum). These sacs provide for additional retention time to complete digestion of complex compounds typically found in plants. Grouse, which dine on hard-to-digest conifer needles, have prominent

This female Spruce Grouse (*opposite top*) devours many spruce and fir needles. Digestion is accomplished with the aid of a large caecum. Gray Kingbirds (*opposite bottom*), like other tyrannid flycatchers, regurgitate pellets of indigestible insect parts. Birds process food rapidly and use their muscular cloacas to excrete both digestive and urinary waste. This Atlantic Puffin (*above*) is doing just that. (*Opposite top and bottom: John Kricher; above: Bruce Hallett*)

caeca. Many groups of birds, especially those that are insectivorous, such as swifts, lack caeca.

The final chamber of a bird's intestinal system is the cloaca, a word derived from the Latin word for "sewer." The muscular cloaca exits at the bird's vent. This single opening allows the passage of intestinal and urinary waste (usually combined in birds into a paste-like substance) as well as reproductive products—semen and eggs. The cloaca is a reptilian cavity dating back to the earliest ancestors of birds. It has been modified in different ways in various bird species.

The solid waste products of birds are evacuated as a paste-like slurry that contains both solid matter and uric acid, the substance that makes up urinary waste (see next paragraph). The cloaca is muscular and can "shoot" the excrement some distance. Many birds evacuate over the edge of a nest, and raptors in particular are known for leaning forward and projectile excreting just prior to taking flight. A few bird species, such as various geese, feed on grasses and produce firm cylindrical droppings, usually green. Owls and hawks routinely regurgitate indigestible fur and bone in the form of pellets. Some insectivorous birds also regurgitate pellets. For example, tyrant flycatchers of various species that consume numerous large insects composed of hard-

to-digest chitin skeletons and big wings routinely regurgitate pellets composed of these substances. Cuckoos devour hairy caterpillars and periodically slough off stomach lining dense with caterpillar hairs, a thick, hairy pellet.

Bird kidneys, lobed, beanlike structures found high in the upper back of the bird's body cavity, function to filter and eliminate toxic nitrogenous waste. The end product is uric acid, a semisolid, concentrated, low-weight form of this waste. Unlike more toxic urea, which is the mammalian form of nitrogenous waste, uric acid need not be dissolved in large volumes of water. Birds lack bladders and they don't consume nearly as much water as mammals. For example, Chipping and other species of sparrows that rely heavily on seeds in their diet may obtain enough metabolic water from the seeds that they need not drink frequently. Uric acid moves through two long ducts (one per kidney) called ureters. These ducts lead directly to the cloaca, where the uric acid mixes with solid waste prior to ejection via the cloacal vent.

Bird kidneys do not concentrate salts, such as sodium chloride, as effectively as do mammalian kidneys. Many birds that normally have low-salt diets of fruits, arthropods, and worms do fine at eliminating excess salt via kidney filtration. But what of shorebirds, gulls, terns, albatrosses, and shearwaters that routinely ingest brackish or salt water? These groups, as well as all others with high intake of salt-concentrated water and food (such as fish), utilize unique salt glands, paired glands on the forward part of a bird's skull, above the eyes. All seabirds rely on salt glands to concentrate and eliminate excess salts. Gulls and sea ducks such as eiders shake their heads from time to time. That is because the salt solution flows out through the nostrils onto the beak. Shaking the head clears the beak of the salt goo. Tube-nosed birds such as shearwaters and storm-petrels have prominent paired tubes on the dorsal side of the bill that allow the salt solution to trickle well down the length of the bill, preventing it from getting into the eyes of the birds. Some of these seabirds sneeze periodically, ridding their beaks of the salt.

THE REPRODUCTIVE SYSTEM

The avian reproductive system changes with the bird's annual cycle (chapter 7). In breeding season, the reproductive system is fully functional. Outside of breeding season the reproductive system shrinks. Both ovaries and testes are reduced in volume during nonbreeding season to but a fraction of the size they attain in breeding season.

Birds have internal copulation. Groups such as the large flightless birds (ratites: ostrich, cassowary, emu, rhea, kiwi) as well as South American tinamous have a penislike organ of intromission. So, too, do waterfowl, the swans, geese, and ducks. Indeed, one odd duck, the South American Lake Duck, has a phallic organ that measures an astonishing length of 17 inches (42.5 cm), which is very close to the total length of the duck! It is unclear what sort of adaptation is represented by such a length.

In the majority of birds, sperm pass from the testes through a duct called the vas deferens to the cloaca. In the cloaca, the sperm concentrate in a chamber, an expansion of the deferens duct. As sperm become more concentrated, the chamber protrudes from the cloaca, forming what is called a cloacal protuberance (CP). Because the CP is outside the body cavity it keeps the sperm at a lower temperature, which is adaptive because sperm are sensitive to heat. When birds copulate the male balances himself on the back of the female, twists his tail such that the CP is forcefully placed against the cloaca of the female bird, and ejaculates sperm into the female's cloaca (chapter 14). Avian sperm are able to remain viable for a considerable number of days following copulation, and in some species, sperm competition occurs when sperm is present from several males (chapter 15).

A female has the potential to produce thousands of eggs in her lifetime, and reproductive potential

The cloacal protuberance is visible in this territorial male Painted Bunting.

The bird egg is a marvel of adaptation. Two newly hatched American Woodcocks are just out of the egg.

does not seem to strongly diminish with age. As with males, the female reproductive system is only fully expressed during the breeding season. Unlike males, which have two testes, females other than those of kiwis and raptors have but one functional ovary, almost always the left one. This reality leads to an occasional sex change in a female bird. Though not expressed, the right side of the female bird's reproductive system may have an undeveloped testis. If the left ovary should suffer damage or disease and become nonfunctional, the right testis may develop and what was once a hen, so to speak, becomes a rooster.

An egg is fertilized soon after it moves from the ovary into the oviduct. Then it undergoes maturation and acquires vital components such as albumen, supportive membranes, and, of course, the shell. The various markings of a bird's egg (if the egg has markings—many don't) are acquired as the egg moves in a spiral pattern in the lower part of the oviduct near the vagina. It is then ejected into the cloaca and laid. This process, from egg fertilization to egg laying, takes about 24 hours. A typical bird lays one egg per day until it reaches a full clutch.

A bird's egg is composed not only of the embryo itself but also extraembryonic membranes (genetically identical to the embryo): the chorion, the amnion, the allantois, and the yolk. The entire complex is housed within a shell that is permeable to gases but not to water. This remarkable structure, called the cleidoic or amniotic egg, was evolved long before birds were. Birds (and mammals) inherited it from their reptilian ancestors, so the answer to the question, "Which came first, the chicken or the egg?" is that the egg did. The extraembryonic membranes have various functions. The chorion lies inside the shell and serves as an overall protection for the embryo and other membranes. The amnion is a fluid-filled sac that surrounds the developing embryo, protecting and buffering it. The allantois collects and stores waste products from the developing embryo. The yolk is rich in fat and protein and serves to nourish the embryo as it completes its development.

All birds lay eggs. Reptiles lay eggs too, but not all of them. Some are live bearers, such as some snakes. Back in the Triassic Period all mammals laid eggs, but only a small number of them do now, namely the echidnas and platypuses of Australia. All other mammals, divided between the groups known as marsupials and placentals, birth live young. So why are there no live-bearing birds?

The answer is likely flight. For embryos to develop in the female's body, it would take days—rather than the 24 hours required for an egg to be ready for laying—and the burden of weight would be too severe for the bird. Eggs predate flight, but the reality of flight prevents birds from becoming live bearers.

4

A PERSPECTIVE ON BIRD BEHAVIOR

WHAT ARE BIRDS THINKING?

Birds are not civilized in the way humans define it; they obey no formalized societal laws. When two birds contest for food, it is doubtful that the loser resents the winner in any way comparable to what might occur in a human interaction. It is highly unlikely that two Laughing Gulls, side by side on a beach, are gossiping about a third Laughing Gull some distance from them. Interpreting animal behavior as one would interpret human behavior is anthropomorphic and often highly misleading. Understanding bird behavior is more challenging in some regards than understanding human behavior because we humans are not birds and thus, to put it bluntly, we do not and cannot think like birds. Is this an unbridgeable gap?

Birds sometimes appear to behave in ways strongly suggestive of human-type thought processes, human emotional expression, and logic. I have watched jays methodically select peanuts from among a mixture of seed types. This observation convinces me that jays are making a conscious choice, as I am when picking out cashews from bowls of mixed nuts. And while Laughing Gulls on a beach may not interact in particularly complex ways (but maybe they do), certainly troops of baboons, chimpanzees, and gorillas do, and we think we understand these animals because they appear to be so much like us. We credit them with having rational thought. But crows and ravens do not resemble humans, and yet, as research has shown, they are comparable to apes in their intellectual abilities.

Although thinking about birds anthropomorphically is a slippery slope, it is reasonable to ask why a bird is performing a particular behavior by relating that action to why we, if we were in the bird's place, might do the same thing. Birds constantly

OPPOSITE: Great Black-backed Gull (*right*) and Herring Gull (*left*) contesting a fish. The Great Black-backed Gull won possession of the fish. Such struggles are common among bird species.

The Common Raven is considered by many to be the most intelligent bird species in North America. *(Ed Harper)*

observe each other, learn as they do so, and are excellent information sources for other birds. Birds think, learn, and remember all manner of things on a daily basis, all of which contributes to their survival. They are most certainly not feathered humans, but they are not dumb automatons either.

In his book *Mind of the Raven,* author and researcher Bernd Heinrich describes numerous examples from his years of study of the Common Raven. Heinrich documents such complex and emotion-laden behaviors as deception, cooperation, individual recognition, play (see "The Mental Bird—Play Behavior," page 45), fear, and, of course, memory and intelligence. For example, young ravens help each other to locate food in winter, a form of cooperative behavior that works well for them because the young birds are able, through the strength of their numbers, to repel dominant adults that would otherwise overwhelm single inexperienced birds and drive them from the food source.

Ravens, members of the Corvidae, the crow and jay family, are considered to be among the world's most intelligent birds; only parrots (family Psittacidae) are comparable. The New Caledonian Crow, once a little-known species, has attained something akin to celebrity status for its demonstrated feats of

Common Ravens (*left*) are sufficiently intelligent that they develop complex behaviors within raven society. They "know" each other. Flocks of ravens (*right*) often circle high for reasons known best to them. *(Left: Ed Harper; right: Van Remsen)*

intelligence and problem solving, most particularly tool making. A comparative summary of ape and corvid intelligence suggests they are on par, the corvids as intelligent as apes. This conclusion was reached by comparing the two groups with regard to nonverbal cognitive characteristics, including causal reasoning, imagination, flexibility, prospection, and complex cognition.

Irene Pepperberg devoted thirty years to working with an African Gray Parrot named Alex. Pepperberg's studies focused on the parrot's ability to learn and use words and, in doing so, revealed his cognitive skills. They also brought to light Alex's engaging personality. He communicated. He learned to know people as individuals, a trait evidenced by virtually all parrots that are treated well in captivity. It is worth realizing at this point that birds, with their faces covered by feathers and their bright but relatively expressionless eyes, are challenging for humans to "read," but it can be done. Birds express themselves emotionally by using subtle and nuanced body postures. It takes time and practice to learn how to "read" a bird, especially birds in the field.

Though corvids and psittacines are recognized as uniquely intelligent, other birds have been shown capable of quickly learning sophisticated tasks. Lab experiments have demonstrated that Rock Pigeons are capable of advanced cognition. The pigeons were able to sort among complex categories of photographs and recognize and recall abstract relationships. When pigeons and preschool children were compared, assigned the same tasks, children performed slightly better, but not by much. Just as with children, pigeons showed individual differences in learning skills.

European Starlings use recursive syntactics in their complex songs and calls. This means that the birds are able to selectively embed specific phrases into their vocalizations. Recursive grammar in humans is routine. For example, you might say, "That's a good bird." Or you might say, "Martha just saw a good bird that Bruce missed," or, "Have you seen a good bird this morning?" The phrase "a good bird" is inserted selectively and conveys different meanings in the context of how it is placed in the sentence. Using recursion is a hallmark of human language (and intelligence), but at least one songbird also does it.

Cognition, explained in more detail later in this chapter, is the ability to acquire knowledge and understanding by using thought and intelligence and to then act on that information. Birds continually demonstrate various behaviors that are best described as cognitive.

Charles Darwin wrote a book with the title *The Expression of the Emotions in Man and Animals* (1872). It was the first scientific effort to place animal behavior in a comparative and contextual light with human behavior, with a focus not so much on intelligence but on commonalities of emotions that reflect commonalities of origin, the deep genetic relationship among all life forms.

Nature's rules are simple: eat, grow, reproduce, die. One bird might displace another, perhaps because it saw the second bird acquire food, perhaps because the displacer is more dominant, perhaps both. It is likely that there was a reason—the bird was acting intentionally. The displacer made a choice to try and displace the other bird. The two birds apparently did not regard each other as equals and one of them bested the other. I cannot pass judgment on the bird that displaced the other bird. As I have told my students countless times, "There is no *morality* in nature, only *reality*." Birds are not little people in feathered suits, no matter how much their behavior may, on occasion, seem suggestive of aspects of human behavior. Birds are birds and they live very close to the physiological edge in a world of nature.

Birds show much individualized behavior, including within species. The study of bird behavior in the field will bring you to a better understanding of the

The European Starling has been shown to be able to use recursive grammar in its vocal communications with other starlings.

bird as a bird. But birders must learn to take avian behavior at face value and not read more into it than is merited. That is a big lesson. If a Sharp-shinned Hawk kills and devours a bird such as a Mourning Dove that had been coming to your bird feeder, you are witnessing an act of predation, commonplace in nature, not a vicious murder. The hawk is being a hawk. It is exhilarating to study bird behavior, but always keep it in context: they are birds.

THE MENTAL BIRD

INSTINCT: Since the mid-twentieth century, ornithologists have recognized that birds are innately programmed to express genetically based instincts that enable them to respond to whatever comes along in their world. For example, a Herring Gull chick instinctively pecks at the red spot on an adult gull's lower mandible. The parent gull then instinctively regurgitates food into the beak of the chick. Gull chicks will even peck at a red spot on a stick even though the stick does not look like mom or dad. The spot triggers a *releaser* in the brain that tells the chick to peck. Releaser signals are commonplace among birds and sometimes make birds appear pretty dumb, blindly responding to the releaser, whatever it may be. But releasers are highly adaptive because they save time that would be devoted to learning something that a bird must know.

Shearwaters and storm-petrels nest in burrows,

visited mostly at night by parent birds returning with food for their nestlings. Downy nestlings within their burrows grow considerably fatter than the parent birds. Once the young have reached peak weight the parents end their reproductive duties, leaving the chicks alone in the burrow. Chicks molt into body and flight feathers and become increasingly slender. Each juvenile must, on its own, leave the nest, find its way to the cliff edge, and instantly "learn" to fly off and fend for itself over the ocean. But they don't learn, they just "know." The complex behavior associated with fledging is enabled by instinct.

Various environmental and social stimuli that birds experience release what are termed *fixed action patterns* that birds perform in response to stimuli. Courtship behavior, as when many species of male ducks perform elaborate body posturing while in the presence of females, is an example. Instinct has a strong genetic basis. Birds show numerous examples of instinctive behavior such as "knowing how" to assemble a nest characteristic of their species, "knowing how" to preen feathers, and "knowing how" to form a pair bond.

LEARNING AND MEMORY: Birds have also always shown a capacity for learning. Birds learn to navigate complex environments, quickly locating bird feeders, for example. Birds move easily from one habitat to

Herring Gulls as well as other gull species exhibit a conspicuous red spot on their lower mandible that acts as a releaser in the process of feeding chicks.

another and experience one situation after another. It is difficult to imagine how they could exist without having a strong learning capacity to develop a spatial sense of place, accompanied by an ability to memorize relevant details of that place. Learning is evident in many of the Passeriformes, the "perching birds," such as thrushes, orioles, blackbirds, wood-warblers, and buntings, which must learn their songs, usually from the male parent (chapter 13). Birds of some species routinely mimic vocalizations of other birds. This is also learning.

Avian learning often connects with instinct, as is illustrated by imprinting. Baby birds have a critical time period after hatching when they imprint on the parent bird and thus learn their caregiver as well as their species. Female ducks look different from males. Newly hatched ducklings see their mothers at the nest but not their fathers. Male ducklings learn the appearance of their mothers sufficiently well to distinguish their species from other species in which the females appear similar. Female ducks are generally instinctually drawn to the posterior coloration of the males of their species. This is augmented by behavioral posturing typical of duck courtship behavior.

Herring Gulls learn to pick out their chicks from numerous others in the colony. Black-legged Kitti-wakes never learn to recognize their offspring and would not know if another chick was switched with theirs (kittiwakes lay only one egg per nesting). Why the difference? Herring Gulls have broad colonies on marshes, and a chick may wander from one area to another in the colony. Parents must relocate it. Kittiwakes nest on narrow ledges of high cliffs and do not wander. Any movement from the nest by the chick is usually fatal. Thus kittiwake parents learn their nest location, not their chick.

Numerous songbird species fail to recognize the addition of a Brown-headed Cowbird egg in their nests, nor do they recognize that the cowbird fledgling is not their own fledgling. This becomes very obvious when a bird such as a Song Sparrow is feeding a cowbird chick that is well over twice its size (chapter 15). Many birds instinctively know the overall shape of a raptor such as a falcon or accipiter. But they must learn how to be alert for such predators and act quickly should one appear.

The essence of learning is memory. The avian brain contains a sophisticated memory bank located in a part of the brain called the hippocampus, discussed more in the next chapter. I stopped putting suet in my suet feeder because I was leaving for a protracted period of time and would not be around to refill it. The previous summer a Gray Catbird that nested in our yard had made frequent visits to the suet feeder. I returned from my trip but had not replenished the suet when catbirds arrived in early May. One newly arrived catbird landed on my deck rail where the suet had hung during the previous spring. It looked up to where the suet basket had been. It dropped down on the deck and inspected the boards where chunks of previously fallen suet had typically accumulated. It flew back up on the deck rail and again inspected the pole that once held the suet feeder. After pausing a bit, it flew off. I have no proof but I believe it was one of the same catbirds from the previous spring and summer. It had learned where the suet was and, upon returning, it remembered the location of what had once been a reliable food source. In between it had presumably migrated to the tropics and back.

Black-legged Kittiwakes never learn to recognize their own offspring because of their extreme nesting conditions, on narrow ledges.

Geese have been studied widely with regard to imprinting and learning. This is a family of Canada Geese. *(Bruce Hallett)*

HABITUATION: Birds learn what to fear and what not to fear, which stimuli are meaningful and which are not. City dwellers know that urban pigeons show little fear of people; if chased by a child, for example, the birds will simply scatter and move a short distance away. The same is true of flocks of Sanderlings and gulls on a beach. When a bird learns that a particular stimulus causes it no harm or does it no good but is, in fact, neutral in nature, it ignores that stimulus. This form of learning is called habituation.

Birds constantly show various levels of habituation. In Florida, it has become common in some places to observe pairs of Sandhill Cranes walking

---

This Ovenbird (*top*) is feeling some form of emotion, as evidenced by its raised head feathers and its alert posture. The raised crest and overall countenance of this Blue Jay (*bottom*) suggest that it is on some form of alert. (*Top: Garry Kessler; bottom: John Kricher*)

with nonchalance on the lawns of housing developments, alert with regard to people and their dogs, but not acting unless specifically threatened. The same is true of White Ibises, once a bird confined to marshes and mangroves. In some Florida neighborhoods they now forage on suburban lawns. Burrowing Owls in Florida have habituated to humanity such that some of them nest on small plots of land (where they are protected) within dense housing developments.

The ability to habituate is of strong survival value to any bird. Young birds learn not to waste energy by reacting to such things as rustling leaves, herds of sheep, sounds of traffic, or thousands of other potential stimuli. We will revisit habituation as it relates to stress in chapter 7.

EMOTION: Emotions often trigger behavior. Birds compete and sometimes fight; it is physiologically hard to fight without emotion. So birds are capable of expressing anger. Birds form pair bonds, but the extent to which there is real emotional attachment between the bonded birds is, in most cases, impossible to say. Some birds, parrot species in particular, appear clearly affectionate toward one another, but whether the behavior can be called affection is open to interpretation. I have watched birds react after one of their nestlings has been taken by a crow. The parent bird or birds appear highly agitated, stressed, and angry. Do they grieve? At the other end of the spectrum, is there such a thing as a happy bird? Should the word *happy* ever apply to a bird?

What we do know is that emotions are part of brain physiology, activated by hormonal release resulting from the bird's perceptions. Fear is an emotion that is vital for survival. Submission, which derives from fear, is adaptive in many instances. Agonistic behavior is fueled by emotion, and birds behave agonistically when, for example, defending territories or competing for access to food. Situation-dependent emotional response is very much a part of every bird's behavioral pattern.

Birds reveal their emotions if you know how to recognize them. Birds don't frown, for example, but watch as a Blue Jay or Steller's Jay calls loudly, raising and lowering its crest. Jays on high alert typically raise their crest (as do other crested birds, such as titmice and cardinals). American Robins, when agitated, flick their wings and tails and emit harsh, repeated call notes. Many other bird species do this too. There is an emotional underpinning to such behaviors. Birds control the positions of most of their feathers and do so in response to hormonal and visual signals associated with behaviors ranging from aggression to courtship.

Bird emotions are sometimes expressed in an odd kind of behavior called *displacement activities*. The

There is little doubt that anger, in the avian sense, is being expressed in the interactions between two Northern Fulmars, one of which, the bird with the open beak, appears dominant over the other.

classic example is of two gulls engaging in agonistic behavior such that they adopt mutually threatening postures. One gull might abruptly begin pulling out tufts of grass with great intensity. The explanation for the grass-pulling is that the gull is conflicted between elevating the threat and retreating, the "fight or flight" response. The grass-pulling behavior is a displacement activity. Many other birds show similar behaviors in conflict situations. Some birds engaged in agonistic interactions suddenly start preening, for example. This happens frequently when groups of birds are mobbing a potential predator. Displacement activities result from conflicting releasers of various innate behavior patterns.

INTELLIGENCE AND COGNITION: Birds exhibit impressive cognitive abilities. I observed a pair of Eastern Bluebirds along a roadside capturing arthropod prey. Moments earlier this particular roadside area was being cut by a gentleman riding on a mower. When he departed, the various weeds and associated plants had been cut and flattened. The bluebirds, neighborhood residents, had been perched in nearby trees, obviously observing the cutting. They soon began feeding. Both birds perched about 6 to 9 feet (2–3 m) above the ground in trees that lined the road. Each would fly down to the roadside, spend a brief moment snatching prey exposed by the cutting, and fly back with a bug in its beak. This sort of

Purple Martins routinely use artificial gourds as nesting sites for their colonies (*left*). More recently some Purple Martins have begun nesting on dock pilings beneath pointed cones that cover the piling (*right*).

behavior represents cognition on the part of the birds. They connected the disturbance of cutting with exposure of prey.

Cognition is the acquisition of information that is processed through a complex combination of the animal's senses and thoughts, eventually to become part of its memory bank, its knowledge, even its culture. The bluebirds learned to associate the cutting of vegetation with prey availability. Ring-billed and Franklin's Gulls likewise learn to associate the cutting of fields with prey exposure, just as Cattle Egrets have learned to follow large animals that inadvertently scare up arthropod prey. Aplomado Falcons are quick to converge on burning fields where insect prey is driven into the air, just as Mississippi Kites and Swallow-tailed Kites converge over fields bountiful with dragonflies, which they capture on the wing.

Intelligence is a term that must be viewed contextually with the species in question. American Goldfinch intelligence is not like American Crow intelligence, just as raccoon intelligence differs from gorilla intelligence, and avian intelligence, in any form, is not like human intelligence. Intelligence, such as it is, has evolved in each bird species in con-

text with its various environments. A Greater Roadrunner in its desert environment is "intelligent" at survival, able to capture snakes and lizards for food. Some roadrunners are better at these tasks than others, and they are the most successful breeders. But a roadrunner would be less competent in a broadleaf forest. The Common Raven has adapted to live in almost all types of environments, ranging from urban areas to tundra to desert to forest. But this breadth of habitats in the range of one species does not mean that a Common Raven adapted to live in northern Alaska would thrive in, say, the southwestern deserts or along the Appalachian Trail. It would have a lot to learn. By contrast, long-distance migrant passerines such as thrushes, orioles, and wood-warblers experience multiple environments in the course of their year and must be sufficiently intelligent to survive in each of them.

The avian brain, briefly described in the next chapter, has the capacity to store enormous amounts of information, which may then be applied to unique situations. There are thousands of examples. Every bird provides them. Purple Martins have learned to recognize clusters of gourds as potential nesting sites for their colony, a relatively recent adaptation dating back at most a few hundred years. More recently, Purple Martins have begun nesting in pointed plastic cones that cover dock pilings meant to discourage birds such as gulls from perching on them. Some Green Herons use various forms of bait to attract fish that they capture and devour. Burrowing Owls sometimes use mammal dung to attract dung beetles, which the owls then capture and eat. Many birds cache food and remember where they have cached most of it (see chapter 12).

A few species, such as the Woodpecker Finch of the Galápagos Islands, use tools to extract food. Seventeen species of birds have been shown either in the field or in laboratory studies to craft tools, usually but not always for use in foraging. This, of course, is but a small percentage of the total number of bird species, which is approximately 10,000 worldwide. Tool use remains uncommon in birds. But consider bird nests. The pioneer American ornithologist Alexander Wilson described in detail how a female Orchard Oriole intricately weaves grasses to form the hanging nest. Constructing an oriole nest would be challenging for any human. Knowing the overall details of nest construction is mostly innate, but assembling nests also requires thought and skill. Birds must find and collect material (chapter 15), which in itself requires some measure of searching and remembering, a form of cognitive behavior.

PLAY BEHAVIOR: Play behavior, which often mimics aspects of adult behavior, is an essential part of human development as well as that of various other mammal species, and play is often associated with learning and intelligence. Humans are familiar with play behavior of dogs and cats. Play demonstrates curiosity, intelligence, and innocence, while having the underlying function of aiding in the development of neural skills used in adulthood. Play is normally seen in juvenile animals that have a protracted period of immaturity. Play aids in learning and is therefore adaptive. Adult animals also play, as seen in cats and dogs (and people).

Birds do not normally have a protracted period of "innocence" as juveniles. Juvenile and immature birds fend for themselves once fully fledged; therefore, because they grow up so fast, most birds do not have the luxury of play during their juvenile lives. More accurately, most are innately familiar with their foraging methods and thus require little practice or training. That said, play is known to occur in some bird species.

Defining play behavior in birds is not as easy as in mammals. There are, however, published accounts that document play behavior in some birds, including such diverse groups as gulls, hummingbirds, and woodpeckers. There are also accounts of play behavior of various species of raptors, mostly dealing with manipulations of objects similarly to how they will handle prey items. Play in species such as Common Ravens suggests that it engenders enjoyment. Juvenile Tree Swallows sometimes carry white feathers in flight, drop them, and then swoop down and grab them before they hit the ground. This appears to be play. In most instances of play behavior in birds, the playing seems related to juvenile birds learning techniques that will be useful to them as adults. In the case of Tree Swallows, they typically incorporate white feathers into their nest structure.

Immature gulls hover and drop objects such as stones and shells. Play? Adult gulls habitually drop clams and other bivalves on hard surfaces such as parking lots, which seems to be a learned behavior. Doing so breaks the shells, making the contents available for the gull to consume. For inexperienced gulls to drop stones and other objects could be interpreted as playing while learning, or perhaps just learning. We humans associate play with having fun. Do birds have fun? As with most types of bird behavior, it is a slippery slope to try and relate play in birds to play in humans or even other mammals.

At Everglades National Park in Florida, a sign at the entrance to the Anhinga Trail at Royal Palm Visitor Center warns tourists of potential damage to their parked autos, damage caused by Black Vultures, which are numerous there. Much as New Zealand Keas (a species of parrot, see below) seem to enjoy pulling rubber linings from vehicle windshields and wipers, apparently Black Vultures have acquired the same habit, at least at Royal Palm. The park supplies tarps free of charge to cover auto windshields and windows and thus prevent damage from vultures.

I watched an adult Black Vulture for several minutes at a spot along the Anhinga Trail. The bird continually pecked and pulled at an abandoned baseball cap that was initially mired in mud. The actions of the bird were similar to what one would observe if the bird was manipulating and tearing at carrion. It was, however, highly doubtful that the vulture regarded the cap as potential food, and it is equally doubtful, at least to me, that it was "practicing" for when it would have access to actual carrion. The bird was an adult. It likely knew perfectly well how to go about devouring a decomposing rabbit or armadillo and did not need to practice on a hat. Was it just having fun? Nervous energy perhaps? Other observers have reported seeing Black Vultures manipulating various objects in a manner suggestive of play.

Play has often been documented for captive parrots. The birds typically express curiosity, interest,

An adult Herring Gull is experienced at dropping bivalves such as this scallop onto hard surfaces to break open the shells. Juvenile and immature Herring Gulls also drop things, often stones, of no nutritional value.

and apparent pleasure in manipulating objects, often repeating actions much as a child does when at play. Parrot play is much less frequently reported for wild parrots. But one parrot species, the Kea of New Zealand, has long been known for a peculiar behavior that many ornithologists have interpreted as play. Keas habitually land on automobiles and, using their large and strongly hooked beak, pull loose the rubber lining around the windshield and other areas of the vehicle. They do not eat the rubber linings or otherwise use them for any adaptive purpose. Some observers have opined that Keas just enjoy doing it, a form of play, and they have done so for many Kea generations. It has become part of Kea culture.

Raven play is expressed by repeated acts such as hanging upside down and flipping over to land on the feet, manipulating sticks in various ways, and dropping sticks in the air and catching them before they hit the ground. Ravens apparently enjoy snow-bathing, a form of play that involves splashing in snow just as the birds would in water. The play also includes sliding down snowbanks, apparently just for the pleasure it affords. Sometimes play is solitary, but with ravens it is usually social. Many examples of aerial play in ravens have been documented, particularly in Bernd Heinrich's studies.

UNEXPLAINED BEHAVIORS: Ornithologists interpret bird behavior with the assumption that behavior is an intentional act directed toward an outcome favorable to the bird doing the behavior. But sometimes behavior can be puzzling and defy easy explanation.

One such example went viral on Twitter and then YouTube. It was a film about Wild Turkeys, taken with a phone camera. Seventeen Wild Turkeys were methodically circling around a road-killed cat in the town of Randolph, near Boston, Massachusetts. The birds were following one another in a wide circle around the deceased cat, a behavior captured on film and easy to find on various websites (search for "Wild Turkeys circle dead cat"). Explaining such behavior is difficult. Opinions (guesses really, and none convincing) ranged from inspecting a potential predator to ritualistic following of the bird in front of the bird in front of the bird. One observer suggested that the group circling behavior represented both curiosity and fear; others called it a "death dance." Clearly the behavior affected the entire flock of turkeys and was intentional, but the reasons for doing it reside within the turkeys.

More cerebral birds, such as American Crows, are known to engage on occasion in what have been called "crow funerals." Crows have been observed to fly near and gather around a recently killed crow, carefully inspecting the corpse. Crow researcher John Marzluff suggests that the birds are attempting to ascertain the danger that might explain the death of a crow. On occasion a crow will actually attempt to copulate with a fresh crow corpse. One possible explanation is that crows recognize each other as individuals and are attempting to discern just who was killed. For birds as intelligent as crows, being drawn to a recent corpse might be stress related,

eliciting conflicting reactions having to do with the realization that the bird is dead. But do crows actually understand what death means?

SUMMARY: WHAT IS BIRD INTELLIGENCE ANYWAY?

Many bird behaviors appear to be innate while others suggest intelligence and learning. Bird behavior is both genetic and learned. Mental problem solving (insight learning) involves thinking, presumably with mental images of the real world in which objects are virtually manipulated. The New Caledonian Crow makes and uses tools to solve unique problems.

An excellent discussion of the evidence buttressing mental problem solving in animals is found in Frans de Waal's book, *Are We Smart Enough to Know How Smart Animals Are?* De Waal also discusses the importance of cultural transmission of learned behavior, as noted earlier with the example of immature gulls observing adult birds dropping clams to break open the shells. Insight learning and cultural transmission are different phenomena, but both represent forms of avian intelligence, even if part of that is simply to watch and learn from another bird. Avian intelligence is real, but much of it is remains enigmatic.

This adult Black Vulture, for whatever reason (known only to it), seems to be playing with a discarded hat. Its actions are similar to how it tears at carrion.

A BIRD'S BRAIN AND SENSES

"BIRD BRAINS"

The familiar expression "bird brain" likely originated because birds sometimes display seemingly naïve behaviors, suggesting to human observers that birds are robotic and stupid. The names "booby" and "gooney bird" (for albatrosses) convey such a view. Hunters know that nonanimated and often crude decoys attract waterfowl. However, they also know that waterfowl soon learn to be wary.

Bird behavior that seems unintelligent to people is often simply our misunderstanding of context. Northern Cardinals (as well as many other species) vigorously and persistently attack their own images, flying and pecking at a car side-view mirror or a window. In context, such behavior is adaptive. Northern Cardinals defend a breeding territory and repel intruders of their species. The closer an intruder is, the more agonistic the behavior. When a male cardinal suddenly sees another male cardinal during breeding season, he becomes reactive, even if the cardinal is an image of himself. How could he know about mirrors?

PROXIMATE CUES, ULTIMATE ADAPTATIONS

Biologists distinguish between proximate causation and ultimate adaptation. This dichotomy is often described as asking "how-type" versus "why-type" questions. Consider the mirror example. A male Baltimore Oriole establishes a territory and courts a female. A pair bond forms. The female is that male's ticket to his genetic future. If another male oriole appears, the territorial male attacks. He doesn't stop and think about it, but instead goes after the intruder, chasing him away. Why? It takes only seconds for birds to mate and the territory holder cannot afford to allow another male to linger in his territory (chapter 14). Instinctively the bird "knows" to go after an intruder.

OPPOSITE: A male Baltimore Oriole beholds himself reflected in a window. His brain tells him that he is looking at another oriole.

The proximate reason for a male oriole to chase another male is that he recognizes his species and the potential threat that another male poses. Early imprinting is likely *how* the bird came to recognize others of his species. But ultimately, the reason *why* he reacted aggressively is because it is adaptive not to have rival males in one's territory. There is no wiggle room here, no negotiation, no compromise. Get him out. Now. So orioles react, albeit mistakenly, in situations in which they see their own images reflected. Birds do not recognize themselves in a mirror. Instead, they see another bird.

Carefully observe how a bird behaves and then try to place the behavior in the context of understanding survival and reproduction. Bird behavior, of necessity, is adaptive, with regard to either immediate survival or to potential reproductive success.

WHAT ACTUALLY IS A BIRD BRAIN?

All vertebrates—fish, amphibians, reptiles, mammals, and birds—have brains that share evolutionary core similarities, composed of a forebrain, midbrain, and hindbrain.

The hindbrain consists of a cerebellum and a medulla oblongata. The cerebellum, large in birds, controls coordinated body movements and balance. The medulla oblongata, often referred to as the brain stem, controls basic physiological processes, such as blood circulation, digestion, and respiration. The medulla functions without conscious thought and is thus involuntary.

The midbrain consists of paired optic lobes. In birds, which have outstanding vision (see "Vision," page 51), the optic lobes are large, bulging laterally on either side of the brain.

The forebrain has paired cerebral hemispheres and is the center of thought, instinct, emotion, and memory. It is where consciousness lies. In birds and mammals, it is the largest part of the brain, filling most of the cranium. However, the forebrain of birds differs in structure from that of mammals. Mammalian forebrains show obvious wavy convolutions throughout their bulbous surface. These deeply convoluted sulci characterize the huge cerebrum of humans as well as apes and dolphins. In mammals, the outer part of the cerebrum, the cerebral cortex or neocortex, is the thought center. This region encompasses all aspects of intelligence, learning, memory, and personality. "You" reside in your cerebral cortex.

In birds the cerebral cortex is thin and smooth. This led to the mistaken belief that birds are ill-equipped for sophisticated thought processes, intelligence, and learning. That is not true because other, more inner regions of the avian cerebrum are well developed. An area called the striatum within the base of the cerebrum is associated with instinctive behavior. Much of the forebrain interior consists of the pallium and the dorsal ventricular ridge (DVR), containing the nidopallium, mesopallium, entopallium, and archipallium. The pallium and associated regions within the cerebrum are the avian equivalent of the mammalian neocortex, where the various components of cognition, learning, and intelligence are located. In corvids (crows, ravens, jays) and parrots, recognized as being extraordinarily intelligent, the forebrain (pallium) is about 80 percent of the total brain volume. In comparison, the forebrain of a chicken is about 53 percent of brain volume and that of a Rock Pigeon about 48 percent. Another area, the hyperpallium, also called the Wulst, bulges from the top of the forebrain and functions to coordinate cognitive and binocular vision.

Yet another prominent part of the forebrain is the hippocampus, essential in converting short-term memory into long-term and in storing memories. In hummingbirds the hippocampus is uniquely large in proportion to that of other birds, perhaps an aid to remembering the specific locations of widely scattered flowers where nectar is available. One would predict that the hippocampus would also be large in species such as sunbirds and sugarbirds, also nectar feeders. The hippocampus is also large in seed-caching birds such as chickadees and various corvids, birds that are able to recall exactly where they've stashed seeds.

In songbirds, areas of the brain that have to do with song production and song learning grow new neurons prior to the bird entering into breeding season. This annual neuronal genesis is unique to birds.

At the base of the forebrain, the hypothalamus and pituitary gland function to direct hormonal secretion and thus coordinate most aspects of a bird's annual cycle (chapter 7).

It is beyond the scope of this book to go into additional detail about the avian brain (but see references). Suffice it to say that while a bird's cerebrum is wired differently from a mammalian cerebrum, it nonetheless performs similar though not identical functions. Birds and mammals both think, but not in the same ways.

BIRD REACTION TIME

Birds are cautious and adapted to react quickly to situations that pose threats. Suppose you come upon a flock of Chipping Sparrows feeding on the ground. Up they swoop into the protection of shrubs or tree branches. You stand there hoping they'll return so you'll get to see them well, but they usually won't, at least not very quickly. The birds sit quietly, biding their time, not inviting risk caused by exposure.

Birds fly in complex three-dimensional habitats, such as forests. It is amazing to see, say, a chickadee or an oriole fly with such ease through a dense array of branches and land safely on the branch of its choice. Indeed, it is wondrous how they make such a choice so suddenly, but they do. Hummingbirds, which can attain flight speeds of up to 35 miles per hour, fly skillfully between clusters of branches when seeking flowers. There are numerous videos on YouTube that show species such as Northern Goshawk flying rapidly and unerringly through

Most bird species have outstanding vision. Note the conspicuously large eye of this Veery, typical of many bird species. The bird inhabits shaded forests and migrates at night.

The importance of vision in birds is shown in this young male Bee Hummingbird, which has disproportionally large eyes for his tiny body size.

narrow openings. (See the episode called "Goshawk Flies Through Tiny Spaces in Slo-Mo!" in the BBC series *The Animal's Guide to Britain.*) Similarly, birds ranging from Great Crested Flycatchers to Hooded Warblers adeptly pursue and snatch flying insects within dense vegetation.

Quick reaction times are the hallmark of many forms of foraging behavior. Swallows and swifts and many passerines such as flycatchers and some wood-warblers capture their food on the wing, nimbly snatching flying insects. Flocks of starlings, blackbirds, and shorebirds are nothing short of amazing as they form dense flocks of hundreds to thousands, flying in close proximity without ever crashing into one another.

The reaction times of birds in all of the above examples surpass what mammals are capable of doing, with the possible exception of bats, which, like birds, have powered flight. Birds could not have such remarkably quick reaction times without the aid of outstanding vision coupled with a brain capable of high-speed processing.

VISION

Birds take in most of their information about their world as we humans do, from sights and sounds. Human beings have outstanding vision, but compared to birds, we are lacking in a number of respects.

Birds' eyes are proportionally larger than mammalian eyes. The bird eye is so large that in the case of owls, the skull and braincase appear to be distorted to accommodate the huge orbits of the eyes. The world's smallest bird species, the Bee Hummingbird, found only in Cuba, has a disproportionally large eye considering its diminutive body size. And the Bee Hummingbird itself is smaller than the eye of an ostrich.

Birds' eyes are not movable within the orbits of the skull. A thin circle of bone, the sclerotic ring,

This Burrowing Owl demonstrates that birds have and use their feathered eyelids to blink.

surrounds avian eyes. But birds have highly flexible necks (chapter 3) and thus they can swivel their heads with ease, allowing their eyes to take in a wide swath around them. You might also have seen owls bob their heads up and down when looking at something. By such movement they mentally triangulate and judge distance to an object. The position of the eyes on the skull also enables birds to have an extraordinary range of vision (described below).

The bird eye has a cornea and lens as well as a variable iris, moved by muscles that regulate the amount of light entering the eye. Bird species have various colored irises, and many birds undergo changes in iris color from juvenile to adult. In some raptors, for example, immature birds have yellow irises and adults have red ones.

Although birds have eyelids and occasionally blink, they mostly rely on a membrane called the nictitating membrane, located below the eyelid, to sweep laterally across the eye, wiping it of debris. This membrane helps protect the eye during foraging movements. Birds of prey use it to protect the

eye during a dive, and some waterbirds use it when diving under water. The nictitating membrane is often translucent but may be opaque. You have likely seen birds with an eye that momentarily appears cloudy. What you saw was the bird's nictitating membrane. While the nictitating membrane is essential to birds, it did not evolve in birds. Nictitating membranes, like sclerotic rings that surround eyes, are present in reptiles, too, and thus the evolution of the nictitating membrane preceded the emergence of birds. Birds' eyes also have lacrimal glands, which produce a tearlike liquid that acts to lubricate the eyes.

Avian eyes have high acuity and high sensitivity. Acuity is the ability to see in sharp detail, even from a distance. Sensitivity is the ability to see with limited light. Diurnal birds in general have high acuity and nocturnal birds have high sensitivity.

Vision happens because sensory cells in the retina detect the image that comes through the cornea and lens of the eye. The image is converted to electrical impulses by photoreceptor cells in the retina,

Ospreys have large eyes (*top*) and the nictitating membrane (*bottom*) is transparent, aiding in capturing fish.

and these impulses are then transferred via the optic nerve to the brain. Ultimately it is really the brain that "sees." There are several kinds of sensory cells in the retina of a bird, just as there are in mammalian retinas. Photoreceptors called cone cells process color vision. Other receptor cells called rods are engaged in low-light vision. When you become dark-acclimated after being outside at night, it is primarily your rod cells that are providing your visual image.

Humans have three kinds of cone cells that differ according to the wavelength of light they respond to. One kind is stimulated by red, another by green, another by blue. When the signals from these cells are coordinated in the brain, color vision is the result. Birds, too, have red, green, and blue photoreceptors, but they also have ultraviolet (UV) receptors. UV light has shorter wavelengths than red, green, or blue light. Seeing in the UV range adds an additional component of color that birds see but humans cannot.

Imagine two Blue Grosbeak males side by side. To us they may look identical, but to other Blue Grosbeaks, particularly females during breeding season, one may shine more brightly than the other because of the intensity with which his feathers reflect UV light. Monochromatic species (in which males and females look alike), such as Cedar Waxwings, may not appear monochromatic when viewed in UV light. If so, the sexes could more easily tell each other apart because of different degrees of UV reflectivity. King Penguins of the sub-Antarctic have vivid orange and yellow colors on their face and breast that apparently reflect in UV light. Perhaps they distinguish each other more easily because of

This male Boat-tailed Grackle demonstrates the use of the nictitating membrane as it sweeps over the eye.

sensitivity to UV wavelengths. Newly developed window glasses embed complex patterning visible only in UV light (see Ornilux, for example). Humans cannot see these patterns (if we did, it would spoil the view out the window), but birds can see them, so they help prevent birds from crashing into a window that they otherwise would not see.

Rod cells are essential for sensitivity in low light. Most birds are diurnal and thus their eyes have more cones than rods. Not so in groups such as owls and nightjars, which are mostly nocturnal. The density of rod cells is high in these birds and they have both strong visual acuity and high sensitivity under low light conditions. Owls cannot see in total darkness because some light must be available to stimulate the rods to work. And, as any birder knows who has observed Snowy Owls and Short-eared Owls, owls do see well in daylight. They have cone receptors, but their color vision is less developed than that of diurnal birds.

The large eyes of nightjars and owls reflect red eyeshine. The eyeshine comes from a structure called the *tapetum lucidum,* a layer covering the back of the eye that acts as a mirror, reflecting light in the retina and at the same time enhancing light by stimulating more sensory cells. Mammals with eyeshine, such as cats and deer, have a similar structure.

Birds have many more light receptors per unit area than humans. Some bird species have ten times the density of photoreceptors as humans. This greatly enhances visual acuity. If you and a Red-tailed Hawk were side by side and someone 20 feet away from you held up a mouse, you would both see the mouse. But if the mouse were to be moved farther away from you, it would become invisible to you far sooner than it would to the Red-tailed Hawk. One study determined that an American Kestrel, a small falcon that hovers over fields to spot prey, is able to see a 2-mm insect (0.08 inches) when it is nearly 60 feet (18 m) away from it. You try seeing a tiny grain of rice when you are about 60 feet away from it.

Some birds have enhanced visual acuity because they have two foveae. A fovea is a pitlike and dense concentration of cone cells in the retina where vision is the sharpest. Humans have a single fovea, and many bird species do too. But terns, raptors, shrikes, hummingbirds, kingfishers, and swallows have two foveae, one of which is called a shallow fovea, the other a deep fovea. The shallow fovea is used in close-up visual acuity. The deep fovea is positioned differently and acts as a sort of telephoto lens, magnifying an image.

Another component unique to the avian eye is the pecten. Attached to the optic nerve, it is a comb-shaped pleated structure dense with blood vessels. Its function is to enhance blood flow to the eye. The avian retina does not have blood vessels throughout, so the pecten has likely assumed the role of bringing oxygen and nutrients to the eye.

Most birds' eyes are positioned laterally on the skull. If your eyes were positioned like those of an American Robin or Rock Pigeon, you would have greatly enhanced peripheral vision and thus a far broader visual field. Your forward field of view would be more limited with regard to binocular vision, but you would still have a field of three-dimensional vision.

Birds with laterally positioned eyes often bob their heads distinctively as they walk. Pigeons especially are known for doing this. The pigeon focuses its eye on its surroundings and holds its head steady as its walking body "catches up." As the body moves, the pigeon constantly repositions its head. This gives it an accurate and detailed view of its surroundings as it is walking. Pigeons placed on a moving treadmill, where the side vision field remains the same, do not bob their heads.

Birds such as waterfowl and shorebirds have eyes positioned relatively high on the skull, aiding in aerial predator detection. Owls and raptors have eyes positioned forward. This provides for an overlapping forward visual field and accurate stereoscopic vision essential in capturing prey from the air. Unique among birds, woodcocks (one species in North America, seven others worldwide) have an immense visual field. Their eyes are large and positioned high on the skull. Woodcocks apparently are able to see for fully 360 degrees on a horizontal plane and 180 degrees on a vertical plane. They see what is behind them, beside them, and in front of them. Imagine what it would be like to be inside a woodcock's brain and see as a woodcock does.

Some birds' eyes are lateralized, meaning one eye focuses on close objects and the other on distant objects. Chickens use their right eye for close vision, left eye for distant vision. Lateralization is adaptive. A feeding bird needs to discern food objects but, at the same time, it must be vigilant for predators (chapter 12). Having an eye that focuses close while the other eye is scanning the sky allows the bird to effectively feed and remain on alert. As is the case with mammals, the optic nerves cross such that what a bird's right eye detects is "seen" by the left side of the bird's brain, and vice versa.

Some birds, such as shorebirds and waterfowl, sleep with one eye open (chapter 9). This suggests that one-half of the bird's brain sleeps while the other remains awake. Diurnal birds typically sleep at night, and nocturnal birds slumber during daylight hours. Almost all bird species are fitful sleepers, awakening often, likely an adaptation to avoid predation.

Finally, birds' eyes are able to detect polarized

American Woodcocks have large eyes positioned to present a unique visual field approaching 360 degrees.

light. When you wear polarizing sunglasses, glare is largely eliminated because you are seeing light vibrating only in a single plane. This sensory ability enables birds to more easily follow the position of the sun, an adaptation for migratory birds (chapter 16).

HEARING

Birds lack pinnae, external ears common to mammals, including bats. The absence of external ears might be considered an adaptation to flight because external ears would add to drag, but reptiles lack pinnae and thus their absence on birds is likely historical. In most bird species, the ear openings, which are behind and below the eye, are covered with feathers, appropriately called auricular feathers. Feathers do not reduce the bird's ability to detect sound waves, and they protect ears in flight and, for waterbirds, when diving.

The ear canal is short, and one bone, the columella, transfers sound-wave vibrations from the middle ear to the inner ear. (Note that humans have three middle ear bones, popularly called the hammer [malleus bone], anvil [incus bone], and stirrup [stapes bone].) The bird's inner ear, the cochlea, is lined with cilia, fine hairlike sensory cells that trans-

fer sound impulses to the brain. As with sight, a bird "hears" with its brain.

The hearing range of birds is similar to that of humans, which is why we hear birds' song and call notes so well. Sound waves are measured by frequency, in units called hertz (Hz). One cycle of sound vibration per second is defined as one hertz. Humans are able to detect sounds within the frequency range of about 20 to 20,000 Hz. Peak human sensitivity to sound occurs at between 2,000 and 4,000 Hz, which coincides with the range of peak sensitivity in most bird species. Human hearing at the high and low ends of the frequency range actually exceeds that of most birds. Bats hear in a range that spans 20,000 to 200,000 Hz, which is why most bat sounds are inaudible to humans and birds. Older birders may lose their sensitivity to high-frequency sounds such as those made by some wood-warblers, waxwings, and gnatcatchers.

Though humans and birds share a similar range of sound sensitivity, birds are better at processing multiple sounds occurring simultaneously. Birds have very discerning ears, another way of saying their brains work at warp speed to process incoming sounds. They identify specific sounds of their species and of various individuals during the dawn chorus of song emitted from dozens of species, all singing loudly and simultaneously (chapter 13). This is equivalent to your hearing a friend speaking some distance from you in a room densely populated by people, all of whom are in animated conversation. This enhanced ability to process sound allows birds to hone in on specific sounds made by potential prey, even in a noisy environment. Barn Owls can capture a mouse in total darkness by hearing it move. Great Gray Owls and Northern Harriers are able to capture voles beneath snow, again by hearing them move. Some woodpeckers are able to hear beetle larvae under bark.

The conspicuous facial disks of many owl species help direct sound waves to their ear openings. Some owls, such as the Great Gray Owl, have asymmetrical ear openings, one opening higher on the skull than the opening on the opposite side. Such positioning aids in triangulating on prey, assuring greater accuracy and increased chance of capture.

Another example of how birds discern sounds better than humans do is found with the call of the Whip-poor-will. Humans typically hear three notes, "Whip-poor-WILL." But if you look at a graphic illustration—a sonogram—you will see that there are actually five notes, not three. The human brain is

Birds' ears are normally covered by feathers, called *auricular feathers*. But some species, such as Wild Turkey, for example, have bare heads with the ear opening visible. Note the ear just behind the eye.

normally not able to discern all of the notes. But Northern Mockingbirds often mimic Whip-poor-will calls, and when the mockingbird's mimicked Whip-poor-will sound is represented as a sonogram, it contains all five notes.

Birds tolerate noise levels that humans find intolerable. Snowy Owls and flocks of Snow Buntings inhabit airports in winter and go about their foraging directly adjacent to active runways, apparently oblivious to the deafening noise created by engines at full throttle as jets take off—the engine noise is equivalent to about 150 decibels, enough to rupture a human eardrum.

Avian researchers are studying how some bird species are altering the volume and frequencies of their songs in noisy environments, such as urban and suburban settings. Studies performed in Europe have shown that Great Tits, European Robins, and Eurasian Blackbirds (a thrush similar to an American Robin) sing at higher frequencies when in urban environments where there is typically loud background sound. Song Sparrows have shown similar patterns in North America.

OLFACTION AND TASTE

Many past students of ornithology have been taught that birds have a poorly developed sense of smell and that their sense of taste is equally deficient. After all, American Crows, as intelligent as they appear to be, have been observed to eat almost anything, including human vomit. Do wood-warblers actually taste those caterpillars they devour during breeding season? Do nestling herons and pelicans savor the regurgitated fish brought by their parents? Perhaps they do. Taste normally acts to ensure food consumption. Perhaps we have underestimated birds. Maybe species such as Tennessee Warblers are stimulated to keep consuming caterpillars because they actually taste good to them. But we do not know this with any certainty.

Some bird species have keenly adapted olfactory abilities as well as highly attuned organs of taste. Birds have olfactory lobes in their forebrains, and in some groups the avian sense of smell is finely adapted to specific functions. Why do shearwaters and storm-petrels seem to come out of nowhere, attracted to boats that toss out fish oil, popularly called chum? And how do they locate their mates, young, and nest burrows among jumbles of rocks when returning at night, in darkness? How do Turkey Vultures manage to find carrion lying on a shaded forest floor when they are flying high above the dense canopy of trees? These are well-known examples of how birds utilize their olfactory abilities.

Tubenoses, the order Procellariiformes, which includes the albatrosses, shearwaters, storm-petrels,

The large facial disk of the Great Gray Owl is an aid in focusing sound, giving the bird the ability to hear prey even when the prey is covered by snow. *(John Grant)*

and fulmars, may have the most highly developed olfactory senses of any order of birds. They do smell their world. They locate swarms of krill (shrimplike crustaceans) in open oceans by detecting the odor of dimethyl sulfide (DMS), a chemical byproduct of the krill. To these oceanic wanderers, the ocean is anything but featureless; it is a patchwork of differing densities of prey readily detectable to birds as they cruise above the waves. They don't *see* differences in the ocean, they *smell* them. Unfortunately, masses of discarded plastic items that now pollute the oceans also emit DMS as they slowly decompose, with the disastrous result of attracting seabirds that might consume items that are impossible to digest and will cause them harm.

Shearwaters and storm-petrels have a musky body odor, used for recognition. I once rescued a beleaguered Leach's Storm-petrel from a boat deck after a severe storm had drenched it to the point that it was too wet and weak to fly. I kept the bird with me overnight in a canvas travel bag. I zipped the bag shut for the night, and by morning the storm-petrel had recovered its strength and alertness. I carried it to the deck and released it. The bird flew in several expanding circles until it got its bearings and then headed off, appearing none the worse for its experience. My hands smelled strongly of storm-petrel musk. My canvas travel bag so reeked of storm-petrel musk that I could never use it again.

Turkey Vultures detect a characteristic odor of decomposition, smelling ethyl mercaptan, the compound that produces the noxious odor emitted at a certain phase of the decomposition process. Turkey Vultures inhabit forested regions where carrion detection purely by vision would be difficult. Black Vultures, which are grassland and savanna species, have not developed the ability to locate prey by olfaction. They often appear to watch Turkey Vul-

tures and follow them to carrion. Two Central and South American vulture species, the Greater Yellow-headed Vulture and King Vulture, both forest dwellers, also are suspected to use olfaction to locate carrion. Old World vultures, all of which are typically found in open woodland and savanna where carrion is easy to locate by sight, have not evidenced use of olfaction in locating food.

Olfaction requires that molecules producing the scent be detected by sense organs after passing through the nostrils. Nerves from sensory olfactory cells transfer the impulses to the olfactory region of the forebrain. Birds have convoluted chambers of thin bone in their beaks, called conchae, containing membranes dense with olfactory sensory cells. Tubenoses, highly reliant on odor, have the largest olfactory bulbs in their brains of any group of birds.

A few kinds of birds lack external nostrils. These groups include the family Sulidae (gannets and boobies) and families Phalacrocoracidae (cormorants) and Anhingidae (Anhingas and darters). Anhingas have small nostrils when hatched, but the nostrils quickly become permanently sealed by the time the birds forage for themselves. Brown Pelicans also lack nostrils on their beak, though they do have tiny nostrils beneath the rhamphotheca, or bill covering. These species all dive forcefully in pursuit of fish, the Sulids and Brown Pelican diving from the air, the cormorants and Anhingas from the surface. As they swiftly pursue prey, the absence of

The Northern Fulmar is an example of a *tubenose*, with well-developed olfactory senses.

Turkey Vultures (*top*) have been shown to have a keen olfactory sense in detecting odors associated with freshly decomposing flesh. Black Vultures (*bottom*) must find prey by sight, but they sometimes follow Turkey Vultures to carrion.

Look carefully at the bill of this Double-crested Cormorant. It lacks nostrils.

external nostrils is thought to protect them from excess water pressure. Consequently, these birds breathe through their mouths and none show evidence of using olfaction in food detection or any other activity.

Some ornithologists have suggested that birds use olfaction as a possible source of information when orienting to specific locations. Experiments have suggested that Rock Pigeons may use odor when orienting to their home roost, and it has been claimed that they create a mental "olfactory map" of their surroundings and use it in homing behavior. The degree to which birds use olfaction in geographic orientation remains controversial.

Sense of taste has historically been considered weak in birds, but that is not necessarily so. Birds have four basic taste buds: bitter, salty, sweet, sour. But birds do not have bountiful taste buds, most having far fewer than mammals. Looking at birds in the field you might make some guesses about how taste would affect birds, how it could be adaptive to have a sense of taste. Grassland birds that rely heavily on small seeds might detect little in the way of taste because they swallow seeds whole. But they still must discern the seeds as edible and taste might matter in that regard. What of waxwings, tanagers, and other fruit-consuming birds? These birds crush

their food and could presumably detect specific tastes. Parrots, which crush fruit before swallowing it, have been shown to have a well-developed taste sense, the taste buds concentrated in the fleshy tongues of the birds. But parrots also crush seeds, some of which may be toxic, so taste may be an important signal in determining edibility of certain seeds. Hummingbirds have been shown clearly to be able to detect the sweet taste of nectar and discern among various nectar concentrations.

In a classic experiment created to test whether birds avoid eating butterflies such as the brightly colored monarch, Blue Jays raised to adulthood without prior experience with monarchs readily devoured the butterfly, only to soon vomit and appear quite miserable for the experience. This was because the butterflies contain the chemical cardiac glycoside, acquired as caterpillars from feeding on milkweed. Cardiac glycosides are powerful drugs and have a distinct taste. Once the Blue Jays learned the unpleasant results of eating a monarch, they avoided it. Brightly colored orange monarchs advertise to their potential predators, birds, that they are best avoided. The birds appear to quickly learn that truth.

The Blue Jay/monarch experiment brings up another point. I mentioned that American Crows have been observed to eat almost anything. Vultures eat carrion. Woodpeckers eat raw suet. None of these delicacies would find its way to a human dinner table, but among humans, many of us have high regard for foods that others of our species studiously choose to avoid, all owing to our different preferences of taste and texture. In a broad sense, birds have adapted to eat food items characteristic of their species. In some cases, such as hummingbirds, taste looms large because it informs the bird of the richness of the food source. In other cases, such as herons, egrets, Anhingas, and cormorants, which ingest live fish whole, taste scarcely matters. It is hard to know if Northern Cardinals or House Finches visiting bird feeders taste differences in the kinds of sunflower and safflower seeds they macerate before swallowing.

Finally, consider the Curve-billed Thrasher, a bird shown to have a real taste for very hot chili peppers—except that they don't, at least not literally. Just the opposite: the thrashers ingest the hot chilies but lack taste receptors for the chemical compound capsaicin that produces the heat sensation. Thus, the birds never know the chilies are hot because to them, they aren't. Mammals taste the capsaicin and avoid hot chili peppers because they are put off by the heat. This distinction is adaptive for the chili and also for the thrasher. Birds are good seed dispersers. They ingest the chilies, the seeds pass unharmed through the bird's digestive system, and out they come in the bird's droppings, distant enough from the parent plant that the seedlings can sprout and grow with less competition for soil nutrients, water, and sunlight. Mammals would likely chew the seeds, destroying them. Birds are far better potential seed dispersers, so chilies and thrashers enjoy a mutualistic interaction in which both benefit from their adaptive relationship.

TOUCH

Birds are sensitive to touch. Birds spend considerable time preening their feathers, coating them with protective oil (see chapter 9). They could hardly do such an action without feeling the feathers as they touch them with their beaks and heads. Beyond that, birds have a type of feather called filoplumes—small, hairlike feathers with a bare rachis, with no vanes except for a tuft at the feather tip, and not all

This male Rose-breasted Grosbeak might be enjoying the taste of the mulberries he is consuming, but the degree to which taste affects food choice is still not clearly known.

The Curve-billed Thrasher is an important disperser of chili seeds. It does not sense the capsaicin that makes the peppers taste hot.

This female Mallard with her beak open reveals the sensitive membranous lining of the lower mandible, allowing the duck to feel and filter her food as she forages.

have that. Filoplumes have sensory touch receptors (chapter 8). The crests of cormorants in breeding season are composed of large filoplumes.

Some birds, including parrots, herons and egrets, albatrosses, and others, engage in allopreening, wherein one bird preens the feathers of another. Just as birds feel their own feathers as they are preening them, they also feel another bird's feathers as they allopreen, as does the bird being preened.

A beak appears to be inert, a solid rhamphotheca covering the bony mandibles. Beaks are often compared to tools: some are like fine forceps, some like heavy pliers, some adapted to tear flesh, some to crush seeds, and some to drill into bark. But beaks do more than that. In many kinds of birds, beaks are internally grooved and dense with sensory cells, an aid in processing food items. Birds must be able to discern food items from nonfood items, and many birds feed in such ways that vision is not helpful in selecting food items. Shorebirds probing in sand and mud cannot possibly see the animals that will become their prey. Ducks diving in a murky pond or bay, probing in the sediments for various animal food items as well as plant material, must rely on a sense of touch to be able to separate a worm or a clam from a pebble or coarse sand.

The major group of touch sensory cells are called Herbst corpuscles, clusters of complex pressure-sensitive cells associated with nerves that signal the brain. Herbst corpuscles are found in the skin, beak, tongue, legs, and feet of birds. When a Black Skimmer skims (chapter 12) its lower mandible in the water, it does not see the fish it will likely snap up; it *feels* it because of the action of the Herbst corpuscles in its mandibles. As you watch shorebirds probing a mudflat, the Herbst corpuscles in their beaks enable them to detect small pressure waves created by the presence of solid items, such as worms, clams, and mussels, informing the birds of specifically where these prey items are buried. This was originally demonstrated for Red Knots, but likely most probing shorebirds are able to detect tiny pressure waves. Many shorebirds time their feeding with the tides, feeding at night as well as during the day. Being able to capture food by touch makes that possible. Whether it's a Roseate Spoonbill sweeping its broad spoonlike bill beneath the water surface or a Common Eider or Surf Scoter diving for mussels, these species are dependent on their arrays of Herbst corpuscles for the sensitivity that allows them to capture food.

Ibises and sandpipers, as well as the kiwis (which occur only in New Zealand), have evolved yet another kind of touch-sensitive cells, called Grandry corpuscles. These are located at the bill tip and are used to detect prey when the birds probe in mud or sediment. Fast reflexes are essential in species such

White Ibises (*top, center*) and Wood Storks (*bottom*) are among the bird species that rely strongly on touch when capturing prey.

as Wood Stork, in which the bird typically forages with its beak partially open, snapping it shut instantly when it detects a fish. Species such as American Woodcock, Wilson's Snipe, and Red Knot exhibit rhynchokinesis (chapter 3), the ability to turn the tip of their upper mandible upward as they probe, an aid in grabbing food once it is detected at the bill tip.

Birds also have cells sensitive to temperature (thermoreceptors) and cells that detect pain (nociceptors). Both thermoreceptors and nociceptors are found throughout the bird's skin and in the beak.

In addition to the senses described above, birds also have remarkable abilities with regard to geographic orientation. This subject is discussed in chapter 16.

6

UNDERSTANDING BIRD DIVERSITY

BIRDS AS SPECIES

To many birders, building a life list is the most important aspect of birding. Listing is fun, competitive, and often conveys some measure of status in the birding community. Whether it be a Christmas Bird Count, a Big Day, a year list, or a global life list, many birders take listing seriously. Why?

A large part of the fascination with birds lies in their collective diversity, and *collective* is the right word. It is fun to watch the list grow, to *collect* new species. For a neophyte birder, thumbing through a bird guide to North American birds can seem both overwhelming and wondrous. There are so many different species of birds in North America, to say nothing of globally. In the East, you see Indigo Buntings. In the West, you find Lazuli Buntings. Varied Buntings inhabit dry southwestern deserts, and Painted Buntings are found in the coastal Southeast and in Texas. Different regions, different buntings. But what really are they, all these names next to pictures in the bird guide? What do they represent? They are species. The number of species is the measure of any list. And there is a good reason for that.

The species is the fundamental unit in natural history. Some argue that genes, which are specific locations on the DNA double helix, are the real fundamental units of biology as they contain the genetic code. But one does not see genes in nature— rather, one sees the result of the interactive workings of genes, namely individual organisms, all organized into various species. Ornithologists continue to struggle to formulate a good working definition of *species*. But field guides to birds really do work because species really do exist. No one doubts this.

There are just over 1,000 species of birds documented to occur (some as extreme rarities) in North

The Lazuli Bunting (male shown here), widely distributed across western North America, is but one of four species of buntings in the genus *Passerina*. (Ed Harper)

America. Learning them presents a challenge at multiple levels. Seeing them all takes appreciable time and effort. Should you wish to be a global birder you have the thrill of potentially encountering over 10,000 species of birds, from the penguins of Antarctica to the trogons of the tropics, to the many species of little brown look-alike cisticolas that challenge birders from Africa to Australia. No one to date has made the claim of actually seeing all of the world's bird species in the wild. That bar

OPPOSITE: The Atlantic Puffin is but one of more than 10,000 bird species that currently inhabit earth. Two other closely related species of puffins, the Horned Puffin and the Tufted Puffin, occur only in the northern Pacific coastal region.

This spectacular male Elegant Trogon (*top*), found in the American Southwest and a prize for any birder, is but one of 43 species of trogons that await the global birder. A Steller's Jay (*bottom*) recognizes others of its species and will normally not mate with other jay species or any other bird species. This is the essence of the species concept. *(Top: Dan Tankersley; bottom: Bruce Hallett)*

is really high, requiring extraordinary effort, time, money, and luck.

The current definition of species, as applied to birds, is based on sexual reproduction. A species is a population or group of populations of birds that recognize each other, interbreed with each other, and do not breed with other species (though sometimes they do and then it becomes complicated, as I will discuss shortly). A Steller's Jay may forage in a ponderosa pine in proximity to a Western Tanager, but the two birds have no mating attraction toward one another. Steller's Jays mate only with other Steller's Jays. Nor will a Steller's Jay mate with a Canada Jay, though these species may cohabit the same area and encounter one another. Jay species are both members of the family Corvidae, but such a grouping is a human construct and is meaningless to the birds. The concept of species is thus not abstract but rather a natural unit that the birds themselves define by with whom they choose to mate. Mate recognition isolates each species from all other species. The jays named above may occur together, but each is reproductively isolated from the other and they apparently know it. This is why the definition of a species is called the *biological species concept* (BSC). Reproductive behavior of birds results in reproductive isolation from other species, including similar and/or closely related species. This is the linchpin of the BSC, the species definition currently favored by most ornithologists.

One clear example of how the BSC works is seen among the complex of tyrant flycatcher species in the genus *Empidonax*. Birders are challenged to identify the various species of Empids, which appear extremely similar to one another. That is the case with a supposed species once called the Traill's Flycatcher, originally named by John James Audubon to honor Dr. Thomas Stewart Traill of Edinburgh. Birders sometimes still use this name but only when they cannot really tell what species the flycatcher they're observing actually is. That is because the Traill's Flycatcher is, in reality, two species, the Alder Flycatcher and the Willow Flycatcher. These species are so similar in appearance that from the time of Audubon until the latter part of the twentieth century, they were not recognized to be separate species. Both species live near freshwater habitat, but the Willow Flycatcher prefers willow thickets while the Alder prefers bogs and wet meadows of birch and alder. Most importantly, the Willow Flycatcher has a kind of wheezy song often described as *fitz-bew*. The Alder Flycatcher sounds similar at a distance, but its simple wheezy song is different, described as *vfee-bee-o*. The short, buzzy, nonmelodic songs are the key to species recognition. Neither species is attracted to the song of the other. The songs and, to a lesser degree, range and habitat differences serve to reproductively isolate populations of one species from the other.

The biological species concept is not 100 percent inviolate, however. Hybrids between species do occur, as in groups such as waterfowl, where various species often share similar bodies of water at precisely the time when they experience the urge to mate. Hybridization is also known in other groups, such as wood-warblers and gulls. Birds are sometimes apt to be ambiguous in going about their reproductive behavior, sending ornithologists mixed signals. There are examples of closely related species that meet at part of their respective ranges and form a zone of hybridization. This is the case, for example, with Black-capped and Carolina Chickadees over part of their range, as it is with Hermit and Townsend's Warblers and with Baltimore and Bullock's Orioles, to take three well-documented examples. Several pairs of species frequently hybridize where their ranges meet in the Great Plains. These include the Rose-breasted and Black-headed Grosbeaks, the Indigo and Lazuli Buntings, Eastern and Spotted Towhees, and the aforementioned Baltimore and Bullock's Orioles.

In the case of all of the pairings named above, the zone of hybridization is relatively narrow and not spreading. This indicates that the hybrids are not as reproductively fit as the parent species and have less of a chance of successfully reproducing. For example, when Bullock's and Baltimore Orioles interbreed, where do the hybrids migrate? The two species have different wintering ranges. Migrants must inherit a combination of migratory directions from both parents and, as a result, may not migrate with great success. Any trait that makes hybrids less apt to reproduce indicates that its parent species are, indeed, separate species.

Assigning species status to birds is constantly a work in progress, and the devil is in the details. To accurately assign species status between similar populations requires documenting reproductive behavior, as was the case with Alder and Willow Flycatchers. Islands or other geographically isolating factors present serious challenges. If two similar species are separated by one being on an island and the other on a mainland, or if mainland populations are physically separated, how do we know whether they would successfully interbreed if brought together?

Geese (*top left*) often hybridize, wild forms such as Canada Goose hybridizing with domestic varieties, as shown here. Buffleheads occasionally hybridize with Common Goldeneyes, as is the case with the duck in the upper right of the photo (*top right*). The Townsend's Warbler (*bottom left*) and Hermit Warbler (*bottom right*) are separate species but are known to hybridize where their breeding areas overlap. (*Top left: Garry Kessler; top right: George Martin; bottom left and right: Ed Harper*)

The widespread California Scrub-Jay (*top*) is recognized to be a different species from the Island Scrub-Jay (*bottom*), whose population is entirely restricted to Santa Cruz Island, California. (*Top: John Kricher; bottom: Dan Tankersley*)

An example is seen when comparing the California Scrub-Jay, Woodhouse's Scrub-Jay, Florida Scrub-Jay, and Island Scrub-Jay (of Santa Cruz Island, California). Each is now considered to be a separate species but not based entirely on whether or not they interbreed. Some, such as the Florida Scrub-Jay and California Scrub-Jay, cannot interbreed because of a lack of geographical overlap. The Island Scrub-Jay is, indeed, isolated on an island. The decision to award species status to each was based on comparing morphology, song, behavior, and genetics (similarity of DNA). The preponderance of that evidence suggested that each should be considered a valid species.

Birders are generally aware of the terms *lumping* and *splitting*. An example of a split recently occurred with the Winter Wren. In North America, it was split into two species now known as the Winter Wren and the Pacific Wren. An additional split was also made. The Winter Wren was once considered to be the same species as a bird called "the Wren" in Great Britain and Europe. That species has been split from the North American species, so where there was once one species with the scientific name of *Troglodytes troglodytes,* there are now three: the Eurasian Wren, *Troglodytes troglodytes;* the Winter Wren, *Troglodytes hiemalis;* and the Pacific Wren, *Troglodytes pacificus.* These wrens look very much alike, but there are differences in range, habitat, and song. They do not interbreed.

Lumping is the opposite of splitting. In past decades birders have worked hard to find and identify Thayer's Gull for their life lists. That has now changed. Thayer's Gull was lumped with Iceland Gull in 2017 because it successfully interbreeds with various populations of Iceland Gulls with no loss of fitness (reproductive potential) in the offspring. So bye-bye Thayer's gull.

In the case of the wrens, what was considered to be one species became three species. In the case of the Thayer's and Iceland Gulls, what were considered to be two species became one. Both cases represent evolution by decree, so to speak. Each year a committee of the American Ornithological Society (AOS) makes decisions based on various lines of evidence presented in proposals about whether to lump or split various species. The AOS maintains an updated and official checklist of all North American bird species, available at the AOS website.

Sometimes changes in the environment result in habitat alteration such that species come into contact and may then hybridize. This has been the case with eastern populations of the Blue-winged and Golden-winged Warblers. The regrowth of young forest has helped Blue-winged Warblers and negatively affected Golden-winged Warblers, which tend to prefer less woody sites. As the two species have come into contact, they have hybridized. Birders know to search for two hybrid examples, the "Brewster's Warbler" and the more uncommon "Lawrence's Warbler." Because of loss of favored habitat, namely early successional fields, plus hybridizing with Blue-wings, Golden-winged Warblers have become rare or extirpated over most of their northeastern range. The two species are genetically similar (as genetically close as two humans selected at random) and the pronounced difference in plumage between them is due to the effects of only a few genes.

As I said, the devil is in the details. Blue-winged and Golden-winged Warblers look distinct, have different songs, occupy somewhat different habitats, and have some differences in geographic range, though they overlap in many areas. They often respond to one another's song. In spite of their plumage differences, they hybridize. The DNA in the nucleus of their cells is extremely similar. But there is also DNA inside the cytoplasm of the cell, called mitochondrial DNA (mt-DNA), which is characteristic of all organisms other than bacteria. Ornithologists have made extensive use of mt-DNA analyses in comparing similar species and, in the case of the Blue-winged and Golden-winged Warblers, the mt-DNA shows areas of difference sufficient to support them being designated as separate species.

Where there was one, now there are three: the Winter Wren (*left*) is now split from the Pacific Wren (*right*). Both are also split from the Eurasian Wren. (*Left: Bruce Hallett; right: Ed Harper*)

The Blue-winged Warbler (*top*) and Golden-winged Warbler (*center*) are considered to be separate species even though they frequently hybridize, resulting in hybrids such as "Brewster's Warbler" (*bottom*). (Top: Dan Tankersley; center: John Kricher; bottom: Garry Kessler)

Normally it is easy to assign species status. No one doubts that a Scarlet Tanager is a species separate from a Summer Tanager, or that a White-throated Sparrow is a species separate from a White-crowned Sparrow. It should also be apparent that when applying the biological species concept to birds, behavior is extremely important since the essence of the BSC is how birds within a population respond to one another with regard to pair bonding and mating. But what does muddy the waters a bit is the reality of subspecies.

BIRDS AS SUBSPECIES

The scientific name of the species is *Junco hyemalis*. Its common name is Dark-eyed Junco, and it occurs throughout North America. In the East it is known as "Slate-colored Junco." Eastern birders know to search for "Oregon Juncos" occasionally appearing among flocks of "Slate-coloreds," just as western birders seek out "Slate-colored Juncos" within flocks of "Oregon Juncos." Then, depending upon location in North America, there are "White-winged Juncos," "Gray-headed Juncos," or "Pink-sided Juncos." These groups are recognizable subspecies of Dark-eyed Junco. Why are these groups, easily separated in the field, not considered full species? According to the biological species concept, they do not meet the criteria for full species because they successfully interbreed where their ranges intersect, with no loss of fitness among the hybrids. Thus, they are designated subspecies, locally adapted populations with distinct appearances but not genetically isolated from one another.

The same is true for Fox Sparrow, easily separated into four subspecies: Red, Slate-colored, Sooty, and Thick-billed. These subspecies differ in range, and the differences reflect local adaptations to habitat. For example, the Sooty subspecies inhabits dark, moist coastal forests in the Pacific Northwest. Its plumage is dark ("sooty") because of concentrated melanin pigment in the feathers. Wet, dark forests present favorable growing conditions for feather bacteria (chapter 9) that could negatively affect feather durability. Enhanced melanin makes feathers more resistant to the degrading effects of these bacteria as well as to wear and tear. This example illustrates the importance of local adaptation, a hallmark of subspecies. In the case of the juncos, differences among subspecies do not appear to be correlated with obvious environmental characteristics, whereas with the Fox Sparrows they do. But what matters is that the subspecies are interfertile.

The subspecies category identifies distinctions among geographically separated populations, some of which reflect local adaptations to habitat. Desert Song Sparrows are pale, for example, while Song Sparrows in the Pacific Northwest are, like Sooty

The Dark-eyed Junco is divided into recognizable subspecies, including the "Oregon" (*top*) and the "Gray-headed" Juncos (*bottom*). (*Top: John Kricher; bottom: Bruce Hallett*)

Like the Dark-eyed Junco, the Fox Sparrow is divided into recognizable subspecies groups. The bird in the *top* photo is from the "Red" group, while the one in the *bottom* photo is from the "Sooty" group. (Top: John Kricher; bottom: Bruce Hallett)

Fox Sparrows, very dark in plumage. But some distinctions are far from obvious and cannot be noted in the field. The Slate-colored Junco is actually part of a three-subspecies group, the subspecies not identifiable in the field; the Oregon Junco is part of a five-subspecies group. There are two subspecies of Gray-headed Junco. So, in the case of the juncos, there are five subspecies groups that, in total, contain 12 subspecies, most of which are not separable in the field. In the case of the Fox Sparrows, there are four subspecies groups with a total of 17 subspecies. The Song Sparrow has long been the champion bird for subspecies diversification. It is divided into six groups that are recognizable in the field but that contain 29 subspecies. Depending on which taxonomic classification is used, Horned Larks have been considered to have between 20 and 40 subspecies. Subspecies are designated by museum examination of specimens. It should not be lost on the reader that much of what subspecies represent are human judgment calls.

When a species is divided into several subspecies, it is termed *polytypic,* meaning "many types." By no means is every species polytypic. There are many monotypic species that have not been separated into subspecies populations. Examples include Northern Mockingbird, American Black Duck, Bald Eagle, and Ring-billed Gull. Subspecies have always presented a conundrum to ornithologists. The "Audubon's Warbler" of the West was lumped with the "Myrtle Warbler" of the East into the Yellow-rumped Warbler; and the "Red-shafted Flicker" of the West was lumped with the "Yellow-shafted Flicker" of the East into the Northern Flicker. These examples represent instances of successful hybridization where ranges meet. So, according to the BSC, they are considered to be single species—for now. Ornithologists are open to re-evaluating whether or not to re-split, for example, Yellow-rumped Warbler.

THE SLIPPERY SLOPE OF SUBSPECIES

Birders have long challenged ornithologists to justify why populations that appear distinct from one another should not merit full species designation. The answer has always been that under the biological species concept they must be lumped because they freely interbreed in some areas with no loss of fitness in the hybrids. But not all ornithologists have a high comfort level with the BSC. Some have advocated other species definitions that would, if adopted, elevate many subspecies to full species.

Some ornithologists prefer a combined morphological and genetic approach to species designation. Morphology is the body form of a given animal, its appearance. Genetics refers to the animal's "bar code," its DNA. Some ornithologists have advocated adopting what is termed the *phylogenetic species concept* (PSC) that would define species based on the most recently evolved genetic and morphological differences. For example, it is likely that the plumage distinctions between the Oregon Junco and Slate-colored Junco (as well as the other major subspecies groups of juncos) evolved as the species spread in North America. The various subspecies are each more recent in origin than the emergence of the species itself. By this line of reasoning, what was once a single species has branched into multiple and more recent phylogenetic species.

Genetic and song analyses of various populations have helped reveal what are called cryptic species, which are species not initially obvious by appearance to ornithologists but which are, in fact, reproductively isolated. We have already discussed two examples, the Winter Wren group and the Traill's Flycatcher pair (Alder and Willow Flycatchers). There are numerous other examples that may result in recognition of more cryptic species. Evidence from morphologic and genetic analysis suggests that Common Ravens in parts of California are distinct from ravens in the rest of North America, making them potentially a cryptic species.

Cryptic species are evident in the Red Crossbills. The species shows variation in bill dimensions and flight calls throughout its broad range in North America. At least 10 different call types of Red Crossbills may represent 10 distinct but cryptic species. Birders now record call types and identify which of the crossbill types they are encountering. In 2017, the American Ornithological Society's North American Classification Committee split the Cassia Crossbill (*Loxia sinesciurus*) of Idaho from the Red Crossbill (*Loxia curvirostra*) but did not opt for other splits within the Red Crossbill complex. The split was largely based on the large bill size and facial muscles of the Cassia Crossbill. The birds feed exclusively on seeds extracted from the cones of lodgepole pines (*Pinus contorta*). In many areas in the West crossbills occur with red squirrels (*Tamiasciurus hudsonicus*). But in one particular region of Idaho, red squirrels are absent and the cones have evolved into a larger size, presumably in response to seed predation by crossbills. That, in turn, served as an evolutionary stimulus, what is termed a *selection pressure,* for the crossbill population to evolve deeper, stronger bills for seed extraction. The Cassia Crossbill shares its range with two other Red Crossbill types and, as you might expect, the Cassia Crossbill has calls and a song different from the other crossbill types. Note that the Cassia Crossbill is not all that cryptic since its large bill dimensions are recognizable in the field.

The Evening Grosbeak, declining in some parts of its range, may be a cryptic species. Like the Red Crossbill, its different call types are associated with different geographic populations. But unlike the case with the Red Crossbill complex, the call types do not appear to be associated with different food preferences, and the various call types mix when winter flocks form. Examples such as the Red Crossbill and Evening Grosbeak complexes are examples of evolution in action.

Molecular analysis of genetic differences among bird populations increasingly suggests that cryptic species exist within well-recognized species. If that is the case, there is the potential to greatly increase the number of recognized species. One group of researchers, led by George F. Barrowclough (see references), has calculated that as many as 18,043 species of birds occur in the world, a huge increase from the current number of approximately 10,000. Extrapolated to North America, it suggests that as

The Cassia Crossbill is a newly recognized species within the Red Crossbill complex. *(Dan Tankersley)*

many as 1,800 species might occur. But that all depends on what we call a *species*. Molecular techniques coupled with morphological comparisons continue to provide data, and ornithologists have made major changes in the arrangement of bird orders and families. Groups such as pigeons and doves, cuckoos, nightjars, swifts, and hummingbirds now precede most waterbird groups in bird lists, and caracaras and falcons are now placed between woodpeckers and parrots.

BIRDS AS POPULATIONS

A biological population is defined as consisting of a single species of organism, such as a population of, for example, House Finches. Anyone who watches House Finches first notices that they are streaky brown and that some, the males, have red on them. Next comes the realization that some of the males are richer in reddish color than others. If the birds are feeding, some birds may aggressively displace others, usurping their perch or feeding area. When watching individual birds closely, some may appear more robust and active than others. Birds in populations react to one another such that dominance relationships and social networks develop within many populations (chapter 11).

Salt marshes along the East Coast have distinct populations of Saltmarsh Sparrows, each popula-

tion isolated to a degree from others because salt marsh distribution along the East Coast is patchy. Painted Bunting populations have a geographic gap between their southeastern range in the coastal Carolinas, Georgia, and Florida and their western breeding range in Texas, Oklahoma, Louisiana, Arkansas, and other midwestern states. The distribution of populations within a species is strongly affected by migratory habits. Some species, such as Sandhill Cranes, Eastern Towhees, and Dark-eyed Juncos, contain both migratory and nonmigratory populations. Individual Ospreys that are in the same breeding population may be widely separated on their wintering ranges, in essence each bird becoming part of a different population during its nonbreeding season.

Both Charles Darwin and Alfred Russel Wallace, the codiscoverers of the theory of natural selection, realized that the local population is the actual unit of evolutionary change among species. Individuals do not evolve. They age, their appearance and vigor may change, but they contain the same genetic material, the same DNA, throughout their lives. Evolution occurs through relative changes in DNA distribution within populations, and such changes occur because some individuals within any population are better at survival and reproduction than others. These individuals will, on average, leave more offspring, carrying a greater share of their genes into the next generation, the process that Darwin and Wallace termed natural selection. Over time a population evolves genetically, perhaps eventually diverging to become a new species.

Red Crossbills, as noted earlier, offer a good example of natural selection. In the Red Crossbill complex it is well established that the bill dimensions of the various types of Red Crossbill, including the newly recognized species, Cassia Crossbill, evolved in response to selection pressures exerted by the various species of coniferous trees upon which the birds extract seeds from cones. Also in play was competition from other seed predators, such as red squirrels.

All bird populations have a sex ratio, the proportion of males relative to females. One would expect this ratio to be pretty close to 50:50. But ornithologists have long recognized that males tend to outnumber females in many (but not all) bird species populations. This is the case in spite of the fact that the hatch rate of males and females tends to be the expected 50:50. Thus, males appear to have longer average survival than females. There is no simple answer as to why, but in many species females devote disproportionally more energy into nesting

and caring for young and are at greater risk while doing so. Females use more energy in the combination of egg production, nest building, brooding, provisioning, and fledging young than is the case with male birds. Females may therefore "age" faster.

Bird species in the temperate zone have a compressed summer breeding season (chapter 15) such that the age structure of the population is strongly skewed to contain numerous juveniles immediately after the nesting season, but the mortality rate of juveniles is high. By the time the next nesting season commences, many of the birds hatched the previous summer have perished. Fewer migrants typically move from the nonbreeding grounds to the breeding grounds than is the case in the other direction (chapter 16).

The highest rate of mortality occurs while birds are either still within the egg, in the nest (or are downy chicks such as turkey poults), or newly fledged. Once a bird is able to be on its own it still has to learn how to survive, by gaining experience. Among adult passerines, an individual is at about the same risk of perishing on any given day throughout its life. Birds typically live far shorter lives in nature than they are physiologically capable of living (chapter 7). Many bird species in zoos or confined to

The Evening Grosbeak has been divided into three subspecies and five call types, but there is not yet sufficient evidence that these distinctions merit recognition of multiple species.

a life of captivity because of injury live far longer lives than is normally the case for free-living members of their species.

HABITAT SELECTION

There is no habitat anywhere in North America or, for that matter, in the vast majority of the world that is bird-free (hurray!). Even in the depth of winter in

Chipping Sparrows (*top*) are common birds in suburban habitats but have a strong propensity to require pine trees for nesting. American Bitterns (*bottom*) are anatomically and behaviorally dependent on being in marshes.

the far North there is some bird presence, often Common Ravens (Christmas Bird Counts throughout the years have shown this clearly). Only by wandering deeply in the central part of the continent of Antarctica would you find a place where no bird ventures. Nonetheless, the fringe of the Antarctic continent abounds with penguins and seabirds. All deserts host numerous bird species, each adapted to live in the harsh desert conditions. Cities have Rock Pigeons, European Starlings, House Sparrows, and much more when you really begin to look. Forests have the highest numbers of bird species, especially tropical forests. Birds have adapted to an immense diversity of global habitat types and are thus essential components of all of Earth's ecosystems.

How do birds select habitats? Birds are innately shaped by their evolutionary histories to recognize ecological features that tell them "This is where I belong." Birds imprint on the habitat in which they have fledged. Habitat is a complex variable encompassing multiple dimensions. One of the first lessons birders are taught is that to find a variety of bird species, they must visit a variety of habitats. That seems obvious. What the bird sees in a habitat, the key elements by which it recognizes its habitat, is known only to the bird—and they *do* know. Chipping Sparrows have been shown to recognize pine trees as essential components of habitat. For a migrating American Bittern the habitat bar is pretty high. They require a particular kind of marsh or wet meadow, structurally complex with sufficient tall grasses, cattails, or reeds to maintain *crypsis,* the ability to conceal themselves from predators by blending into the environment—essential for any bittern. During migration and during winter, bitterns are able to use both brackish and freshwater marshes (they nest in freshwater marshes). But at any time of year they don't take well to open fields, woodlands, or golf courses. Imagine a bittern migrating through the night and, upon dawn, having to locate just the right habitat, a marsh, before landing. But they do—at least, most do.

Sometimes birds provide a hint of how effectively they recognize habitat, and it often seems logical to birders. For example, when the water level in a reservoir or other body of fresh water occasionally drops to the point that mud is exposed, shorebirds of various species seem to magically appear, foraging on the mud. Not magic at all. These birds, as they overfly the area, key in on the sight of exposed mud near water and know that food is likely available. I once came upon an Ovenbird during its spring migratory period that was, of all places, at Logan International Airport in Boston. The bird was outside the departure terminal walking around, as Ovenbirds do, in a tiny patch of planted trees (meant to be decorative) that bordered the terminal

The Kirtland's Warbler has rigid habitat requirements and is thus a specialized species. It requires jack pine forests of a specific age for nesting.

building. The bird was in an alien environment (at a huge airport) but it managed to locate a tiny patch of habitat where it more or less could behave normally and attempt to forage. Some years ago, there was a Boreal Owl that spent a few days in downtown Boston, observed by many. It was roosting in a small hemlock tree just outside of a private home along a major street. It fed on rats and mice at night and returned to the roosting tree to spend the day. The tree, a needle-leaved tree, resembled the kind of tree in which it would normally roost if it were in its usual boreal habitat. The owl, like the Ovenbird at Logan Airport, recognized and responded to a familiar habitat characteristic.

Some birds utilize a wide range of habitats, while others are confined to a very narrow habitat type, a case of specialization. Species such as American Crows and American Robins are ecological generalists, particularly good at adapting to human habitats, such as to parks and yards, but also at home in less humanly affected habitats. At the other end of the spectrum are specialist species such as Kirtland's Warbler, which requires recently burned jack pine forests of a specific age in which to nest, and is essentially confined to areas within the Lower Peninsula of Michigan plus a few other small areas nearby. The species does not occupy a jack pine

tract until about six years after the tract has burned. It will occupy it only until the trees are about 15 years old, and then the warbler abandons the tract. The management plan to sustain Kirtland's Warbler now includes a pine planting regimen with small openings set between pine trees, mimicking the fire-produced openings in early successional jack pine stands that the nesting warblers prefer. Small openings among the pines are key references that Kirtland's Warblers must see in order to know to nest there. The Kirtland's Warbler spends the non-breeding season on several islands in the Bahamas, occupying early successional scrubby habitats.

Habitat selection is a deeply and genetically de-termined component of birds' behavior, but many species do adapt as habitats change. Pileated Wood-peckers, denizens of closed forests, are becoming increasingly more adapted to living in proximity to suburban residences, especially in the Southeast. As habitat loss increases and habitat diversity declines, it is inevitable that some bird species will follow, declining and risking extinction. Many already have. Others will broaden their habitat selections, perhaps as American Robins apparently did during forest clearance in the East. Habitat selection, with all its dimensions, is an essential requirement that must be met by any bird species.

THE ANNUAL CYCLE OF BIRDS

THE BIRD'S YEAR

A bird experiences its year far differently from how humans do. Birds change physiologically throughout the year as they respond to certain shared needs, such as the need to molt and grow new feathers (chapter 9), the need to reproduce and raise offspring (chapters 13–15), and, in many cases, the need to migrate from one region to another (chapter 16). These behavioral changes are seasonally driven by and ultimately encoded in the birds' genes. Humans do not change physiologically in relation to seasonal changes, at least not nearly to the degree that birds do (some humans do develop depression at low light levels, typical of the depth of a northern winter). Humans do not have a breeding season. For humans, the most obvious rhythmic physiological change occurs when our sleep cycle is disrupted by events such as long-duration air travel, the cause of jet lag. Our 24-hour biological clock then requires gradual resetting. Human physiology is attuned to what are termed circadian rhythms, in which our cycles of appetite, activity patterns, and need for rest and sleep are based on daily rather than seasonal changes. The word *circadian* means "about day" (from the Latin words *circa dies*). Birds, too, exhibit circadian rhythms, but for birds, there are additional strong—indeed essential—annual rhythms, termed the annual cycle.

The annual cycle of an American Robin typifies that of many bird species in the temperate zone. Let's begin in January, when the calendar year gets its start. Robins in winter typically associate in flocks that roam in search of fruiting shrubs and trees. They roost together, sometimes with European Starlings and other species, in dense trees such as conifers. They vocalize, communicating with call notes, used in maintaining flock cohesion as well as other functions, but in winter they sing only on rare

occasion. During winter robins tolerate close physical proximity to one another, forming large nightly roosts. But as winter wanes and spring approaches, robins begin to react to hormonal changes and become edgy to migrate, first the males, then the females. These hormonal changes are triggered by the effect of increased day length on the bird's brain. Light is the variable that is most important in coordinating the annual cycle.

Male robins systematically move northward to breeding areas, following the 37°F isotherm, which means that they pace their rate of migration to generally avoid areas colder than 37° (2.7°C) (see chapter 10). The rate of migration is influenced by local weather conditions as well as ambient temperature. Warm southerly winds stimulate migration while storms or sudden drops in temperature stop migration until conditions moderate. Males become increasingly aggressive toward one another as their testosterone levels and testes size increase and the urge to establish a territory asserts itself. Song begins. In some passerines, the brain will have grown additional nerve cells in areas that coordinate various aspects of singing behavior. Female robins begin their migration following the males, experiencing the same urge to migrate but with their timing later than that of the males.

Males ultimately return to their breeding areas, often to exactly the same territories they occupied the previous season (chapter 14). By then they have become carnivorous, seeking worms and other forms of animal protein on lawns and in woodlands. Males defend a territory from other male robins, though unlike many birds, robin territories often have considerable overlap, and males, even during breeding season, often roost together at night. Males sing vigorously throughout much of the day (chapter 13) and chase other males, sometimes physically fighting in defense of their chosen territory. Once territorial boundaries are established, male robins become more tolerant of one another and share feeding areas, such as broad lawns.

When females arrive, males engage in courtship

OPPOSITE: This "dancing" Sandhill Crane is responding to physiologically induced hormonal urges to maintain its pair bond, part of its annual cycle.

behavior. The female selects a particular male, and they form a pair bond. The female constructs a nest, aided by the male bird, who brings nest materials. The female lays the egg clutch.

Robins hatch as altricial, featherless young with eyes closed (chapter 15). Before hatching, both sexes incubate the clutch (usually three or four eggs, occasionally more). They also take turns incubating the newly hatched nestlings, and both sexes bring lots of food to the nestlings, an energy-consuming task. Parents and sometimes neighboring robins vigorously defend nestlings from predators, often putting themselves at risk. Many nestlings fail to survive owing to a variety of causes, ranging from nest parasites to nest predation. The young that do survive soon fledge, leave the nest, and must learn to fend for themselves. The male bird assists in provisioning fledglings while, in most areas throughout the range, the female begins a second brood. Double brooding is common in robins as well as other species but by no means all bird species. Nestlings and fledglings grow rapidly and are soon on their own.

Once nesting is over, parent birds begin what is called prebasic molt. This process takes several weeks and results in a gradual and orderly replacement of all feathers. Fledged young, which have speckled breasts (called juvenile plumage), also soon undergo a prebasic molt and become funda-

mentally indistinguishable from older robins. Robins form flocks and begin their fall migration to their wintering areas.

That is the annual cycle of the American Robin. Note that annual cycles differ in details, not only among species but also among individual populations within a species. The annual cycle described here for the American Robin provides a representative example of the typical bird's year.

WHAT GOVERNS THE ANNUAL CYCLE?

The annual cycle is hormonally controlled, driven by a genetically based biological clock housed in the hypothalamus of the bird's brain. In response to an external cue, typically a change in the length of days, the hypothalamus sends a signal to the anterior pituitary gland located at the base of the hypothalamus. The pituitary, in turn, sends chemical signals in the form of various hormones, carried in the bloodstream, to target organs such as the gonads and the cortex of the adrenal glands. These glands, in turn, secrete specific hormones into the bloodstream that act on targeted organs and trigger various behaviors. The hormones are classed as *steroidal* because they are derived from the compound cholesterol. Hormones and hormonal concentrations are recognized by various target organs that provide chemical feedback to the pituitary, creating a series of complex biochemical feedback loops that regu-

This male American Robin experiences numerous changes in behavior over the course of his hormonally driven annual cycle, which is completely typical of most bird species.

The Razorbill is one of many bird species that undergo obvious coordinated plumage changes over the course of the annual cycle. Breeding plumage (*top*) is distinct from nonbreeding plumage (*bottom*).

late the bird's homeostasis while also stimulating essential seasonal changes in behavior. Hormones stimulate the urge to migrate and gonad development as well as manage stress levels. A dozen different types of hormones ranging from estrogen and testosterone to thyroxine and adrenocorticotropic hormone (ACTH) combine to control various aspects of the daily and annual cycles.

The bird's internal clock is genetically encoded. In a controlled environment where light and temperature are constant, birds still go through an annual cycle, but the cycle will drift (in calendar days) from what it would be if the birds were in nature. The same is true of daily rhythms. Circadian (daily) and circannual (seasonal) rhythms, though both part of the animal's genetic code, must each be set by some external environmental cue. The German word *zeitgeber* has been adopted by ornithologists to define the environmental cues that set biological clocks. For birds, the major *zeitgeber* is light, and the recurring cycle of light and dark is called a *photoperiod*. Day length varies seasonally with latitude and longitude, except directly on the equator, where there are 12 hours of light and 12 hours of darkness throughout the year. Thus day length serves as the dominant cue to coordinate birds' biological clocks,

Flocking behavior, evident in Snow Buntings (*top*) and Dunlin (*bottom*), represents a collective adaptation to potential predation. Birds reduce individual risk by flocking, and flocks take to the air quickly when any member of the flock detects potential danger.

both daily and seasonally. For example, in spring, an increase in day length by as little as five minutes per day is sufficient to stimulate increased testes size.

It would be reasonable to assume that birds measure light using photoreceptor cells in their eyes, which are, of course, connected to the brain via the optic nerve (chapter 5). However, the part of the brain most sensitive to photoperiod, the part that actually measures light and thus sets and engages the biological clock, is deep within the brain, very ancient, tracing its evolutionary roots to a structure called the pineal gland. This tiny organ is situated within the brain's hypothalamus. The pineal gland secretes the chemical melatonin, a regulator of circadian and annual cycles. In birds, ambient light penetrates through the feathers, skin, and thinly boned skull to stimulate a group of photoreceptor cells in the hypothalamus that measure the light and thus coordinate the annual cycle.

NATURAL STRESS AND HOW BIRDS COPE

Gulls resting idly on a beach or crows nonchalantly perched in a tall tree may be, for want of a better term, relaxed. Nonetheless, birds typically encounter numerous daily potential stresses. In addition to stimulating gonadal development, the secretion of hormones (adrenocorticotropic hormone, or ACTH) by the pituitary gland acts to stimulate hormonal production by the adrenal cortex, the gland that secretes adrenaline. Corticosteroid hormones stimulate and coordinate the fight-or-flight response (this happens in humans too). Perception of risk is expressed emotionally as fear and/or rage and causes increased heart rate and blood pressure, moving blood flow to the brain and away from digestive organs (including stimulating defecation as part of the fear response) and providing for a rapid reaction time. The bird may either respond with aggression or try to escape.

Stresses in nature are diverse and commonplace. Prolonged activities such as breeding, while normal in the bird's year, raise levels of stress that remain high throughout the breeding cycle. Weather events such as severe late-winter storms raise corticosteroid levels, a direct measure of stress, which may cause birds to stop breeding. In many cases stress requires a bird to make a split-second decision, quite often the decision to escape. Hormonal influences act through chemical feedback loops such that when the stress attenuates or lessens, the hormonal concentration stimulating the stress response is reduced and the bird's behavior changes.

I emphasize again that stress is normal for birds. It abounds in their world. Birds have, in the course of their evolutionary trajectory, adapted to stress, habituating to stimuli that hold no threat to them. Many examples will be evident in the following chapters.

This female Yellow Warbler located her nest very close to a boardwalk with dense human traffic, and, presumably, she adapted to that reality. She remained tightly on the nest as dozens of people walked past.

FEATHERS AND FLIGHT

TYPES OF FEATHERS

The earliest feathers appeared around 140 million years ago on small, bipedal, terrestrial, carnivorous, nonflying dinosaurs (chapter 3). Dinosaurs, all of them, are part of a group of reptiles called archosaurs (along with crocodilians, ancient and extinct pterosaurs, and modern birds) and presumably had color vision, at least in the red through blue visual spectrum. Therefore, feathers, which were mostly hairlike but with tufts of barbs, did contain pigments of various colors, initially byproducts of metabolism. The colors imparted on the nascent feathers could and likely would have been useful in signaling (and recent work by paleontologists has been able to demonstrate coloration patterns on dinosaur feathers). As is the case in modern birds, early dinosaur feathers presumably were attached to muscles that could raise and lower them, as a Northern Cardinal does with its crest. Once feathers evolved greater structural complexity, they began to diversify into categories such as down, contour, and other kinds of feathers, and acquired additional functions of providing warmth and flight.

Feathers form in follicles in the epidermis of the skin and are composed of an embedded tip, the calamus, and an exposed central shaft, the rachis. Vanes extend from either side of the rachis. The two major categories of feather structure are down and contour feathers. Down provides insulation and serves to preserve body heat. All birds have down feathers worn, rather like underwear, below their visible contour feathers and sometimes as the lower part of the contour feather. The insulating capacity of down was essential in the evolution of warm-bloodedness (endothermy) (see chapter 3). As might be expected, birds that inhabit cold environments, species such as Snow Bunting, have a

OPPOSITE: Flight requires numerous adaptations. Though highly diverse, birds are quite similar in characteristics enabling powered flight. Here, probably quite by chance, a Snowy Egret flies just above a Yellow-rumped Warbler.

thick layer of down beneath their contour feathers. Waterfowl, such as various eider species, also have dense undercoats of down. Birds in warmer areas possess less down.

The vanes of down feathers lack the structures present on contour and flight feathers that stiffen the vane and allow it to hold its shape. Instead, the vanes of down feathers are fluffy, a structure essential to heat retention. Down is lightweight, and its insulating capacity is impressively high. Down is the first kind of feather to form on a hatchling bird and comprises the juvenile feathering for species of birds that hatch precocious young, such as waterfowl and gallinaceous birds. It makes baby goslings, ducklings, and chickens look fluffy and cute. Because a down feather looks plume-like, down is referred to by ornithologists as a plumulaceous feather.

Contour feathers provide additional insulation, but their primary function is to shape the bird, provide coloration, and support flight. Contour or pennaceous feathers are what you see when you look at a bird. The vanes are structured such that the component barbs interlock. This is accomplished as follows. Barbs extend like branches from the sides of the central rachis. Each barb has a central shaft, the ramus. Like little branches off a tree limb, numerous barbules extend from the ramus, and these tiny structures have even tinier hooklets called barbicels. Like Velcro, the barbicels interlock, stiffening the vane, making a structure that is lightweight but strong. Preening (chapter 9) is how birds groom their feathers to keep the vanes properly arranged as well as oiled. Anyone who holds a contour feather can demonstrate how feather integrity is restored if the barbs become separated. Just smooth the feather between your fingers and watch how the vane shape comes back together. You just preened the feather.

Typical contour feathers are pennaceous on the outer (distal) part of the feather and plumulaceous on the inner (proximate) part of the feather so that the feather provides for both warmth and shape. Some pennaceous feathers also have an aftershaft of down that lies against the bird's skin. The combination of pennaceous and plumulaceous feathers contributes to heat retention, enabling endothermy.

In most bird species, feathers emerge along distinctive lines called feather tracts. These are easily seen in the skin of a plucked bird such as a chicken or turkey. Feather tracts are also obvious on altricial baby birds as the ensheathed and growing feathers begin to emerge. The presence of feather tracts allows feathers to grow in rows such that they overlap.

Wing feathers are called remiges (singular: remix). Primary feathers characterize the outer wing and secondary feathers compose the inner wing. Primaries supply both lift and thrust while secondaries essentially provide lift (chapter 3). The entire wing serves effectively as both an airfoil and a propeller. Normally there are ten primaries, but some birds have eleven and some nine. A few birds, such as storks, grebes, and flamingos, have twelve. The number of secondaries is variable. Birds with short wings, such as hummingbirds, have as few as six secondaries, whereas birds with long wings, such as seabirds, have many more. Some albatrosses, for example, have up to forty secondaries.

The leading edge of flight feathers, especially the primaries, is much narrower than the trailing edge. In addition, in some birds, particularly large soaring species such as hawks, eagles, and vultures, the outmost part of the outermost primaries is noticeably narrower. These are called emarginated primaries. The emarginated primaries aid in precision flying, as when soaring. Birds have muscles that enable them to manipulate each of the primaries as well as numerous filoplumes (see below) to gather information about the positions of these feathers.

Several short feathers are located on the "thumb" joint of the wing, and together they form the alula (occasionally called the *bastard wing*). The alula is visible when birds land. Although exactly how the alula functions is not entirely clear, we do know that it enhances control of landing by acting as a slot in the wing to break up airflow, helping with the controlled stall that is, in fact, the act of landing.

Groups of contour feathers called wing coverts are found on both the upper and lower wing. One group partly overlaps the primaries and another the secondaries. Feathers called tertials are found on the upper wing area and act to cover part of the primaries and secondaries of the folded wing. Some ornithologists also refer to the innermost secondaries as tertials. The coverts of the upper secondaries are in three rows: lesser, median, and greater. One set of coverts, the primary coverts, overlaps the primaries. Finally, a set of feathers termed scapulars covers the upper area of the closed wing in the shoulder region.

OPPOSITE: Examples of feather types: (*top*) Wild Turkey breast feathers showing plumulaceous base and pennaceous tip. Normally only the tips show. (*center*) A Wild Turkey primary wing feather showing the narrow leading edge and wider trailing edge. (*bottom*) The alulas are shown on the upper wing at the wrist on this Roseate Spoonbill as it lands. See text for details.

Tail feathers, called rectrices (singular rectrix), serve many functions including acting as a rudder in flight to help the bird turn and also as a brake when landing. There are twelve rectrices on most bird species. The two central tail feathers are attached to the tip of the vertebral column, the pygostyle. The others are embedded in a muscular area of flesh that surrounds the pygostyle. Birders sometimes notice "tailless" birds. Birds do not molt all tail feathers simultaneously, so the loss of all rectrices is more likely related to a near-miss with a predator. Birds are able to fly if they should lose their rectrices but they obviously fly more effectively with a full set of tail feathers.

BODY SHAPE

All flying birds exhibit an aerodynamic, streamlined body shape. Whether you are looking at a goose, a gull, a grouse, a grebe, or a grackle you are seeing an animal with basically an elliptical shape from head to tail, particularly when in flight. The body of a bird is its "fuselage," smoothly contoured with overlapping feathers, allowing air to stream over and under the bird while it's in flight. Once on the ground, some bird species present a less aerodynamic appearance, particularly species such as long-legged and long-necked egrets and herons. Flightless birds, such as ostriches, are not aerodynamically shaped because there is no selection pressure to maintain such a shape.

SEMIPLUMES, FILOPLUMES, BRISTLES, AND POWDER DOWN

Variable and widely distributed, semiplumes combine plumulaceous and pennaceous structure, providing both insulation and shape. Most are not obvious, except for the long breeding plumes of various egrets and herons. It was the quest for these unique plumes, used in women's fashions, and the subsequent catastrophic decline of egrets that helped stimulate the Audubon conservation movement of the late nineteenth century.

Filoplumes are inconspicuous sensory feathers that resemble hairs. A filoplume has a long rachis without vanes except toward the tip, where short vanes occur. Filoplumes are mixed among flight feathers and are numerous around head and neck, informing the bird of feather positions. In general,

ALULA

Some birds are hatched alert and covered with down, such as this Canada Goose gosling (*top*) being observed by a turtle. Other birds are hatched naked and need to grow all of their feathers. These baby robins (*bottom*) show the feather sheaths that contain the newly growing feathers. (*Top: Jeffory A. Jones; bottom: Leonard Kessler*)

Birds are aerodynamically shaped, forming an airfoil that supports the demands of flight. This is an immature Western Gull.

filoplumes go unnoticed. However, the tufts of feathers on the head and face of cormorants during breeding season, such as the "double crests" of Double-crested Cormorants, are composed of filoplumes. The delicate plumes that male Anhingas sport on their face and neck during breeding season are also filoplumes. Passerine birds sometimes show filoplumes protruding from the feathering on the nape of the neck, and a plucked bird usually has what appear to be tiny hairs scattered around. These hairs are actually filoplumes. Perhaps the most elaborate filoplumes are the decorative long white "whisker" feathers of the Whiskered Auklet. These feathers are essential to the birds as they navigate their way into darkened rock crevices where they nest. The touch-sensitive filoplume whiskers orient the bird as it moves through darkened passageways among the rocks.

The coarse hairlike feathers around beaks on species such as nightjars, flycatchers, some wood-warblers, and other birds are called rictal bristles. Their function is still being investigated, but evidence suggests that they help protect the bird's eyes while it is maneuvering as it forages for flying insects. Some bird species, particularly birds of prey, have bristle feathers that resemble eyelashes, and various bristles characterize the facial area of many bird species. The unique Bristle-thighed Curlew is named for (often hard to see) bristles on its knees.

Powder down feathers occur among contour feathers of some bird species, particularly herons and egrets. These feathers slough off a powdery substance that permeates the feathers, but its function remains unclear.

PIGMENTS, COLORATION, AND FEATHER PATTERNING

The most widely distributed feather pigment in birds is melanin, responsible for the tans, browns, beiges, and black that characterize many bird species. Brighter colors—the yellows, oranges, and reds—are provided by carotenoid pigments. Melanins are made metabolically, derived mostly from protein, but carotenoids must be obtained from the bird's diet and subsequently formulated into feathers as part of the bird's metabolism. The intensity of oranges and reds in birds is related to how much dietary carotenoid they obtain.

Some birds develop abnormal pigmentation by ingesting various fruits, particularly those of some invasive plant species. Since about 1950, individual Cedar Waxwings have been seen with orange, rather than yellow, terminal tail bands. The orange pigmentation results from ingestion of pigment from Morrow's honeysuckle (*Lonicera morrowii*), native to Asia. Thus far there is no indication that the color difference in the terminal tail band has any effect on fitness. Likewise, some Northern Flickers of the eastern "Yellow-shafted" subspecies *auratus* occasionally have red rather than yellow in some wing feathers. The pigment rhodoxanthin, found in alien honeysuckles, appears to be responsible. Flickers regularly feed on honeysuckle fruits from Tatarian (*L. tatarica*) and Morrow's honey-suckle in late summer before they molt into new wing and tail feathers. Because of the red coloration on flicker feathers, it was originally presumed that such birds represented some hybrid introgression with western "Red-shafted" (subspecies *cafer*) Northern Flickers.

Another common abnormality in many bird species is leucism, which refers to the presence of white plumage in places where feathers should be darker. The condition is a mutation in which melanin cells were not deposited where they should have been and thus the feather develops no pigmentation, instead being white. Sometimes leucistic birds are mostly white, mistaken for being albinos or so-called "partial albinos." Albinism is known in birds but is rarer than leucism and is all or nothing, never partial.

The displaying Snowy Egret (*top*) shows its unique aigrettes, its breeding plumes. Whiskered Auklets (*bottom left*) show their long filoplume "whiskers." Tan filoplumes are evident on this male Anhinga (*bottom right*). (*Top and bottom right: John Kricher; bottom left: Debi Shearwater*)

Albinism is characterized by no pigmentation whatsoever, including in the irises of the eyes. Both leucism and albinism reduce survivorship because feathers are weaker and more subject to breakage without melanin and, of course, the birds are more exposed and easier for predators to find.

Pigments absorb and reflect light at specific wavelengths. Melanins absorb most wavelengths in the visible light spectrum, making the bird appear dark and, in some cases, black. Carotenoids absorb at selective wavelengths, reflecting wavelengths that produce the orange of orioles and the scarlet of a Scarlet Tanager. Birds that are mostly white have feathers that reflect at all wavelengths, making them appear white.

Besides pigments, bird coloration is often dependent on the fine structure of feathers as well as the position of the bird with regard to the sun, as is the case with iridescence. Birds that appear blue, such as various jays and bluebirds, contain no blue pigment. Their feathers are fundamentally gray but are structured such that light is refracted, scattering blue wavelengths, making the birds appear blue. But if you hold a blue feather up against a light source, it will appear gray. On the other hand, if you hold a red feather from a cardinal up against a light source, it will appear red because it has red pigment. There is no "blue pigment" in any bird species; blues are structural colors.

Many birds, such as grackles, Wild Turkeys, various ducks, and hummingbirds, have iridescent plumage on all or part of their bodies. Iridescence makes feathers appear to emit a metallic sheen. The structure of the feathers reflects light from different angles. Barbules of iridescent feathers are positioned such that the combination of keratin, melanin, and air spaces acts to make the feather appear different depending on the angle at which light strikes it. As

Melanin pigments are the most common form of bird pigmentation. This female Brewer's Blackbird (*top left*) has an abundance of melanin throughout her feathers. The male Scarlet Tanager (*top right*) has abundant melanin in his wing and tail feathers but his scarlet coloration is due to carotenoid pigments. This Common Gallinule (*bottom left*) is an example of leucism, its feathering white due to a mutation that resulted in no melanin being deposited in the feathers. Blue Jays (*bottom right*) and other blue-colored birds lack blue pigment and the blue coloration is entirely structural. (*Top left, top right, bottom right: John Kricher; bottom left: Bruce Hallett*)

Iridescence is often widely dispersed in bird plumages, as in Wild Turkeys (*top*), for example. Many ducks have various patterns of iridescence, as in the head of a drake Mallard (*center*). The gorget of this male Ruby-throated Hummingbird (*bottom*) is showing well.

the bird moves, the light's angle changes and so does the iridescence. As a hummingbird moves its head, the throat gorget that might first appear black suddenly glows with sparkling color . . . until the hummingbird again moves its head and the iridescence is no longer evident.

Many bird species have cryptic patterning (crypsis). Examples abound and are hardly surprising given that birds are subject to strong predation. Such coloration aids in protection from predators. Some patterns, such as those found in nightjars, various owls, snipe, woodcocks, and sapsuckers, permit birds to effectively blend in with their background environment. Some plumage patterning, called disruptive coloration, aids in breaking up the body shape outline of the bird, rendering it less obvious. Breast barring and other forms of bold patterning on species such as Killdeer and Horned Larks that inhabit open grassy areas are examples. Many shorebird species that forage on open mudflats or rocky coastlines are amazingly cryptic, using a combination of crypsis and disruptive plumage patterning. Even Ruddy Turnstones, so obvious with their harlequin-like patterning when seen closely, become difficult to discern when feeding on a rocky shoreline. Many female birds are cryptic, while many male passerines molt into a much more cryptic plumage in late summer.

Species that are noncryptic when seen in the open become far more cryptic when one is trying to locate them in foliage. Male Northern Cardinals and Summer Tanagers and both Eastern and Western Bluebirds are much more difficult to see in shaded foliage because reflections of reds and blues are absorbed by the foliage. Downy and Hairy Woodpeckers, so obviously black and white when seen at feeders and in the open, appear cryptic when foraging in shaded trees. Sapsuckers, which spend much time methodically pecking rows of holes, are cryptic when seen from behind.

Countershading is also common in many bird species. Countershading makes the bird dark above but white below. Think of a species like the Razorbill or the Horned Grebe. When seen from below, which is how potential prey might detect a foraging Razorbill or grebe, the bird is pale, white, hidden by the reflection of the sky above the water. But seen from above, as by a potential predator, the bird is dark, its plumage melding into the darkness of the water. In addition, being dark helps absorb light and warms the bird. Sea ducks such as scoters are not countershaded, but there would be no advantage to being so. Scoters feed on mollusks attached to rocks. As another example, many species of wood-warblers are olive-green above and various degrees of yellow below, a form of countershading when seen in the shade of foliage.

Examples of cryptic coloration: (*Top left*) Western Screech-Owl (*Ed Harper*); (*top right*) female Ring-necked Pheasant; (*center right*) three Ruddy Turnstones (in basic plumage) are cryptic as they forage among the rocks; (*bottom*) Canada Warbler.

Countershading is common particularly among seabirds. These Common Murres provide but one of many possible examples.

Many groups of bird species show considerable similarity in feather patterning. There are two reasons why this is so. One is that the bird species may be evolutionarily close relatives, such as species within various genera of sparrows, gulls, or shorebirds, for example. Recently evolved species such as the Gray-cheeked and Bicknell's Thrushes resemble one another closely because they have only recently become separate species.

Another reason for similarity is evolutionary convergence. In cases of convergence, birds are not closely related but nonetheless wear similar plumage patterning. A classic example is seen when comparing North American meadowlarks with African longclaws. Meadowlarks are in the blackbird family, Icteridae. Longclaws, which do not occur in North America, are in the family Motacillidae (to which

wagtails and pipits belong). A comparison of meadowlark plumage with that of the East African Yellow-throated Longclaw reveals stunning similarity. The two are about the same size, similarly shaped, and have yellow breasts with a black V. Both have complex brown mottled plumage on the back, wings, and tail. Both have white outer tail feathers, conspicuous in flight. And most importantly, they inhabit similar habitats, albeit on different continents. The plumage pattern represents how natural selection adapts species with similar ecologies and habitats to look alike, because looking as they do confers fitness and higher survivorship in an open, grassy habitat.

Feather patterns frequently serve as signaling devices. White outer tail feathers, evident on numerous passerines, such as juncos, longspurs, and

The Eastern Meadowlark (*top*) (and Western too) bears a striking resemblance to the Yellow-throated Longclaw (*bottom*) of East Africa. Both occupy similar habitats.

(*Top*) The Northern Flicker (Yellow-shafted subspecies shown) displays a prominent white rump as it flies. (*Bottom*) The Yellow-rumped Warbler (myrtle subspecies shown) has a prominent yellow rump as well as white outer tail feathers.

Birds use wings to balance when they copulate, as shown by this pair of Laughing Gulls.

Horned Larks, act to provide flock cohesion when the birds are together in the air. The same is true of conspicuous rump feathering, such as white rumps on flickers, as well as wing striping on many shorebird species, and colored outer tail bands, such as are seen on waxwings and Eastern Kingbird.

Sexual dimorphism is a common feathering pattern. More accurately termed sexual dichromatic plumage, it means that female and male birds do not have the same plumage pattern, particularly when in breeding plumage. Think of male Western Tanagers and Rose-breasted Grosbeaks, for example. Most species of birds are not sexually dichromatic. But many are dichromatic, including most waterfowl and many passerines. Examples of sexual dichromatic plumage range from dramatic, as is seen with Northern Cardinal, Red-winged Blackbird, Orchard Oriole, Painted Bunting, Bobolink, and most duck species, to far less obvious, as is seen in kinglets, various woodpeckers, kingfishers, and American Robins. Sexual dichromatic plumage results from an evolutionary process called sexual selection (chapter 14).

THE WING: AIRFOIL AND PROPELLER

Wings are modified pectoral appendages—the equivalent of our hand, wrist, forearm, upper arm, and shoulder girdle. Obviously wings function for flight, but they have other functions, such as swimming, aggression, courtship, and balancing during copulation.

The shoulder girdle consists of three bones, each of which articulates with the upper arm bone (called, as in humans, the humerus). The scapula is a long, thin, blade-shaped bone that extends from its point of articulation with the coracoid bone parallel to the vertebral column but is not attached to the vertebral column (thus, like in humans, the scapula is able to move in relation to the vertebral column). The scapula is thin and elongate because birds have little in the way of back musculature. Scapulas in mammals are much larger because mammals have powerful back muscles. The thick, dense coracoid bone articulates with the humerus and the large breastbone, better called the sternum, which contains a prominent hatchet-shaped keel to which the dense flight muscles are attached. The thin and flexible clavicles fuse to become the furcula or V-shaped wishbone that articulates with the coracoid and humerus. The wishbone flexes in flight in many birds.

The humerus is the innermost wing bone, closest to the body. It is shorter than the two forearm bones, but is thick, well-muscled, and it is masked within the contour feathering such that it is not obviously visible when the bird is flying, though it is very much in use. If you are fond of eating chicken wings, you may know the humerus as the "drumette." The forearm bones, the radius and ulna, form the visibly obvious part of the inner wing when the wing is in use. Again, if you eat chicken wings, you'll recognize this as the part of the wing that has two slender bones and thinner meat. Secondary feathers extend from peglike attachments on the ulna bone of the forearm. The secondary feathers form much of the wing's airfoil, providing a convex upper surface and concave lower surface (see page 89, photo).

The bird's wrist and "hand" have few bones, a result of evolutionary loss and fusion. It is on the wrist and three-fingered hand that the large and obvious outer wing feathers, called primaries, attach. Primary feathers act as both airfoils and propellers to provide thrust as the bird flaps its wings. Because primaries are long, the overall wing appears much larger than the skeleton that supports it.

The hatchetlike sternum is dominated by a bony keel (carina). The sternum articulates with the rib cage and the coracoid bone. Birds have two large flight muscles on the sternum, comprising 40 percent of body weight. These are the muscles that you devour as "breast meat." The larger of the two major flight muscles is also the outermost, facing you as you look at the breast meat. It is called the pectoralis or sometimes the pectoralis major. The innermost of the two flight muscles is the supracoracoideus, sometimes called the pectoralis minor.

The huge pectoralis enables the downstroke of the bird's wing, the major power stroke that provides thrust. The supracoracoideus primarily enables the lifting of the wing, the recovery stroke. How can two muscles lying side by side on the ventral side, or "bottom," of the bird provide both the power stroke and the recovery stroke? Obviously, the muscles must work alternately, the pectoralis contracting (downstroke) while the supracoracoideus rests, and vice versa. Both muscles have tendons that attach to the humerus, pulling it downward in the case of the pectoralis and upward in the case of the supracoracoideus. (Note that tendons connect muscle with bone, and ligaments connect bone with bone.) In the case of the supracoracoideus, there is a uniquely long tendon that runs upward through a small opening where the bones of the coracoid, furcula, and scapula meet to a point of insertion on the upper (dorsal) part of the humerus. Thus, when the supracoracoideus contracts, the tendon acts like a pulley that raises the wing—quite amazing and effective.

One might ask why the big flight muscles, and most especially the supracoracoideus, are beneath the bird, on its ventral side. It seems odd. Well, you guessed it: flight. The muscles are massive—and thus heavy—and positioning them beneath and central to the bird distributes the weight to achieve

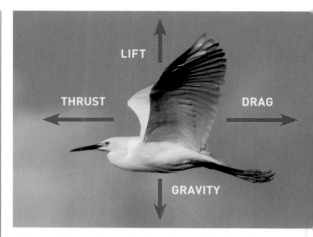

A Snowy Egret shows the four forces involved in flight.

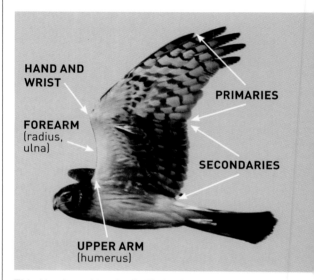

This Northern Harrier in flight shows the location of the skeletal parts of the wing as well as the primary and secondary feathers.

a favorable center of gravity, critical to sustained flight. The bird is neither front-heavy nor back-heavy when airborne.

BASICS OF FLIGHT, CAPSULE VERSION

Powered flight requires an object heavier than air to overcome gravity by use of force and to have a body shape suitable for doing so. A bird's streamlined body shape and wing structure represents an airfoil. The wing is convex on its upper side and concave below. The air has longer to travel over the top of the wing and so lift is created beneath the wing. It is this reality, Bernoulli's principle, that accounts for the physics of flight. Aircraft wings are similarly structured.

But in order to activate Bernoulli's principle, the bird must have sufficient power to take off and maintain suitable airspeed, called thrust. While flying, it must not only maintain thrust but also must minimize drag, which adds to the energy required for flight. Most air passengers are aware that airliners raise their landing gear immediately as they rise from the runway. Birds do the same with their legs and feet. Drag is a big deal when fighting gravity. The bird must also maintain a suitable angle of attack while in flight. The angle of attack is evident in cruising airliners in that the nose of the aircraft is tipped upward when the plane is in level flight. Too high or too low an angle of attack will create a disruption of airflow over the wing, breaking up the airfoil and resulting in a stall.

GETTING AIRBORNE: HOW TO TAKE OFF

Because birds weigh so little, getting airborne is easy for most species. A typical passerine merely raises its wings and jumps (legs are long and feet are large), pushing off from wherever it is perched. As it jumps, the powerful downstroke of the wing, facilitated by the big pectoralis major muscle of the breast, will take it aloft. It then repositions its legs, reducing drag, either pulling them close to the body or extending them behind. As the bird becomes airborne, it adjusts its body and wings to obtain a favorable angle of attack, gaining sufficient thrust to maintain lift and minimize drag. It raises the alula at takeoff to aid in this effort. Off the bird goes, fully airborne.

If you watch large birds, such as herons and egrets, just prior to taking flight, you can observe them bending their knees (you won't actually see the knee, but the bird's body is lowered) and flexing at the ankles. As the bird leaps up from the ground, it then rapidly extends its legs. Large hawks and owls appear to pump their legs as they jump when taking off, adding momentum to gain flight. Many waterfowl species, such as dabbling ducks (Mallards, pintails, etc.), become airborne by using their webbed feet and powerful leg muscles to leap up from the water as they raise their wings and begin to flap.

Heavier birds as well as birds with legs positioned more posteriorly have a more difficult time becoming airborne. Diving ducks must run a few steps along the water as they flap their wings to obtain sufficient force to become airborne. Sea ducks such as eiders and scoters, the heaviest of the ducks, must run a considerable distance along the water, flapping their wings to get into the air. The same is true for loons, grebes, and shearwaters; loons and grebes sometimes become stranded in ponds and lakes that have iced to the point that there is not ample area for the bird to take off. Birds such as gulls, geese, and swans normally take a few running steps as they become airborne. And albatrosses, some of which are among the largest birds, appear to strain to get airborne. If on land, an albatross needs considerable distance to gain flight—a long runway, so to speak. On the water, it must do the same, running with wings open, gradually gaining sufficient lift to clear the water.

A Green Heron (*top*) leaps into the air as it launches; a male Blackburnian Warbler (*bottom*) leaps into flight.

Seabirds such as this Cory's Shearwater (*top*) have a high aspect ratio, optimal for their dynamic soaring flight patterns. Albatrosses, such as this Laysan Albatross (*bottom left*), are able to lock their wings in place, making soaring virtually effortless for them. Soaring hawks (*bottom right*) (immature Red-tailed Hawk shown) exhibit a high wing loading and low aspect ratio. *(Top and bottom right: John Kricher; bottom left: Ed Harper)*

Both Brown Pelican (*top*) and Common Eiders (*bottom*) typically fly low over water to gain an advantage in lift.

WING SHAPES AND SIZES: HOW FLYING CHANGES WITH HABITAT

While all birds' wings share a common anatomy, wing proportions and shapes vary among groups of birds because, like the structure of aircraft wings, they are adapted for different forms of flying. Two measurements are essential in understanding wing function. One is called wing loading, which reflects the weight or body mass of the bird relative to the total wing surface area. A bird with higher wing loading has a smaller wing area relative to its mass. Heavy birds with relatively short wings, such as Wild Turkeys, have to really flap to become airborne, and then it is a struggle to stay in flight. Waterfowl also exhibit high wing loading, which is why duck species are constantly beating their wings as they fly. Ducks have relatively short wings in proportion to their body weight, and thus high wing loading, but with rapid flapping they remain comfortably airborne.

The other essential wing measurement is the aspect ratio, the length of the wing relative to its width. Seabirds such as shearwaters and albatrosses have high aspect ratios, as do coastal species such as terns and many shorebirds. Swallows, swifts, and falcons also have a high aspect ratio. Turkeys, grouse, eagles, and buteo hawks have low aspect ratios. Most species of birds have moderate to low wing loading and moderate aspect ratios. As might be expected, long-distance migrant birds ranging from passerines to shorebirds and terns tend to have wings with a high aspect ratio, characterized by proportionally long primary feathers that help provide strong propellers for sustained flight. This is also true of swallows, swifts, and terns, which require high levels of maneuverability in their foraging behaviors. Short-distance migrant bird species and nonmigrant species such as House Sparrows have comparatively shortened wings with more rounded shape and lower aspect ratios.

Many species of birds use various forms of soaring flight. Soaring conserves energy. Birds that soar over land, such as storks, vultures, and buteo hawks, benefit from a low aspect ratio because the added wing area allows use of warm air currents (thermals) that rise up from the land surface. Seabirds, the albatrosses, shearwaters, and petrels, exhibit dynamic soaring. They continually turn and bank, changing altitude over the water, dropping very near the surface of the sea, where the friction between the wind and the waves is highest, and then quickly rising, banking as the wind adds higher velocity with altitude. Albatrosses are unique among dynamic soaring birds because they are able to lock their wings at the shoulder joint such that they need not hold them in place when soaring. The wings become fixed in position until the bird unlocks them when it decides to flap.

Sea ducks, cormorants, pelicans, and shorebirds tend to fly just above the water surface, a behavior called skimming. This behavior is related to the physics of airflow near water, called the ground effect. If the bird is within a wing span of the water, it benefits from added lift because at low elevation, friction forces air to flow proportionally faster over its upper wing and adds lift. Ground effect occurs on land, but given that birds must be within a wing length of the ground for it to happen, there are too many obstacles on land to make it effective for birds.

Flying in a V formation, common in large birds such as waterfowl and cormorants, is beneficial to all but the lead bird. The birds take advantage of a wingtip vortex created by the upstroke of the wings of the bird ahead of it, reducing drag, saving energy. The efficacy of formation soaring is not advantageous in small birds because they are too small to create a sufficiently strong wingtip vortex that could be used by a trailing bird, which is why you do not see flocks of Cedar Waxwings or Common Grackles flying in V formation.

You might wonder how birds figure out the physics of ground effect and formation soaring. The answer is that they don't, at least in any cerebral way. It is simple trial and error. Birds fly in air, a more complex environment than we ground-dwelling humans might appreciate. Birds adapt quickly to fly in a manner that saves energy by simply feeling the air as they move.

Some species, such as goldfinches and woodpeckers, undulate in flight, flapping to gain altitude and then gliding as they lose altitude. This kind of flying also saves some energy, rather like coasting on a bicycle between bouts of pedaling.

Wing shape is an evolutionary compromise. While a narrow wing may be useful in long-distance migration, a shorter and rounded wing is advantageous for maneuvering among trees in a forest environment. Thus, the wings of thrushes are not shaped like the wings of swallows, even though both groups of birds have species that are long-distance migrants. The precision of natural selection with regard to numerous bird characteristics, from beak dimensions to wing characteristics, is remarkable. For example, a population of Cliff Swallows in Nebraska

Snow Geese typically fly in V formation but are constantly adjusting the order of birds as the flocks move along.

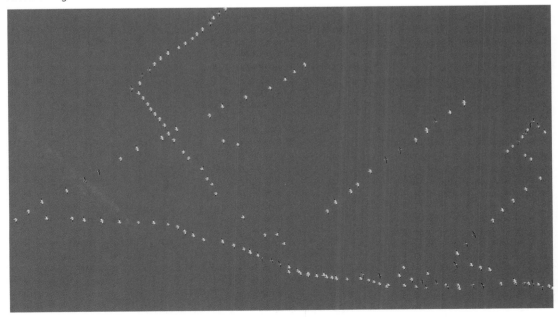

The wings of alcids such as this Razorbill, shown diving, represent a compromise to functioning both in air and underwater.

has evolved shorter wings in recent years. These birds are from colonies that nest under highway bridges and overpasses where fast-moving traffic sometimes kills swallows. The road-killed swallows that were collected over nearly three decades showed that the dead birds had wings just slightly longer than birds typical of colonies today, and that over the three decades, wing length has decreased by several millimeters, about the length of a rice grain. Decreasing wing length allows for more maneuverability, enhancing the chances that the swallows could avoid collisions with vehicles. Today there are fewer road-killed Cliff Swallows from these Nebraska colonies than in the past . . . and the birds' wings are just a bit shorter.

Some species of diving birds provide examples of how wings are adapted both for flight and for swimming. Eiders partly open their wings during a dive. Alcids actively flap their wings underwater in pursuit of prey. The narrow alcid wings work well underwater but require that the birds flap vigorously and continually while airborne because of high wing loading. Diving birds such as cormorants and loons with long- to moderate-distance migration do not use wings underwater, relying instead on their powerful legs and feet to propel them.

PIGMENTS IN THE PRIMARIES

The majority of bird species, including white birds such as gulls and terns, have dark wingtips. Like all other feathers, wing feathers—and particularly outer primaries—are subject to wear. Melanin pigment helps feathers in the wingtips resist abrasion far better than white feathers would.

Some bird species have all-white primaries, such as some egrets, Ivory Gull, and some other northern gulls. Egrets are generally short-distance migrants, not making long, nonstop flights. Ivory Gulls, residents of the high Arctic, are largely nonmigratory, though a few appear annually in the lower 48 states. Adult Ivory Gulls, with all-white primary feathers, are encountered much less frequently than immature Ivory Gulls, which are more apt to disperse and which do have black primary and secondary tips. White-winged gulls such as Iceland, Glaucous, and Glaucous-winged are generally short-distance migrants.

FINAL APPROACH: HOW TO LAND

Landing is fundamentally a controlled stall. Stalling occurs when the angle of attack breaks up the flow of air over the bird and ceases to support lift. The bird must slow, alter its wing angle, add drag, and use its wings and tail both to brake and as a rudder as it lands. Ultimately the bird works with gravity, the wings closing with landing. Birds do this so skillfully that the process appears seamless. YouTube videos showing birds such as Northern Goshawk and Eagle Owl in slow motion in the process of landing are amazing to watch.

Video has demonstrated that the bird never moves its head as it approaches its chosen perch. The head, after all, knows where the bird is intending to land and is making multiple decisions associated with the landing. The body, however, moves a lot. Wingbeats slow as the wings angle upward, creating the impending stall. The tail spreads, acting as both rudder and brake. As the bird reaches its landing point, it fully spreads its wings, the alula on each wing visible. Legs and feet drop, adding drag, and at the moment of impact, legs swing forward and feet grasp the branch or ground.

OPPOSITE TOP: Black primary feather tips characterize many bird species and are particularly evident
in coastal species such as this immature Royal
Tern. Note the distribution of black melanin
pigment elsewhere on the bird's secondaries and
tail feathers.

OPPOSITE BOTTOM: An American White Pelican coming
in for a landing.

Some birds landing under difficult circumstances,
such as herons coming to dense trees at a nesting
colony, hover as they land. Birds landing on water,
such as ducks, cormorants, and pelicans, open their
webbed toes on impact, the feet effectively breaking
the momentum of flight.

Seconds before the bird lands, the airfoil collapses
and turbulence over the wing causes the feathers to
vibrate as the bird completes its landing. Birds such
as woodpeckers and nuthatches, as well as other
cavity-nesting birds that land on vertical surfaces,
do so by swooping upward toward the tree trunk,
spreading tail and wings, projecting legs outward,
and grasping the bark with the feet. Birds such as
bluebirds, House Wrens, and chickadees fly directly
to nest boxes or nesting cavities, land at the cavity
entrance, and require no external perches.

HOVERING: WHAT YOUR AIRLINER CANNOT DO

Many species routinely hover, sometimes briefly
during landing, sometimes as they forage. American Kestrel, Red-tailed Hawk, Northern Harrier,
Osprey, Short-eared Owl, and Belted Kingfisher all
hover as part of foraging behavior. Many passerine
species (chickadees, kinglets, various wood-warblers) momentarily hover as they glean food from
foliage. During hovering birds alter the angle of
their wings, using both shoulder and wrist. Hovering requires more energy than sustained flight, but
in short bursts it is effective, enhancing agility.

Hummingbirds are legendary hoverers. Their
unique wing structure enables them to hover for
extended periods as well as to fly backward, something no other group of birds is able to do. As hummingbirds beat their wings (at around 50 beats per
second) they gain lift by both the downstroke (as all
birds do) as well as the upstroke, which in other
birds is the recovery stroke. Not surprisingly, the
pectoralis muscle of hummingbirds is proportionally large but so is the supracoracoideus, the muscle
beneath the pectoralis that raises the wing. Most of
a hummingbird's wing is composed of primary
feathers. By moving its flexible shoulder joint and
wrist, a hummingbird is able to change the angle of
attack to permit hovering, backward flight, and various degrees of speed for forward flight.

The Great Gray Owl (*top*) is one of numerous
birds of prey that typically hover as they seek
prey. (*John Grant*). Hummingbirds are remarkable fliers by any measure, as this male Black-
throated Mango (*bottom, from Panama*) is
demonstrating.

Slow-motion analysis of hummingbird flight
reveals that the wingtips inscribe a figure eight as
the bird flies or hovers. In the case of hovering, as
the shoulder joint rotates, the underside of the wing
actually points upward on the recovery stroke. This
amounts to maintaining lift while canceling direction, allowing indefinite hovering. But by altering
the angle of the shoulder joint and thus the angle of
attack, the hummingbird may move forward or
backward, and with considerable velocity. Hummingbirds are really cool.

MAINTENANCE BEHAVIOR OF BIRDS

ANNUAL MOLT CYCLE

Feathers wear and must be replaced. Flying birds frequently reveal missing wing or tail feathers, often a clear indication of ongoing molt. Birds sometimes appear bedraggled, and some may even develop baldness on the head as old feathers drop out and new ones slowly emerge. Several factors contribute to feather degradation. Feathers abrade from particles encountered while airborne and from contact with components of a bird's habitat, such as scraping against vegetation and when birds move into and out of nests. Feathers are also damaged by feather mites, feather lice, and bacteria. The need for annual feather replacement should be no surprise.

Feather replacement requires energy for new feather growth and must be well coordinated. Careful timing is essential. Ideally, feathers should be molted and replaced at a time of year when it is least energetically costly to the bird. But not all birds have the same annual cycle or the same ecology, so timing and pattern of molt varies among species. Molt is a universal necessity among birds, but different molt strategies have evolved independently. Even within the same species, molt timing and pattern may vary geographically.

Molt is genetically based and hormonally controlled, part of the annual cycle. Ornithologists identify four major molt patterns, described below, termed *molt strategies*. A logical time to initiate molt is after breeding season. Young birds are fledged and the parent birds have to provide only for themselves. In addition, feathers are abraded from the efforts of nesting and, in many cases, migration looms in the near future. Some species undergo what is called molt migration (chapter 16), in which the majority of molt happens away from the breeding grounds.

Ornithologists have historically used terms like

Two Boat-tailed Grackles, an adult (*right*) and a juvenile (*left*), show heavy molt.

This juvenile Brown Pelican shows obvious molt on its flight feathers.

OPPOSITE: This Snowy Egret is preening its underwing feathers. Preening of feathers is essential, and birds spend time daily engaged in such maintenance behavior.

A Brown Pelican engages in preening, an essential daily function for birds.

This male Northern Cardinal is due for a new set of feathers, a work in progress, as the photo demonstrates. *(Jeffory A. Jones)*

pre-nuptial and post-nuptial for the molts into breeding and winter plumages. These terms are misleading, particularly when applied outside of North America, as in the tropics, for example. Thus, a system of terminology to more accurately describe molt was developed in 1959 by Philip Humphrey and Kenneth Parkes, now referred to as the Humphrey-Parkes (H-P) system, and universally adopted by ornithologists.

Most species molt all of their feathers at least once per year. Exceptions are some larger species, such as hawks and eagles, that molt flight feathers over a period longer than a year because they have large primary feathers that require time to grow and that last longer. Molt typically begins immediately after breeding. At this time, adult birds molt into what, under the H-P system, is called basic plumage. The molt itself is referred to as the *prebasic* molt, and it involves gradual replacement of wing and tail feathers as well as all body feathers. While molt occurs, plumage is frequently scruffy in appearance, and a few species, such as Northern Cardinals and Blue Jays, may lose virtually all of their head feathers for a brief time in late summer before the new feathers appear.

Once breeding is completed, ducks molt all of their flight feathers and thus are rendered flightless for a period of weeks while the new feathering develops. During this *eclipse* plumage phase, males are hard to distinguish from females, and the birds are confined to marshes, living furtive lives until new flight feathering is in place.

In some species, prebasic molt is the only molt. Ruby-crowned Kinglets, Gray Catbirds, and Song Sparrows are examples. Each goes through a prebasic molt after breeding. Molt is complex, however, and for both the Ruby-crowned Kinglet and the Gray Catbird it has been reported that they do on occasion have partial *prealternate* molts involving

A male American Goldfinch in what was once called nuptial plumage, now called alternate plumage.

A female American Goldfinch looks quite different. Such a difference in plumage between males and females is a common pattern in many breeding birds.

Male American Goldfinch (*middle*) transitioning to alternate plumage. Note that only the body feathers are in molt—not the wing and tail feathers, which were molted after breeding season.

A male Northern Cardinal with a naked head. Many people see this manifestation of molt when the birds come to feeders and mistakenly assume the bird is ill or has a parasitic infection. His appearance is common for that stage of the molt process. *(Van Remsen)*

only some body feathers. These molts do not affect the appearance of the birds: they look the same before and after the prealternate molt, if it occurs at all. Much of what is known about patterns of molt comes from bird-banding studies in which birds in hand are studied in minute detail.

In many shorebirds, both sexes molt into brighter *breeding* plumage during spring migration. Dunlin, Red Knots, and Sanderlings are examples. These species carry their basic plumages into winter and then begin a prealternate molt, culminating in alternate plumage for breeding season. Usually, but not always, the prealternate molt will not include flight feathers, only body feathers.

The first plumage after fledging is termed *juvenile* plumage. Think of the breast spotting in birds such as American Robins and Eastern Bluebirds and the streaked breasts and grayish bodies of Cedar Waxwings. Many species—but not all—wear this plum-

age for a short period. Some birds wear distinctive juvenile plumage well into their first winter. Cedar Waxwing is an example, as well as some shorebird species, such as Least Sandpiper and Sanderling. Juvenile-plumaged Roseate Spoonbills are paler than adults, and juvenile Wood Storks have yellow bills, though otherwise they look like adult birds. Juvenile Double-crested Cormorants have pale upper breasts. There are many other examples. When birds eventually molt out of juvenile plumage, they undergo what is called a *preformative* molt, the molt that will take them into basic plumage. Gulls, various raptors, and some waterfowl require several years to attain adult plumage. These species molt from juvenile to various immature plumages.

Gulls require anywhere from two to four years to transition from juvenile to adult plumage. It is easy to see varying-aged gulls wherever they assemble.

Gray Catbirds sometimes have a partial prealternate molt. Birders will not notice this in the field. Much of molt analysis must be accomplished in the hand, by bird banders.

Ruddy Turnstone in basic plumage (*left*) contrasts strongly with alternate plumage (*right*), though both are cryptic when birds are observed foraging because the plumage patterning breaks up the body outline of the birds.

A Cedar Waxwing in juvenile plumage (*opposite top*). The White Ibis (*opposite bottom*) is undergoing a slow, progressive molt from juvenile to adult plumage, and still has a ways to go. Juvenile Wood Storks (*above*) look much like adults but have yellow bills.

For example, in its first autumn post-hatching, a Western Gull is basic brown, in its juvenile plumage. It moves through a gradual molt to become a first-cycle (or first-winter) plumaged gull, quite similar to its juvenile plumage but somewhat paler. In subsequent winters, it will become a second-cycle, third-cycle, and finally fourth-cycle or adult-plumaged Western Gull. Thus, it requires four years for a Western Gull to attain full adult plumage. But why use the term *cycle*? Molts in gulls (and many other kinds of birds) are variable, complex, and overlapping, and thus plumages vary within age classes. The term *cycle* refers not just to age or time of year, but mostly to variable plumage patterning among individuals that may be the same age. A second-cycle Herring Gull often does not look exactly like another second-cycle Herring Gull because plumage change progresses somewhat differently among individuals.

Fundamentally there are four kinds of molt strategies. All bird species fall into one of the four.

The Simple Basic Strategy characterizes albatrosses, petrels, New World vultures, some hawks, some gulls, swifts, and the Barn Owl. There is but one annual molt and no preformative molt. Barn Owls look like Barn Owls from the time they fledge, though as nestlings they wear downy plumage. This

molt is, indeed, simple. But only a few kinds of birds show it.

The Complex Basic Strategy is common to many families of birds, including most owls (other than Barn Owl), kingfishers, tyrant flycatchers, and many other passerines. It is similar to the previous one in that birds have but one prebasic molt and wear their basic plumage throughout the year. But the difference is that they also have a preformative molt with a formative plumage in their first cycle. Often this formative plumage may be subtle and of short duration.

The Simple Alternate Strategy includes molt in loons, pelicans, ibises, some gull species, and some alcids. There is a juvenile or first basic plumage followed by a first alternate plumage. That is followed by annual basic and alternate plumages.

The Complex Alternate Strategy is exhibited by ducks, many shorebirds, some gulls, and many passerines, including wood-warblers, tanagers, sparrows, and cardinals. Juvenile plumage is followed by a preformative molt, formative plumage, then a first alternate plumage, followed the next year by definitive basic and definitive alternate plumage.

The standard pattern seen in molting among all strategies is that flight feathers are molted once a

OPPOSITE TOP: A first-cycle Western Gull with the parent bird behind it.

OPPOSITE BOTTOM: The Bobolink (shown here), a long-distance migrant passerine, and the Franklin's Gull, also a long-distance migrant, are the only two North American bird species to undergo two complete molts annually.

year, but body feathers are usually molted twice annually. There are two exceptions, the Franklin's Gull and the Bobolink. These species, each of which is a bird of open areas and each of which endures a very long-distance migration, undergo two complete molts per year.

Molt occurs in an orderly manner. Birds must molt feathers of the wings and tail, and these feathers are generally large, requiring more time to grow to full size. Molting and regrowth of flight feathers requires temporal coordination. As you observe soaring hawks, eagles, vultures, and other large birds, you might notice that the right wing is missing some primaries, maybe two to three. What you will also see is that the left wing is missing those same feathers. Primary feathers are numbered from the inside to the outside of the wing. If primaries 6, 7, and 8 are missing or small on the left wing, primaries 6, 7, and 8 will be missing or small on the right wing. The same is true with the secondaries and with the tail feathers. There is symmetry in feather loss during molt, enabling birds to still fly effectively with fewer full-sized feathers.

PLUMAGE CHANGES NOT INVOLVING MOLT

In late summer into winter, European Starlings have heavily spotted breasts and dark beaks, but as winter turns to spring the spotting disappears and the beak turns yellow. The beak color change is hormonally related to the annual cycle, as the birds are coming into breeding mode. The loss of breast spotting is due to feather abrasion, not molt. A European Starling in spring is wearing exactly the same feathering as it was in fall.

Another example is the Snow Bunting. Breeding males are strikingly patterned in crisply defined black and white. After the prebasic molt in late summer, their plumage is a combination of rich brown, white, and black. During the winter, males gradually change into breeding colors, but they do so entirely by feather abrasion, not molt. The rusty brown feather tips gradually wear off, exposing white or black, depending upon where the feathers are on the bird's body. Snow Bunting flocks observed in late winter and early spring often contain males that are obviously transitioning into their breeding appearance.

The male House Sparrow has a primarily gray

This Little Blue Heron shows symmetrical feather replacement as it gradually molts from immature all-white plumage to its adult blue plumage.

upper breast after prebasic molt in late summer that gradually becomes black (this is variable among males—see chapter 14) by breeding season. The black appears as the gray feather tips gradually abrade during winter, exposing black-pigmented feathers below.

Some species look odd as they transition in plumage, perhaps none more than the Brown-headed Cowbird, especially males. They take on a patchy appearance, variable from one individual to another.

ECTOPARASITES: UNPLEASANT REALITIES OF BEING A BIRD

Mites . . . lice . . . ticks . . . fleas . . . Birds have them all and then some. An ectoparasite, as distinguished from an endoparasite (such as a tapeworm), is a parasitic animal that lives on the skin or feather surface, consuming some of the host animal (blood or feather) for its own nutrition. Birds commonly encounter and accumulate ectoparasites. Birds' nests are sometimes replete with them, so much so that baby birds can be at risk of suffering too much blood loss. People who tend bluebird nest boxes, for example, frequently encounter thousands of pinhead-sized mites when they remove an infested nest.

Mites are tiny arachnids (related to spiders) that sometimes densely infest skin and feathers. Bird banders assess levels of parasite load by spreading wing feathers, holding them up to light, and counting the black spots of feather mites embedded in the feather vanes. Lice are tiny insects that roam the surface of birds, consuming blood. An entire louse

OPPOSITE TOP LEFT: A male Snow Bunting in November after prebasic molt.

OPPOSITE TOP RIGHT: A male Snow Bunting on breeding grounds in June. The black feathering results from the buffy feathers wearing such that only black remains. The same is true of the all-white head, where buffy feather tips have gradually worn away. *(Ed Harper)*

OPPOSITE BOTTOM: Brown-headed Cowbirds transition from juvenile to adult plumage and often pose identification challenges with their patchy plumage.

order, Mallophaga, is confined exclusively to birds, and many lice species are specific to various bird groups and even species. Many kinds of ticks, including those that carry Lyme disease, are found on birds. Add to that a diversity of fleas as well as a particularly daunting-looking group of creatures named hippoboscid flies. These are commonly called louse flies and are dorsoventrally flattened, somewhat resembling little crabs, their shape enabling them to remain very close to the bird's skin, over which they freely roam, sheltered within a forest of feathers. It is indeed a bit creepy to be inspecting a bird in the hand, part its feathers (especially around the head), and come upon one of these most unpleasant-looking insects.

Taken together, ectoparasites represent a major selection pressure. Birds cannot escape them. Parasites move easily from bird to bird by direct contact and infestation of bird nests. They are acquired from grasses and other components of habitats. Any bird is endowed with its own thriving populations of various ectoparasite species, a diverse community of annoying animals. Something must be done about this reality. Here are the things that birds do.

PREENING

It must feel odd to be a bird, with thousands of these large epidermal objects called feathers poking out of your skin, all of which require precise arrangement. Feathers are felt and controlled through the nervous system using signals from filoplumes. Birds employ subepidermal muscles to fluff and smooth feathers, erect crests, spread tail and wings, and facilitate display plumages (such as the fanned tails of grouse and turkey). Feathers require daily attention, which becomes a major part of daily activity. This attention to feathers is called preening.

Mostly preening entails smoothing the feathers by running the vanes through the bird's bill, and often coating them with oil. The oil is obtained from the uropygial gland, which is located on the rump of the bird. Almost all bird species have uropygial

(*Top*) A male Northern Shoveler preens. (*Center*) A juvenile western Red-tailed Hawk preens. (*Bottom*) A Canada Goose preens after having rubbed its bill against the uropygial gland.

glands. The oil functions to enhance the robustness of feathers and helps repel ectoparasites. Uropygial oil may also help protect against feather bacteria. It was once assumed that preening oil was essential in making feathers water resistant but water resistance is also attributable in part to the structural integrity of the contour feathers. Nonetheless, numerous birds liberally apply uropygial oil as they preen.

Groups of birds ranging from blackbirds to herons often preen together, making preening a group activity. Perhaps when a few begin preening they stimulate others to join.

Waterfowl body feathers are dense and are sufficiently water resistant that their bodies remain dry whether the birds are in rain or diving for food. Water typically beads on the feathers of well-oiled birds. Preening oil is lost when the birds come in contact with pollutants, such as oil leaking or spilling from a ship. Should this occur, the bird becomes vulnerable to cold because its feathers no longer repel water, the cold water gets to the bird's skin, and it rapidly begins to lose heat. Birds react to foreign oil on their feathers by intensively preening, and they may be harmed from inadvertently ingesting the foreign oil.

If you have the time and the patience, next time you observe a bird beginning to preen, carefully watch it throughout the process and observe the thoroughness with which a bird keeps its plumage intact.

BATHING

Bathing is a regular maintenance activity that most birds practice, exceptions being mostly those birds found in water-limited environments such as deserts. Human-provided birdbaths, including heated birdbaths in winter, offer a considerable aid to the welfare of birds. Bathing is stereotyped in birds, and most species use the same mannerisms to accomplish the task. They dip their heads in the water, allowing the water to reach the upper back, and they use their wings to vigorously splash water over themselves. They fully submerge and spread the tail feathers. They fluff their feathers so that the water is able to reach their skin.

For land birds, most bathing occurs at a pond edge or stream, and sometimes where water collects, such as in parking lot potholes. Typically, birds that are bathing stop frequently and look up and around, presumably aware that bathing makes them more vulnerable to predators. Birds periodically pause while bathing, wings spread in the water for a few moments, and then resume the process. Some groups of birds have unique ways of bathing.

A male Boat-tailed Grackle engages in a vigorous bath, typical of how most passerines bathe.

A male Painted Bunting demonstrates how wet birds make themselves when bathing. *(Garry Kessler)*

Bushtits engage in group bathing.

A House Sparrow dust-bathes.

Swifts and swallows bathe while flying in rain or by momentarily dipping into a pond. Hummingbirds are known to bathe in water trapped on the surface of a leaf.

Many waterbirds bathe regularly, either in fresh or salt water, depending on habitat. Waterfowl are frequent bathers, as are cormorants and shorebirds. Herons and egrets are known to bathe, but the behavior is less frequently observed.

Once bathing is completed, a bird shakes off excess water somewhat like a dog does. The bird vigorously shakes its body, droplets of water scattering widely. This behavior is always followed by extensive preening.

Bathing is thought to aid in removing old oil from feathers, allowing birds to then replace it with freshly secreted oil. It is possible that bathing may help reduce ectoparasites on the bird as well as bacteria that may be on the skin surface or feathers. Birds bathe on cold winter days, so exposure of skin to cold water is apparently not an issue. Birds have warm bodies, and thus their feathers, even when wet, are still insulating. Cold water on a bird likely warms and evaporates quickly. They do not stay wet for very long.

DUST-BATHING

Many land bird species engage in dust-bathing, or simply dusting. The behavior is well-known for gallinaceous birds, but many other birds dust-bathe, especially in arid areas. A bird about to dust-bathe typically excavates a small depression in fine, dry sand or dust and settles into it, using its feet and body motions to make the depression and using its wings to spread dust all over its body.

Mannerisms used in dust-bathing are similar to those used in water-bathing but not identical. The bird flaps its wings vigorously into the dust and tosses dust liberally over its body. It fluffs its body feathers so that dust may make direct contact with skin. The process is not rushed. Birds frequently pause active dusting but remain in the depression they made, wings partially spread, and then reinitiate the process. In some cases, such as with House Sparrows, groups of birds dust-bathe together.

Dusting must be adaptive because it does involve

risk (the bird is on the ground, potentially exposed to predators), and it requires time and energy. But many bird species do it. Dusting may aid in absorbing old feather oil as well as ridding the bird of some of its ectoparasite load and, in particular, its bacteria. Feather-degrading bacteria (see below) may have a significant impact on feather integrity, so reducing bacterial impact increases the bird's fitness.

FEATHER-DEGRADING BACTERIA

All animals host millions of bacteria on their surface, and birds are no exceptions. Various kinds of bacteria have been identified on bird skins and feathers. To what extent might bacteria negatively affect birds? Studies pioneered by Edward H. Burtt Jr. demonstrated that feather bacteria weaken feathers, but melanin pigment contained in feathers has been shown to aid in retarding the effects of these bacteria by strengthening the feather.

Many marsh birds, species such as Swamp Sparrow, Seaside Sparrow, the Belding's subspecies of Savannah Sparrow, and various subspecies of Clapper Rail, have generally dark plumage. Studies by A. M. Peele and others comparing a dark coastal subspecies of Swamp Sparrow with an inland population, lighter in plumage, demonstrated that more feather-degrading bacteria were found on the feath-

ers of the coastal subspecies, the one with darker feathers. This result suggests that dark feathers are adaptive in habitats that harbor denser concentrations of bacteria.

In addition, birds that inhabit dark and moist forests of the Pacific Northwest, species such as the Pacific Wren, the *fuliginosa* subspecies of (Sooty) Fox Sparrow, and the Sooty Grouse are all darkly plumaged. Of course, these color patterns could aid in providing crypsis. But they also could serve to repel feather-degrading bacteria by concentrating melanin in the feathers. Moist, dark habitats provide ideal conditions for bacterial growth, more so than drier habitats. Biologists have long noticed that warm-blooded animals, and birds in particular, are darker in humid, dark environments and have termed the generalization Gloger's rule.

HEAD-SCRATCHING

Birds frequently scratch their heads when they preen, but they also scratch independently of preening. Birds use their flexible necks to reach most of their body surface with their bills, but they cannot use their bills on their own heads. Therefore, scratching with the feet is the only way a bird is able to touch its head and neck area.

Birds head-scratch in one of two ways. Some but

This Clapper Rail is highly subject to feather-degrading bacteria that inhabit marshes, but its heavily melanistic feathering provides protection.

Northern Cardinals head-scratch over their wing.

Shorebirds such as this Marbled Godwit head-scratch under their wing.

by no means all passerines reach over their wing to scratch the neck and head. This is also true of plovers, kingfishers, hummingbirds, swifts, and nightjars. But most groups of birds reach up under the wing to access the neck and head. The method of head-scratching may vary within bird families. Some wood-warbler species scratch over the wing, some under the wing, but the choice appears to be species specific. The next time you watch a bird scratch its head, try and note which method of head-scratching it uses.

SUNNING

You may come upon a bird that is on the ground, its wings and tail spread, in full sunlight, appearing almost mesmerized, "in a zone." You might walk right up on it before it suddenly realizes you are there and flies away. What you have witnessed is sunning behavior. Many species do it. Vultures are often seen in trees with their wings spread fully, facing the sun, or sometimes facing away from the sun. Herons and egrets sometimes adopt an odd posture of standing in the sun with partially open wings, resembling solar panels as they face toward the sun.

Sunning may be related to thermoregulation, warming the bird, possibly stimulating ectoparasites to move, allowing the bird to more effectively preen and remove parasites. The sunlight may facilitate release of vitamin D from the bird's preen oil, and the bird may then ingest it during preening. Vitamin D is essential to folate metabolism, a critical component of reproduction. It has been suggested that sunning "feels good" to the bird, and maybe it does. In any case, they do it.

ANTING AND OTHER SEEMINGLY ODD BEHAVIORS

Some birds, particularly passerines, habitually lie down on an active ant nest, wings and tail feathers spread. Such behavior, infrequently observed, has puzzled ornithologists. The explanation may relate to the fact that many ants contain formic acid, an acrid chemical that repels ectoparasites such as lice. Formic acid extracted from ants has been demonstrated to be a strong repellent to bird lice.

Birds may exhibit one of two forms of anting behavior, passive or active. In passive anting, the bird lies prostrate on the ant nest and allows the ants to scurry around on it. In active anting, the bird picks up ants in its beak and crushes them, subsequently rubbing them over its feathers and skin. Blue Jays are known to do this. Doing so gets the formic acid directly onto the skin. It has been suggested that birds preferentially select ants high in formic acid over ants that primarily defend themselves by stinging. You likely won't find a bird anting on a fire ant nest.

This Yellow-crowned Night-Heron (on Genovesa Island, Galapagos) is warming up in the early morning, wings spread to absorb sunlight.

Birds have also been observed rubbing still-hot embers from fires on themselves and sometimes even substances as strange as mothballs. The only explanation suggested for this behavior is that it could be repellent to lice and other ectoparasites.

DRINKING

Birds are somewhat awkward drinkers. Most species put the head down toward the water source, open the beak, scoop up some water, then raise the head, tilt it back, and allow the water to trickle down the esophagus. Scoop, raise head, tilt, swallow. Doves and pigeons are an exception to this pattern. They are able to suck up water directly, without having to repeatedly raise their heads to swallow. Birds such as swifts and swallows, understandably enough, skim over ponds and drink while airborne. Hummingbirds, unsurprisingly, take in water as they do nectar.

Birds do not usually need to consume as much water as mammals do. Recall that they have no bladders and utilize uric acid as their nitrogenous waste compound. Water is not nearly as necessary as in mammals, which produce urine, a toxic compound that requires dilution. Many bird species, particularly those such as waxwings that feed heavily on

An immature White-crowned Sparrow attempts to drink from a dripping sprinkler. Not so easy, but the bird manages.

fruit, obtain virtually all of their water needs from the food they ingest, though waxwings do drink when the opportunity presents itself. Some arid land bird species obtain all of their water from metabolism of seeds. Recall also from chapter 3 that seabirds and gulls rely on salt glands around their nasal areas to eliminate excess salts, enabling them to drink seawater.

LOAFING

Birds use a lot of energy daily, so it is important for them to conserve energy whenever possible. One form of behavior that conserves energy is idle loafing. Many bird species, particularly larger species, conspicuously loaf. Gulls stand or sit idle on a beach. Groups of shorebirds bunch up and remain idle at high tide. Herons and egrets stand idle on a dock, presumably awaiting a change in the tide that will trigger their foraging behavior. Flocks of geese idle on grassy fields or on the water. Rafts of ducks idle. Crows and starlings idle on utility wires. Even among small passerines it is common to find a bird merely perched, doing nothing, for minutes at a time, especially in the middle of the day. If a bird is sufficiently fed, it is under no pressure to actively seek food.

An immature Great Black-backed Gull scoops up water and swallows it by raising its head.

These Black Skimmers, Royal Terns, and Sandwich Terns are engaging in typical loafing behavior.

Blackbirds, such as these Common Grackles, routinely form large roosts during migration and in winter.

Some observers may not regard loafing as real behavior, but it is. The birds are choosing to do it. Unfortunately for some birds prone to regular loafing, such as gulls, terns, and cormorants, more and more docks and marinas are placing cones or wire arrangements atop dock pilings and other structures specifically to repel birds from choosing such perches to loaf on.

ROOSTING AND SLEEPING

The term *roost* applies to a behavior in which a bird goes to a specific place to rest and sleep. Many bird species roost singly, but many roost in flocks, sometimes large flocks. Individual birds often move their roosts. But some species, such as Downy Woodpeckers and Black-capped Chickadees, roost singly in cavities and use the same cavity repeatedly. At the opposite end of the spectrum, herons, gulls, crows, starlings, and blackbirds typically gather nightly at roosts that may number hundreds if not thousands of birds (see photo of grackles above).

Ground birds such as Wild Turkeys and grouse ascend into trees to spend the night. In the cold weather of the northern winter, birds congregate in various forms of roosts that collectively help maintain warmth. Eastern Bluebirds cluster densely into tree cavities or bird boxes. Golden-crowned Kinglets bunch tightly together on a protected tree branch. In warm areas, most passerines find a pro-

tected branch and sleep there. It is likely that birds are not deep sleepers but are instead fitful, awakening often, living as they do in an uncertain world where predation is never far off.

Some species, particularly crows, gulls, and blackbirds, form large nocturnal roosts. In the case of crow roosts, the birds—often in the high thousands—converge at staging areas in late afternoon and fly to the chosen roost when dusk arrives (chapter 11). The roost is usually in a pine grove or another suitable habitat. The birds disperse in many directions at dawn. Crows are particularly fond of roosting near or within towns and cities.

Roosting, whether it be robins in winter or crows or starlings, provides protection from predators as there are so many eyes awake while others are asleep. Roosting may also play a strong role in communication among birds because the roost may serve in part as an information center, where birds seeking promising foraging areas learn to follow birds that appear particularly robust and healthy. There is little hard data to support exactly why roosting in large numbers is important to these species but it seemingly must be.

Nocturnal birds such as night-herons and owls are also sometimes drawn together to roost in close proximity, in these cases during the day. Long-eared Owls frequently converge on the same pine grove, sometimes with Short-eared Owls. Night-herons

A Clark's Grebe in resting/sleeping posture.

A Black-crowned Night-Heron sleeps with beak tucked among breast feathers.

Western Willets rest, each of them on one leg. This is very common among birds.

breed in colonies, but even outside breeding season they gather in diurnal roosts.

Birds, like mammals, must sleep. And there the similarity between mammalian and avian sleep ends. Imagine being able to sleep with just half of your brain, while the other half—say the right half—keeps your left eye open and alert. Your right brain is awake even as your left cerebral hemisphere is sound asleep. Mounting evidence suggests that some birds do this. A bird may sleep with one eye open and one closed. Measurements of brain function indicate that the brain is both asleep and awake, the cerebral hemispheres alternating between wakefulness and sleep.

This pattern makes sense, especially for long-distance migrant birds that feed much of the day and fly much of the night. They have to sleep sometime, but when? Species such as Blackpoll Warbler and Bar-tailed Godwit fly for several days and nights, nonstop, as part of their fall migration (see chapter 16). Do they sleep on the wing? If their brain hemispheres switch on a regular basis, these birds could sleep and be awake simultaneously. The Common Swift, a Eurasian species, has been documented to remain aloft for up to ten months, maybe more, without landing. In order to remain physiologically healthy these birds must have some means of sleeping while airborne, and hemispheric switching in the brain would accomplish that.

Like mammals, many species of birds that are not required to be active on a 24-hour cycle are capable of sleeping through the night, birds such as passerines, for example. Robins tucked into a thick shrub close both eyes and go to sleep. Nuthatches sleep through the night in their tree cavities while screech-owls sleep through the day in theirs.

When resting or sleeping, many birds habitually draw the foot of one leg up and tuck it into their feathers, presumably to minimize heat loss. They balance easily and often hop on one leg to adjust position among the flock if they are roosting with others. This is common behavior among birds and is discussed more in the next chapter.

I have also been asked about birds that tuck their head under their wing to sleep. Birds do not do this. Birds have highly flexible necks. Many typically turn their heads backward and nestle into their back feathers, beak covered by feathers, maintaining warmth. Some birds sleep with a different posture, their head turned downward, their beak nestled within the warmth of their upper breast feathers.

TEMPERATURE AND BIRD BEHAVIOR

THERMAL TOLERANCE ZONE

Birds are endothermic animals with high and nearly constant body temperature. They are adapted to maintain physiological homeostasis within a broad but nonetheless limited range of ambient temperatures. Each bird species has a genetically determined thermal neutral zone (TNZ), a temperature range at which the bird functions normally, maintaining homeostasis. At one end is the lower critical temperature (LCT), below which the animal will begin to fail physiologically. It cannot survive unless it warms up. Likewise, there is an upper critical temperature (UCT) at the other end of the TNZ, where the bird will not survive unless it cools down. Birds are able to survive for a time above or below their TNZ but only temporarily. Protracted excess heat or cold beyond the TNZ will cause mortality.

Like many bird species, the Eastern Phoebe is limited by cold temperature in winter. Phoebes migrate south for two interrelated reasons. One is that insect numbers in northern regions of the bird's range are staggeringly low in winter, making adequate daily caloric intake impossible. Another is that it becomes cold and stays cold in the bird's northern range, pushing it beyond its TNZ. Phoebes are able to survive at or above about 25°F (or −4°C) but if the temperature drops below that and remains so, the bird will perish. Well over half of bird species that live in the temperate zone show a similar pattern.

In spring, the migration of the Eastern Phoebe, like that of many other bird species, follows an isotherm line that coincides with its temperature tolerance. For the Eastern Phoebe, the 25°F isotherm determines its overall rate of migration north with the spring. Phoebes begin to appear when local temperatures reach 25°F and above, so they are generally an early spring migrant. Should birds arrive

Eastern Phoebes are among many bird species whose distributions are limited by isotherms. Phoebes require ambient temperature equal to or greater than 25°F (−4°C).

OPPOSITE: A Rough-legged Hawk takes off. Its fully feathered tarsi (part of the foot) account for its common name and also keep its feet warm in cold weather.

back from wintering areas under favorable temperatures but then experience an unusual and protracted cold spell, many will perish either directly from cold or from its effects, such as limited food availability. This event is relatively frequent with species such as Tree Swallows and Purple Martins, which depend on aerial insects for food. Some birds, such as blackbird flocks, may "reverse migrate" some distance south in spring if subjected to a protracted cold spell. Once the cold spell concludes, the birds return.

This Carolina Wren is cold, as shown by its fluffed feathers as it attempts to trap and retain its body heat. Carolina Wrens are quite sensitive to protracted cold.

Data from both eBird and the Christmas Bird Count (CBC) show that Carolina Wrens in the Northeast were hit hard by the extraordinary cold winter of 2015, when Boston, Massachusetts, received a record-setting 110 inches of snow and temperatures on occasion dropped below 0°F. Mortality was high. The Carolina Wren population remained depressed for two years. Carolina Wrens are relatively recent colonizers of New England, and winters probably push them near their lower critical temperature for survival. They are frequent ground foragers as well, and thus the combination of snow cover and bitter cold was more than many could tolerate.

The Snow Bunting is the passerine bird species with perhaps the lowest lower critical temperature. Snow Buntings are known to burrow under snow on severely cold and windy nights. The snow keeps the birds "warm" at 32°F (0°C). That may seem cold, but not to a Snow Bunting. The lower critical temperature for the species is between −40°F and −60°F (−40°C to −51°C). Snow Buntings breed north of 52° north latitude, the most northern breeders among passerines, and have become well adapted to enduring cold.

COPING WITH COLD STRESS

Some of the classic literature about birds contains some fanciful prose, such as "When winter winds rage in the forest and snow thickens the air, Chickadee, merry and unafraid, hustles about amid the storm . . ." (E. H. Forbush, *Portraits and Habits of Our Birds,* National Association of Audubon Societies, 1921). Merry? Unafraid? Who knows? Doubtful though, since if birds don't eat regularly, especially in cold weather as "snow thickens," they will quickly perish. Be that as it may, many birds function well in cold weather, including in snow. Feathers are effective snowsuits. The insulating capacity of feathers enables birds to endure cold environments. Birds have also evolved several widespread behavior patterns and other adaptations that enhance their abilities to cope with severe exposure to cold.

Wind chill enhances the effects of cold, so birds typically face into the wind on cold days. This orientation helps keep the warm air generated by body

Snow Buntings regard snow as sufficiently warm to shelter beneath it. *(Garry Kessler)*

Three Sanderlings and two Ruddy Turnstones shelter behind a rock from a brisk wind. Four other shorebirds are directly in the path of the wind.

heat well within their feathers and against their bodies. But you might also be aware that birds usually face into the wind on other days. It should be obvious that they do this in large part to avoid having their feathers ruffled, which would necessitate extensive preening. Birds typically fluff their feathers in cold weather, trapping air and warming it with body heat, making birds appear bigger on cold days.

Birds lose heat from exposed body parts such as feet, beak, and facial skin. Minimizing such exposure helps conserve heat. It is common in cold weather to notice sparrows and cardinals squatting close to the ground, their feet covered by feathers as they forage. Open-field birds such as larks, longspurs, and Snow Buntings appear to creep along as they forage, their lower legs and feet sheltered beneath their belly feathers. Woodpeckers hug a tree trunk in cold weather, their body feathers fluffed right up against the bark, covering their feet.

It is important to understand that the feet and lower legs of birds are uniquely structured. Birds rarely experience frostbite, though it is possible. The feet of birds have little in the way of flesh or muscles. The tissue is composed of hardened, tough scales, the toes of the feet moving in response to muscle

Snow Buntings keep their bodies low, feathers fluffed and covering their feet, as they shuffle along foraging on a cold day.

This Downy Woodpecker is cold, as evidenced by the puffed feathers protecting its legs and feet.

contractions generated close to the bird's body. The muscles from the tibiotarsus (lower leg bone) connect to toes by slender tendons running along the leg and foot to the toes. The arteries that take blood to the legs and toes are interwoven among the veins that return the blood to the body. This means that warm arterial blood from the body core flows directly alongside cool venous blood from the toes. The result is that heat is exchanged, the warmth of arterial blood transferring to the venous blood and so back up into the bird's body, thus retaining heat in the interior of the bird, saving tremendously on potential heat loss (though there is some).

This unique anatomy is called a countercurrent blood flow system. Because of this system, the feet of a gull or duck standing on ice are comfortable and in no jeopardy of frostbite even though they may be many degrees cooler than the bird's body core temperature. The same is true of waterbirds such as grebes and ducks whose feet are immersed in water that may be near 32°F (0°C). Adaptation to cold is most obvious in species that experience it regularly. Species in tropical regions cannot tolerate frostbitten feet probably because they have a less evolved countercurrent blood flow system. A vagrant immature Great Black Hawk, a tropical species that wandered as far as eastern Maine, succumbed to frostbite on its feet in January of 2019.

On a severely cold day, a crow will typically squat, covering its legs and feet.

A Rock Pigeon on a very cold, windy day in a New England winter.

Birds exhibit various behaviors to conserve body heat. Many species tuck one leg under their belly feathers while standing on the other leg, a behavior that also occurs even when it is not cold. By reducing leg and foot exposure from two to one, birds significantly reduce potential heat loss. As mentioned earlier, birds also squat over their legs and feet, cov-ering them with belly feathers. Many waterfowl and gulls do this as they idle. Birds that roost in trees during cold weather typically squat over their legs and feet, covering them with feathers.

As mentioned in the previous chapter, sleeping birds typically nestle their beak into their back feathers, minimizing heat loss from the beak. That is what a flexible neck does for you. Snowy owls close their eyes, sheltering them beneath well-feathered eyelids. Some bird species inhabiting cold areas have feathers that extend down their legs and cover their feet, adding to heat retention. Ptarmigans are an example, with feet and toes feathered. Spruce, Dusky, and Sooty Grouse have feathered tarsi (upper part of the foot), as does the Rough-legged Hawk. Snowy Owls also have feathered tarsi, as do most owls, even species not found in the far north.

Many bird species are physiologically able to acclimate to cold weather. They do this in several ways. Some species grow more body feathers in the cold of winter, particularly breast down. Some passerines in cold regions may almost double their body feather count in winter. This sort of acclimation is also common in mammals. Many northern mammals, for example, grow much thicker coats in the winter. Some species exhibit increased metabolic rates that generate greater internal heat. Chickadees do this, but when metabolic rates increase, they must feed more frequently.

Redpolls, crossbills, and Evening Grosbeak offer a unique example of coping with short winter days

This male Gadwall is standing on ice, but is likely not uncomfortable because his efficient countercurrent blood circulation is conserving his body heat.

and long winter nights. These species have pouches, called esophageal diverticula, extending from their esophagus. The birds ingest and swallow whole seeds that are then stored in the esophageal pouches. When the birds go to roost, they regurgitate the seeds, process them in their beaks, and swallow them for nourishment.

The daily equation birds must solve in severely cold weather or during heavy winter storms is when to keep foraging and when to go to roost and wait it out. Roosting—becoming inactive, finding protection from wind, conserving energy—is a huge factor in survival. In severe blizzards or ice storms birds are sometimes best served by sheltering in place, remaining at roost in dense plant cover, since they would expend more energy foraging than they would obtain, given the high cost of heat loss. Prolonged roosting in severe weather is known among many species.

Birds also cluster together in cold weather. Many species of birds cohabit tree cavities or huddle closely together in shrubbery. Starlings align shoulder-to-shoulder on utility wires. Huge crow roosts may facilitate heat retention by the sheer numbers of birds as well as their propensity to roost in very close proximity on cold nights.

This Common Eider is enduring a cold winter day with beak firmly protected by feathers.

Snowy Owls are well protected from cold and wind with heavily feathered faces, including eyelids, and feathers covering not only the tarsi but also the toes (a characteristic of other northern owl species).

White-tailed Ptarmigan in winter are fully white, quite cryptic, their tarsi and toes fully feathered. (Ed Harper)

Common Redpolls have sufficient space in their esophagus to store seeds for later processing and digestion.

The Rufous Hummingbird's entire lifestyle depends on it being able to enter torpor and tolerate nightly low temperatures. *(Ed Harper)*

TORPIDITY IN HUMMINGBIRDS AND OTHER BIRDS

Hummingbirds are extremists among birds, living as close to the physiological edge as is possible for an animal with such a high metabolic rate. Hummingbirds use more energy per gram than any other group of birds. They must eat almost constantly to maintain their extraordinarily high metabolic rates. A hummingbird could starve if deprived of food for mere hours. Cold nights present a particularly harsh challenge. But hummingbirds have adapted. Although their daytime body temperature is about

104°F, for some species, body temperature drops much lower at night when the birds roost, as does metabolic rate, resulting in a state called torpor.

Torpor is rather like a form of nightly hibernation. It occurs in most species of hummingbirds, including the many that inhabit cold higher elevations of the Andes Mountains in South America, as well as hummingbirds found in North America, such as the Rufous, the Broad-tailed, and the Ruby-throated. Preparing to enter a state of physiological torpor, a hummingbird begins by finding shelter and going to sleep. During sleep the bird's metabolic rate drops precipitously, usually by up to 95 percent. Its body temperature may drop to only 50°F. Its rate of breathing slows and its heart rate decreases by 90 percent. The bird maintains this reduced metabolism until the sun begins to warm it in the morning. It wakes and, once its body temperature reaches about 86°F (30°C), it resumes activity. Birders rarely observe torpor in hummingbirds because the birds are well concealed. When the bird resumes its morning activity all appears utterly normal, because it is.

Torpor is also known in various species of swifts, close evolutionary relatives of hummingbirds. One caprimulgid species, the Common Poorwill, a bird of Southwestern desert habitats, will enter protracted periods of torpor that last for weeks, closely resembling true hibernation.

HABITUAL WING SPREADING

When vultures roost overnight their body temperature decreases somewhat. Upon waking, they typically compensate for this loss by spreading their wings widely as they directly face the sun. The wings act as solar absorbers, helping raise the body temperature of the bird. Anhingas also use wing spreading to warm up. Anhingas have little in the way of dense insulation, plus they have long necks that lose heat. After diving for prey and emerging wet, they often assume a pose with their wings widely spread and backs to the sun, which warms the bird and dries its feathers. Many cormorant species engage in wing spreading, mostly to dry their wings. Cormorants' feathers are not as waterproof as those of other birds, and thus drying the feathers is an

Turkey Vultures typically spread their wings as they face the sun in the early hours of the day, warming up. The birds' dark coloration and increased surface area (when wings are spread) combine to absorb heat. The birds do not fly until it is sufficiently warm to produce thermal currents that the birds utilize in soaring.

An Anhinga with wings spread, a typical resting posture as the bird dries and warms.

adaptive behavior. It is also adaptive for the cormorant *not* to have strongly water-repellent feathers. This means they have less natural buoyancy so that when they dive they are more easily able to remain underwater in pursuit of fish. Storks and pelicans also habitually spread their wings, perhaps to warm up, perhaps to cool down, perhaps to help cope with wing parasites.

COPING WITH HEAT STRESS

The most obvious way to reduce heat stress is to seek shade, which birds do when the air temperature begins to stress their thermal limits. When you observe birds on particularly hot days, notice how they tend to gravitate, if possible, to shady areas. Robins and grackles still feed on lawns, but if the temperature is high they show a preference for shady areas over sunny areas. Some birds idle in shade, avoiding sunlight. If shade is not available, birds may orient to minimize body exposure to direct sunlight, often facing the sun. By facing the sun, a narrower profile of the bird is exposed to direct sunlight, keeping the bird somewhat cooler.

Birds such as gulls and terns brooding eggs in unshaded colonies move during the course of the day so that they face the sun. Many bird species become less active in the heat of the day, remaining in shade to reduce heat stress.

How might birds lose body heat? Humans accomplish this through sweating. We have multitudes of sweat glands covering our skin surface. Evaporation of water via sweat cools us because of the sizable amount of heat transferred as water evaporates. Birds lack sweat glands, and for good reason. Losing heat through secreting water would dampen the bird's body feathers, particularly down, countering the feathers' insulating properties. So what options do birds have?

One is panting—opening their beaks, exposing tissues that line the mouth, and vibrating their throat muscles. This behavior requires an expenditure of energy. In desert environments, it is common during the heat of the day to observe birds, ranging from roadrunners and quail to wrens and thrashers, idling in the shade of desert shrubs, mouths open, panting. On sunny, hot beaches, gulls idle with their mouths open as well as fly with open mouths. When birds are coping with high ambient heat, their bills wide open, more blood is transferred to the beaks and heat is lost.

Species such as Anhinga, cormorants, frigatebirds, pelicans, and boobies, as well as various gallinaceous birds such as turkeys, exhibit a somewhat similar behavior called gular fluttering. They perch with mouths open, vibrating the throat skin, transferring heat from bird to air. Roadrunners also exhibit this behavior when in hot deserts.

Birds also use their feathers to aid in cooling. They may smooth them, not allowing air to heat up within the feathers. But sometimes they fluff their

This Great Egret (*top*) is trying to stay as cool as possible by panting (its mouth open), by opening its wings, and by remaining sheltered in shade. Note that the Laughing Gulls (*bottom*) shown here are clustered behind the dock piling in the shade on a very hot afternoon in coastal Georgia.

feathers, exposing skin. Though they cannot sweat, exposing skin surface does aid in cooling, especially if there is a breeze.

Gulls, terns, and other coastline birds stand in shallow water on particularly hot days. The water, cooler than air, aids in drawing heat from the legs. Birds also bathe more frequently when it is hot. The white color of gulls and terns is adaptive in hot environments. White reflects heat while dark colors absorb heat. Thus, being white reduces the rate of heat gain from sun exposure.

Some birds partially open their wings to expose the undersides, offering an additional opportunity for heat loss. This behavior is typically seen among colonial species such as herons and egrets as the birds are tending nests on sunny, hot days.

Vultures and storks sometimes resort to what, in scientific parlance, is called urohydrosis: they excrete directly on their legs. The mixture of wet uric acid and fecal matter presumably enhances evaporative water loss from the legs, cooling the birds. It isn't pretty, but apparently it does help. In any case, they do it.

What are the effects of heat stress? Idle time spent trying to cool down compromises foraging time and may also compromise foraging location and efficacy. Loggerhead Shrikes have been observed to forage more in shade during hot days but capture less insect food than when they forage in the open. Heat stress increases metabolism as birds try to lose heat, burning calories that will soon need replacing. Protracted heat stress could drive a bird beyond

Loggerhead Shrikes in the South are more successful foragers when foraging in sunlit fields but must compromise on very hot days by foraging in more shaded areas.

This male American Robin is likely from a northern population, perhaps Newfoundland, and has migrated to Massachusetts to winter. He is somewhat larger in body size than robins that nested in southern New England but still must cope with the cold, as he has by fluffing his feathers.

the limit of its upper critical temperature and exert physiological effects similar to what we humans describe as heatstroke.

BODY SIZE AND CLIMATE

Eastern Gray Squirrels are noticeably larger in New England than in Georgia. So are Pileated Woodpeckers. If you are watching a Common Raven in California it will weigh somewhere around 1.76 to 2.09 pounds (800–950 gm). But if you are watching a raven of the subspecies *kamtschaticus* in northern Alaska, its weight may be a hefty 3.22 pounds (1,460 gm). Similarly, the various subspecies of Song Sparrows found in the Southwest are smaller compared with those farther north, particularly the *maxima* subspecies found in Alaska.

These and other warm-blooded (endothermic) animals demonstrate clinal variation in body size. As latitude changes from south to north, body size increases. This pattern of latitudinal variation is sufficiently generalized that it has been termed Bergmann's rule, after Carl Bergmann, the German biologist who first described the pattern.

The explanation for the cline in body size has to do with average air temperature. As the climate becomes cooler, animals benefit from being larger because they have proportionally greater body mass than surface area, compared with smaller animals of the same species. Put simply by way of analogy, a large cube of ice melts more slowly than a small cube because a small ice cube has proportionally greater surface area for melting relative to its volume. Bergmann suggested that larger animals retain heat better than smaller ones because they have proportionally greater volume (heat-producing area) relative to surface area (heat-losing area). This relationship between surface area and volume is easily expressed mathematically.

Some bird species generally appear to demonstrate Bergmann's rule but other variables such as average humidity also affect body size.

Regional variation in body size is likely to be genetic, though it need not be. Experiments in which Red-winged Blackbird eggs were switched between Florida and Minnesota showed that the birds hatched in Minnesota from eggs laid in Florida grew to the same larger body size as birds hatched from Minnesota eggs. This result shows that regional environmental conditions affect body size. In a way, this still supports Bergmann's rule because in the colder region the bird grows larger, but that result, in this case, is not due to genetics but rather to environment.

Allen's rule, named for its discoverer, Joel Allen, describes another geographic pattern in which body proportions vary with latitude. As in Bergmann's rule, temperature is the critical variable. Allen's rule states that extremities are larger in warm environments and smaller in cold areas. Think of Kit Foxes in southwestern deserts compared with Arctic Foxes in the tundra of Alaska and Canada. Desert-dwelling foxes have proportionally long legs, long snouts, and large ears, each useful in losing excess body heat. Arctic foxes have short and stubby legs, compressed muzzles, and very small ears, all useful in reducing excess heat loss.

If Allen's rule applies to birds, it would be predicted that birds in more northern latitudes would have shorter beaks and shorter legs because such proportions would reduce heat loss. For example, the beaks of Hoary Redpolls and Common Redpolls differ, the Hoary's being somewhat smaller. Hoary Redpolls remain farther north than Common Redpolls in winter, at higher latitudes where it is very cold. Ptarmigans have short legs with feathering all the way to their toes. Their beaks are proportionally short. Rough-legged Hawks exhibit the same pattern: proportionally small feet, small beaks, feathered tarsi. The Ivory Gull, a species that rarely leaves the high Arctic, has a short beak and short legs. Canada Jays have proportionally smaller beaks than jays of more southern latitudes. Snow Buntings have proportionally short beaks considering their overall body size. Most high-latitude bird species are migratory and do not remain at high latitudes during the coldest times of year, essentially escaping the selection pressures for Allen's rule.

It has been shown that species of sparrows that breed in salt marshes (Song, Saltmarsh, Acadian Nelson's, and Savannah) have a larger beak surface

Beak size and internal structure in various subspecies of Savannah Sparrow vary, becoming greater for those that inhabit areas with warm and dry summers. This Savannah Sparrow of average beak dimensions was photographed in the Central Valley of California.

area that correlates with mean maximum summer temperature. This is an apparent example of Allen's rule in that higher temperature selects for greater bill proportions and an implied greater ability for heat loss. Correlated with bill size pattern, the internal components of the bill, called respiratory conchae (part of the sinus cavities), are also larger, at least in Song Sparrows, which inhabit areas with warm and dry summers. Bill shape both internally and externally aids in conserving moisture and in cooling.

Both Bergmann's and Allen's rules point strongly to environmental temperature as a significant variable in shaping evolutionary patterns. And this brings up the topic of climate change.

CLIMATE CHANGE, BIRD DISTRIBUTION, AND THE FUTURE

Temperature is not the only important environmental variable affecting bird distribution, but it weighs heavily. The thermal neutral zone for any bird population is a product of its evolutionary history. Climate change results in long-term changes in average ambient temperature as well as in patterns of weather, such as increasingly severe storms. Many bird species are strongly dependent on a particular habitat type, and when that habitat is affected by climate change, the bird will be affected as well. Though temperature changes may affect birds directly, changes in regional temperatures, as they impact habitats, will affect birds indirectly.

As Earth continues to warm during the present century, there will inevitably be numerous large-scale and small-scale alterations in habitats as well as temporal changes in such variables as time of spring emergence of leaves and insects. These changes are occurring now and are measurable and well documented. All bird species in North America, and indeed the world, will be in some way affected.

A thorough summary of the potential impact of climate change on birds is found with the National Audubon Society's Birds and Climate Report, titled *Birds and Climate Change* (2015), available online at climate.audubon.org. This report examines 588 North American bird species and concludes that 314 species stand to lose greater than 50 percent of their current range by 2080. The details of the report are beyond the scope of this chapter, but it is easy to summarize some of the salient points.

Given that a bird species is adapted to a particular habitat and that habitat is significantly changing, it is likely that the species will then be less adapted. There are two ways in which a population may cope with significant deterioration or alteration of its habitat and living requirements, including its food sources, its shelter, or its comfort level with regard to ambient temperature. One way is through natural selection, evolving to keep pace with the changes. Those less genetically adapted and thus less able to cope with the changing conditions perish at greater rates or will leave fewer offspring than those lucky enough to have the genes enabling them to endure, and thus the survivors leave proportionally more offspring. The population evolves to become better adapted to the changing conditions. This form of change, at the evolutionary level, will perhaps happen to some degree, at least in some bird populations. But because environmental changes now occurring are happening so rapidly it will be difficult for most populations to undergo significant genetic adaptation. They will be reduced to being threatened and then endangered, with extinction a possibility. Historically, throughout geologic times of major climate shifts, extinction rates have increased, often dramatically.

A more likely result of habitat alteration driven by climate change is for birds to locally adapt by adjusting breeding times. Some species now breed earlier, presumably due to climate change. This is an example of local adaptation to climate shift. In a broader sense, distribution of whole ecosystems will likely be affected as Earth warms; deciduous oak forests of the Northeast will move farther north. Carolina Chickadees will eventually replace Black-capped Chickadees in Massachusetts, and Black-caps will remain among the oaks as oak forest pushes the boreal spruce/fir forest farther north, compressing the range of Boreal Chickadee. Ecologists know, however, that ecological effects brought about by rapid climate change are far more varied and less predictable.

Climate change tends to alter plant and animal communities as each species' population responds in its own unique way. Some habitats will shrink dramatically, reducing populations of resident bird species. Groups of species will be brought together that have no long-term history, what ecologists call no-analog communities. Novel interactions among species will result. In general, many more populations will decline than will increase. The global biodiversity of birds and other organisms seems destined to decline.

Many regions in North America show clear changes in bird communities that correlate with warming temperatures, though additional factors also contribute. In New England over the past half-century there has been an influx of bird species historically more typical of the southeastern states. These species include Northern Cardinal, Tufted Titmouse, Carolina Wren, Northern Mockingbird, Red-bellied Woodpecker, White-eyed Vireo, Orchard Oriole, Blue Grosbeak, Acadian Flycatcher, Turkey Vulture, and Black Vulture. For some of these species, their northward influx was likely

These Barn Swallows were photographed in Amazonia, along the Marañón River, in March. They have no obvious cues to changes or trends in ambient temperature on their temperate breeding grounds.

aided by increased regrowth of woodlands as well as the increasing popularity of bird feeding. But average temperature has increased over the same time period as the range expansions. Temperature may not be the only factor in changing bird distribution, but it is a major factor.

Climate change alters food webs in many ways, often not predictable. In some areas, species such as Tree Swallows are returning to breed a week or more earlier in spring than was once considered normal. Tree Swallow return coincides with the earlier emergence of leaves and the concomitant insect flush that accompanies leaf-out. Tree Swallows, at least some populations of them, are behaviorally adapting to the realities of shifting climate baselines. But long-distance migrants such as Barn Swallows, which winter in Amazonia, still return at their historical times, likely because they have no way to detect the changes happening thousands of miles to the north. Long-distance passerine migrants, the many flycatchers, thrushes, orioles,

wood-warblers, grosbeaks, and tanagers, all have the potential to be negatively affected in comparison with resident species and short-distance migrants. Climate changes will produce a few winners and many losers among bird species. Returning to breed when there is less food to be had will greatly increase reproductive stress levels and likely reduce successful reproduction for many species.

Climate change is not merely a matter of warming temperature. Warming affects complex weather patterns, including sea-level rise and changing patterns of precipitation and wind. Several authoritative analyses are available online that describe the complexity of climate change as it affects bird abundance and distribution. See, for example, Marra et al. (2014), cited in the references.

Although some species appear to be adapting by adjusting their migratory timing, some are not exhibiting such adaptation, and species ranging from Red-necked Grebe and Black Tern to Eastern Whip-poor-will, Worm-eating Warbler, and Rusty

Blackbird are considered vulnerable. One of the most vulnerable species is Ivory Gull, a species dependent on Arctic pack ice because, as a frequent scavenger, it seeks carrion such as polar bear kills, as well as feces from polar bears, walruses, and seals, mostly found on pack ice. As ice cover annually shrinks, Ivory Gulls are losing essential habitat. Predicted levels of rising sea waters in the present century also endangers coastal species, such as Seaside, Nelson's, and Saltmarsh Sparrows. These species will continue to nest as usual but will become subject to increasingly high tides as sea level rises. Such high tides drown the nestling birds.

There are data to suggest that some bird species are becoming smaller in body size as the climate warms. A study by Van Buskirk et al. examining body mass and wing chord length showed that body mass has decreased since 1961 in long-distance neotropical migrants such as Rose-breasted Grosbeak, a 3.3 percent decrease, and Scarlet Tanager, a 2.3 percent decrease. Such a trend would be predicted by Bergmann's rule and it may or may not be genetically based.

In summary, climate is a significant variable affecting every bird species, every population. The continuing and accelerating effects of global climate change require diligent documentation. Birders who regularly report their observations on eBird and who participate in state bird atlas projects are contributing in a positive way to monitoring what will be unprecedented changes in bird communities as the century proceeds.

The Ivory Gull is highly vulnerable to current trends in global warming and resultant habitat alteration.

11

SOCIAL BEHAVIOR

BIRDS IN A SOCIAL CONTEXT

I was passing a field of corn stubble and noticed four American Crows foraging in the middle of the field, near one another, heads down, presumably focused on searching for corn kernels. The foraging crows were exposed to potential predators such as Cooper's Hawks and foxes, which are common in the area. But also in the middle of the cornfield was a tall utility pole. Perched on top of the pole was a single American Crow, the sentinel for the group. While its colleagues, each of whom it "knew," foraged, it was on guard, the lookout. That's what crows do.

Crows, as with corvids in general, are among the most social of birds. Social behavior is at the very essence of what it is to be a crow. Crows in winter form immense roosts that clearly reflect their intense social nature. Before roosting, the birds methodically assemble at a staging area, hundreds and often thousands gathering. As darkness approaches, the birds, almost in unison, noisily fly to the nightly roosting location. Much is yet to be learned about the dynamics and interactions of crows as they occupy these huge roosts, but the roosts likely offer strong protection for the crows and may serve as essential information centers for birds that daily range very widely in search of food.

Social behavior is commonplace among birds, but it varies widely in form and complexity. Mating, territorial and nesting behavior, flocking, group foraging, and migrating together represent forms of social behavior. Sociality often requires birds to remember

ABOVE: These Semipalmated Sandpipers are all foraging on a sandy beach. Flocking protects each of them (to a large degree) from predation, but they likely have no social bonds other than species recognition.

OPPOSITE: The Florida Scrub-Jay is one of the most social bird species, with complex group relationships.

The crows shown here are staging for their nightly flight to their winter roost. Crows converge from all directions at staging areas prior to flying to their roosting site. (Craig Gibson)

After staging, crows move at dusk to roost, often in pines but sometimes in deciduous trees. (Craig Gibson)

and recognize one from another as individuals. What we see as a small flock of a dozen Cedar Waxwings, none very different from another, may, to the waxwings, be a social group in which they know each other. We are well aware that each of us humans is unique. So are birds.

What does a bird actually think of its fellow flock mates? The answer is complex.

Herring Gulls exhibit individuality as to how they react in an unusual situation. The famed ethologist Niko Tinbergen liked to manipulate birds to learn about how they think and learn. In one experiment, he and his students moved clutches of three eggs from Herring Gull nests and placed them in plain view about one foot from each nest. Each gull had several choices. It could try and move the eggs back into its nest, and on occasion this happened. It could ignore the nest and incubate the eggs where it found them, but the gulls did not do that. Or it could ignore the eggs and sit on its nest as though it was incubating the eggs. The gulls often did that. An obvious conclusion was that the gull's brain apparently orients more to the nest than to the eggs.

But there was more. Tinbergen learned that not all gulls reacted alike. As Tinbergen wrote, "The tests do not always give exactly the same results. Most gulls do not choose rigidly either the nest-site or the eggs, but they hesitate, looking from one to the other, and even sitting down on each alternately. Also, one bird may be more attracted by the nest-site, while another may be more inclined to choose the eggs. This may depend on innate differences between the individuals, or on the differences in the external situation in which the tests are taken in the different cases, or in different histories of the birds involved." (From *The Herring Gull's World,* pages 145–146.)

What Tinbergen's gull studies showed was that the birds were acting as individuals in their behaviors, and the reaction of any one bird was not necessarily that of another. Tinbergen had no way of knowing if the differences were innate or represented a spectrum of what, for want of a better word, could be deemed intelligence among the birds. Bear in mind, Tinbergen was working with Herring Gulls. Had American Crows been subjected to similar manipulations, they would probably perform differently because they are more intelligent than gulls. But any population of American Crows undoubtedly contains a range of individual learning abilities among the birds.

Natural selection has shaped birds' bodies, but it has also shaped their minds and thus their behaviors. Behavior—the perceptions, reactions, and choices a bird makes daily—is as essential to survival as flying. There is not much leeway in nature for a

Outside of breeding season, this Brown Creeper lives a largely solitary life, not interacting with many other Brown Creepers. It mostly sees other species such as chickadees and nuthatches.

The pugnacious Eastern Kingbird loses its aggressive behavior toward others of its species when it forms large migratory flocks in late summer.

bird that has below-average flying skills or below-average intelligence for its species. That is likely why instinct is such a prevalent component of avian behavior. Recall that during Tinbergen's studies, birds were thought to behave in fundamentally innate ways, somewhat like avian automatons. That is clearly untrue. We know they learn. But given the reality of their fast lives and environmental challenges, slow-witted birds would seem to have dubious futures. Thus the range of what we might call intelligence is likely narrow for birds within each species. Individual birds within a flock are probably more similar in their range of mental prowess than would be the case with a randomly chosen group of humans or chimps.

There is a broad range of sociality among bird species. Some are rather nonsocial, highly individualistic. Dippers do not form intraspecific flocks, nor do Brown Creepers. Indeed, in the course of a typical day, Brown Creepers routinely encounter more chickadees than they do other creepers because they associate with groups of chickadees in mixed foraging flocks. So it may be said that American Dippers and Brown Creepers are not social species, though one might suggest that creepers are perhaps interspecifically social. And it should be noted that

creepers sometimes appear to migrate together, as they are found in close proximity following a night of migration. Perhaps that is as social as they get.

What about Eastern Kingbirds? They are the antithesis of sociality during breeding season, among the most pugnacious defenders of their territories, attacking other kingbirds and numerous other species. But once breeding has concluded, Eastern Kingbirds undergo a personality change and form migratory flocks that remain together during the nonbreeding season in South America. In the tropics flocks of Eastern Kingbirds move around in search of fruiting plants, sometimes intimidating larger fruit-eating species such as toucans by their sheer numbers. Thus kingbirds become social during the nonbreeding season. It is to the advantage of each member of the flock to do so. Eastern Kingbirds are a great example of "changes in latitude, changes in attitude."

Large foraging flocks, no matter what the species, gather for a common purpose and tolerate one another, but generally do not know one another and are minimally interactive. If birds in flocks are interactive at all, the interactions are almost always agonistic in nature (see below). Many bird species, particularly oceanic seabirds, cormorants, herons and egrets, spoonbills, Wood Storks, alcids, gulls and terns, and swallows, nest colonially, a form of sociality discussed in chapter 15. But very few bird species exhibit the level of complexity of social behavior seen in some mammals, such as prairie dogs, cetaceans, wolves, monkeys, baboons, and apes. Among the few that come close are species such as Groove-billed Ani, Brown-headed Nuthatch, Western Bluebird, Red-cockaded Woodpecker, Acorn Woodpecker, Harris's Hawk, and various corvids, particularly Florida Scrub-Jay. In most of these species the social components of behavior are based on reproductive interactions, particularly having helpers at the nest. The birds in these social groups know each other as individuals and interact within that context. Some are discussed below.

SOCIALLY COOPERATIVE BIRD SPECIES: EXAMPLES

Red-cockaded Woodpeckers are specialist species that inhabit southeastern old-growth longleaf pine (*Pinus palustris*) forests. Historically such forests have experienced frequent fires and thus are rela-tively open with an understory of palmetto and grasses. The woodpeckers excavate cavities in living pine trees, particularly those infected with red-heart fungus (*Phellinus pini*). The fungus damages the wood in the trees to the degree that it facilitates woodpecker excavation, but the excavation is still slow and time consuming. The tree secretes sticky resin as the birds excavate, and the dripping resin from the live tree helps protect the cavity from such predators as rat snakes (which are skilled tree climbers).

Red-cockaded Woodpecker is endangered because of the historic loss of habitat over the species' range. Careful wildlife management, particularly at military bases such as Fort Stewart near Savannah, Georgia, or Fort Polk in Leesville, Louisiana, has successfully resulted in habitat restoration and repopulation of Red-cockaded Woodpeckers. These woodpeckers form groups composed of a breeding pair plus up to four nest helpers. The helpers, who aid in feeding young, are usually young males bred in the territory held by the breeding pair. Females disperse more widely than males. Groups are highly

A Red-cockaded Woodpecker in a longleaf pine tree. The species lives in social groups that forage in close proximity and engage in cooperative nesting behavior. (*Jerome A. Jackson*)

Artificial nest cavities have aided in the reproductive success of Red-cockaded Woodpeckers.

Acorn Woodpeckers at one of their granary trees.

social, interactive, and remain together, defending territory, foraging, and moving through the forest as a noisy group.

Acorn Woodpeckers are more complexly social than Red-cockaded Woodpeckers, but the basis of their sociality is similar, namely cooperative nesting, including helpers. Acorn Woodpeckers store thousands of acorns in holes drilled in dead trees called granary trees. The acorns serve as a winter food source but are vulnerable and require protection. Acorn Woodpeckers defend their acorn granaries by organizing into territorial groups of anywhere from 7 to 15 birds. Pairs nest in close proximity, and sometimes two females will lay eggs in the same nest cavity. This may seem socially cooperative, but it sometimes results in competition between females and dumping of eggs by one female into another's nest. There are social bonds as well as social tensions among individuals. Females often mate with more than one male. Usually unmated birds act as helpers, feeding the young. The groups consist of related birds, but females typically disperse, joining other groups. The unique cooperative and territorial system of Acorn Woodpeckers likely developed many woodpecker generations ago, based largely on storing acorns, a behavior also observed in other less social woodpecker species, such as Red-headed and Lewis's. The need to defend

This wet Florida Scrub-Jay is a member of a highly social group based on defense of limited territories and aiding in nest success.

The scrub habitat of the Florida Scrub-Jay has been severely fragmented, significantly reducing the population of the species.

a uniquely large store of acorns is perhaps the impetus for the further evolution of the Acorn Woodpecker sociality. Storage of acorns is not a sufficient cause to make Acorn Woodpeckers as social as they have become. Other woodpecker species such as Red-headed Woodpecker also store prodigious numbers of acorns but do not develop such social tendencies as are exhibited by Acorn Woodpeckers.

Florida Scrub-Jays inhabit fragmented patches of open scrub in central Florida. Once their habitat was far broader and contiguous, but that is no longer the case and their overall population is greatly diminished, only 10 percent of what it once was. Because they are relatively long-lived and real estate is scarce, dispersal of individuals to establish new territories is limited. Territories remain in families. Within many jay species some pairs have helpers at their nests, usually offspring from the previous year. The commitment to helping varies among individuals. In the case of Florida Scrub-Jays a dominant nesting pair may have up to six helpers, all offspring from not only the previous year but also earlier years. The helpers feed offspring, act as sentinels, and vigorously help defend the territory. Females tend to disperse to other territories and are welcomed into other groups, but male dispersal happens far less. Should the dominant nesting male perish, one of his sons will take over the territory. It

is thus in the best interests of male offspring to stay with dad. Their options are limited.

COOPERATIVE SEARCHING AND HUNTING BEHAVIOR VERSUS GROUP FORAGING

Resource distribution is often patchy in nature. Species such as waxwings and crossbills wander widely in search of suitable food plants. Various cone crop species, particularly pinyon pine in western states, provide selection pressures for sociality in some bird species.

The Pinyon Jay, a specialist species, has a social structure based on annual variability of pinyon pine cone crops from one location to another. Nomadic flocks of Pinyon Jays in the American West roam in search of clusters of pinyon pines bountiful with cones. The jays are social throughout the year, forming nesting colonies. The birds are effective dispersers of pinyon pine seeds, and the ecology of pinyon pine is thus dependent upon them.

Pinyon pines periodically have bumper cone crops. The nomadic movements of the jay flocks help them locate areas where the pines are dense with cones, and the jays go to work. A flock of about 250 Pinyon Jays may cache up to 30,000 pine seeds daily. Over several months this rate of caching results in some four million seeds being cached. The jays often cache the seeds at long distances from the trees themselves, and they typically place them in areas conducive to seed germination, acting as essential seed dispersers. In return, the larders of energy-rich seeds enable the jays to breed earlier and successfully. Pinyon pines and Pinyon Jays are thus locked into a mutual interdependency in which social behavior is required of the jays.

Many birds routinely hunt together in flocks, swarming over fields or mudflats or skimming in the air for insects, but this is not cooperative hunting. It is group foraging. In the Southeast, fields where dragonflies are massing may attract multiple Mississippi Kites and Swallow-tailed Kites. The kites are drawn to the food source, but they are not interacting to enhance capture rate. They are just good at spotting a food source, usually by observing each other's behavior, a very common event in the lives of birds. Two or more Ospreys often locate fish schools more successfully than individuals hunting alone. This is perhaps one reason why pairs of Ospreys have little aversion to nesting in close proximity with other pairs. But Ospreys do not cooperate in capturing prey, only in finding it.

American White Pelicans routinely forage by grouping together and forcing schools of fish into confined areas where the pelicans are able to scoop them up. This coordinated behavior of pelicans appears purposeful and cooperative. Aplomado Falcons also hunt cooperatively, coordinating their attacks. Peregrine pairs sometimes appear to hunt in a cooperative manner, as do eagles.

The Harris's Hawk does on occasion become a true cooperative hunter. Groups of three or more Harris's Hawks act in concert to capture ground squirrels and rabbits. Some of the hawks force the potential prey animal into the open, where it may be taken by one of the group. Others in the hunting group share in devouring the prey. Harris's Hawks are facultative cooperative hunters, also hunting as individuals, usually for small prey.

Common Ravens have been observed to work in teams of two or more to steal prey from raptors. Ravens have been widely reported to combine their efforts in killing dying or injured animals.

PECK ORDER AND DOMINANCE HIERARCHIES

It has long been well known that chickens in a flock form a dominance hierarchy. The common term for dominance hierarchy has thus become *peck order* or *pecking order*, a term now even applied to human social organization. Chickens in a flock assert their dominance by pecking other chickens. This could quickly become dicey—lots of annoyed and bleeding chickens—but chickens soon learn that some birds, which they recognize as individuals, are best not to peck because they react strongly and win the

Two Pinyon Jays atop a pinyon pine. *(Ed Harper)*

Pinyon Jays, like this one on a cedar, move in nomadic colonies, always in search of a robust crop of pinyon pine cones. *(Ed Harper)*

pecking contest. By exploratory and persistent pecking, the peck order emerges.

In chickens, the dominance hierarchy is among females since several females typically associate as a flock, with a single rooster. (And yes, the rooster is dominant over the hens, all of them.) The dominant hen not only is free from being pecked by others but also gets to choose where the flock feeds and gets to feed first. Studies of chicken dominance hierarchies have shown them to be complex and nonlinear. Dominance hierarchies also frequently change

American White Pelicans come together to force fish into a high concentration, and then they scoop up the fish, a form of cooperative hunting.

Common Ravens, here following a coyote, cooperate to ensure access to carrion. *(Ed Harper)*

Harris's Hawks are true cooperative hunters. *(Ed Harper)*

within the flock. In order to establish a dominance hierarchy, it is obviously necessary to recognize and remember each member of the group. Thus dominance hierarchies tend to break down or not be established at all in large flocks.

During much of the year, Black-capped Chickadee flocks are organized in complex dominance hierarchies. Chickadee societies are being studied using applied social networking theory (see below). Birders sometimes notice agonistic interactions among chickadees at bird feeders when one chickadee displaces another. But in chickadees, the dominance hierarchy is typically expressed by pairs, not individuals. A dominant pair occupy a top position in the flock, which usually numbers about 8 to 12 birds. However, it is also true that males generally are dominant over females except when males are in their first year or are just joining the flock. The longer a bird is a flock member, the more likely it is to move up in dominance, so in chickadee society seniority counts. Chickadee flocks typically include some low-ranking birds called *floaters,* a term also applied to nonterritorial, unpaired birds during breeding season (chapter 14). Floaters attempt to attain acceptance into a social flock. If they succeed they begin as low-ranking birds in the dominance hierarchy. Should a higher-ranking bird perish, a floater may move into its spot.

Dominance hierarchies are fluid in foraging flocks of juncos, sparrows, Horned Larks, and Snow Buntings. That is because these flocks are normally rather large, frequently augmented with new members, making it impossible for all birds to recognize one another sufficiently to sustain a tightly structured dominance hierarchy. There is thus much squabbling, displacement, and momentary fights among flock members as birds forage in close proximity. In junco flocks adult males are generally dominant to females and first-winter birds.

Dominance hierarchies form because they maintain order and reduce energy use per bird. In this way, even a low-ranking bird is better off than it would be if it were to be constantly contesting with other birds.

AGONISTIC BEHAVIOR AND FIGHTS

Agonistic or hostile behavioral displays are widespread in avian sociology, far more than cooperative or benevolent behavior. Birds that fight all the time would risk life and limb as well as waste time and energy. By developing ritualized agonistic social behaviors that signal an individual bird's potential for dominance or, on the contrary, its willingness not to engage (appeasement), overall risk of injury is reduced and physical combat reduced or altogether avoided. Dominance hierarchies are adaptive. Even

A female Boat-tailed Grackle defends her access to the trash bag, clearly threatening the other female.

A male Ruby-crowned Kinglet with crest erect, signaling that he is agitated. *(Jeffory A. Jones)*

These two displaying male Wild Turkeys are breaking off the contest. Notice that the bird on the right is looking down, a clear appeasement gesture.

subordinate birds benefit because social knowledge of their status saves them from continued bullying and harassment.

Threat displays usually have the following in common. The bird attempts to look larger, raising its head or opening its wings or acting in an aggressive manner, often accompanied by vocalizations. Many birds raise their crests or erect their head feathers when agitated. Aggressive display includes brandishing head and wings prominently, commonly seen in threat displays of owls. White-breasted Nuthatches have a threat display in which they momentarily bow and lunge with wings and tail spread, often observed when the birds are in close proximity to other birds at feeders. The bird doing the threat stares directly at its would-be antagonist, often with its head forward. Eye contact is essential in a threat display.

Appeasement displays function to turn off the potential aggression. Like a dog or wolf with its tail between its legs, the appeaser signals that it will not engage further, and usually the tension subsides.

Appeasement displays in birds involve the appeaser looking away, breaking eye contact. It will often put its head down and sometimes lower its body, making it appear smaller. Crouching is common as a form of appeasement.

Ritualistic displays of aggression and appeasement do not always result in a peaceful outcome. Unsurprisingly, actual fights related to territoriality or contesting for food do occur. To attain dominance in a threat display, the bird must communicate to the other bird that it is, indeed, the stronger and more aggressive of the two. But if the birds are equal in that regard the contest may not be settled without at least some physical contact. The fights are usually of short duration and do not result in serious injury. Often a fight involves only momentary contact and then it is over. Two goldfinches at a feeder get a bit too close to one another and aggressively make beak contact only to quickly break off. But sometimes fights become protracted, as when gulls contest for food, for example. Still, injuries are uncommon. Sometimes an individual bird will

These two Great Blue Herons are about to avoid a fight. The bird on the left is attempting to retreat as the bird on the right has asserted its dominance.

chase another for a considerable distance before breaking off. This should not be surprising as the fight may have ensued when an interloper bird entered another's territory. The interloper has to go.

Instances of interspecific aggression are also common. Common Ravens are often attacked and harassed by American Crows. Ravens are larger than crows but crows gang up on ravens. Usually at least three crows will be involved. If the raven is perched, the crows surround and vocally harass it,

Two Semipalmated Plovers in a confrontation. The one on the right is showing an appeasement display. *(Kevin T. Karlson)*

OPPOSITE TOP: Interspecific aggression is common among birds, as is the case here with a Great Shearwater and a Herring Gull competing for access to chum. Concentrated food sources force aggressive behavior among birds.

OPPOSITE BOTTOM: These two Great Black-backed Gulls were contesting for access to chum tossed out during a seabird cruise. It is not uncommon for gulls to aggressively grab at wings of other gulls.

approaching the larger bird closely until the raven flies. Crows then pursue the raven, loudly *cawing* as they do. Why do that? One study by Freeman and Miller suggested three possible reasons, each logical from a crow's standpoint. First, just as crows are nest predators of smaller birds, Common Ravens prey on crows' nests. Having a raven pair in the neighborhood is potentially dangerous to crow reproduction. Driving a would-be nest predator away is therefore adaptive. Second, when several crows are involved, harassing a raven is not very risky to the individual crows. The raven flies away, seemingly just to be rid of the crows. Third, though crows harass ravens more during breeding season, they also go after ravens when not breeding. However, this is the season when food can be in short supply, and ravens do compete with crows for food.

Thus there is just no time of year when American Crows want to share space with Common Ravens. The aggression shown toward ravens by crows is entirely logical.

Sometimes aggression among birds is difficult to explain in adaptive terms, particularly interspecific aggression. A wintering Common Loon at Fresh Pond Reservoir in Cambridge, Massachusetts, was reported by J. B. Miller to have acted aggressively toward Common Goldeneyes, Hooded Mergansers, and Greater Scaups. The loon would assume an aggressive posture with its neck flattened against the water, bill facing the duck, threatening the various species of ducks in turn. The ducks moved away from the loon, which was larger than any of the waterfowl. The loon's behavior was puzzling because the birds that it chased are not serious competitors and posed no threat to the loon. When on their breeding lakes, Common Loons routinely threaten other birds that may pose a danger to their young. Loons are fiercely territorial, sometimes fighting to the death with other loons (chapter 14). Perhaps this loon was still affected by a high testosterone level and was unable to be other than aggressive.

BIRDS AND SOCIAL NETWORK THEORY

Social network theory (SNT) was invented to analyze how human social behavior works to form complex social networks, structuring relationships and loyalties. It has been incorporated into animal behavior studies and has been applied extensively in mammal and bird community analysis. Individuals are termed *actors* and function as *nodes* in the network. *Ties* are the actual interactions between individuals (diads), and the degree of connectivity among ties represents relationships among the various actors, the actual social structure. The complexity of the social network is visualized as a diagram depicting the nodes connected by weblike lines that represent the ties.

The essence of SNT rests in the reality that in human societies various individuals size one another up and form alliances of varying strengths—alliances subject to change, a common proclivity of humans. Alliances typically change over time and with the addition or loss of specific individuals or with changes in connectivity between individuals. SNT does not apply if individuals are unable to discern each other as individuals and form specific and meaningful relationships.

An example of SNT applied in the bird world would be how a flock realigns after a dominant individual is lost or as new individuals attempt to join the society. Social networks are not simple dominance hierarchies. Different individuals react differently to the removal of other individuals, or to the addition of new individuals. The very fact that SNT has been successfully applied to the analysis of some avian communities, such as chickadee society, speaks to the reality that some bird societies are structured by complex group interrelationships.

In Oxford, England, at Wytham Woods studies of Great Tit and Eurasian Blue Tit populations by Josh Firth and colleagues focused on how social networks among Great Tits change when individuals are removed from the group (simulating loss as a result of predation). In the experiments individuals were removed and subsequently returned. As more

Two male Mallards are engaged in a fight of brief duration. Neither is likely to suffer any injury. (*Bruce Hallett*)

The Black-capped Chickadee leads a complex social life as part of a changing social network. The birds in this photo interact on a daily basis and form a social network.

of the birds were "lost" (captured and held for a few days in an aviary) the remaining individuals altered their social connectivity. Relationships changed throughout the flock. Many studies of birds now employ SNT in assessing the dynamics and structure of both intraspecific flocks and interspecific flocks (such as stable mixed foraging flocks). SNT has shown that birds are not species-specific in their social networking. For example, where they overlap, Black-capped and Carolina Chickadees may form diads within a social network.

Observing and analyzing social network structure requires sophisticated techniques to mark and track each individual bird plus subsequent analysis of numerical data using computer-generated algorithms. Thus it is beyond the reach of birders observing behavior in the field. But when you see a flock of chickadees, crows, jays, or juncos, or a mixed-species foraging flock of chickadees, nuthatches, woodpeckers, and kinglets, realize that they may be bonded together in complex social understanding that changes with circumstances and experience with each other. Such social complexity is a mark of avian intelligence and cognition.

This Common Loon in basic plumage in winter may exhibit aggressive behavior toward other birds in close proximity, including ducks.

PROVISIONING AND PROTECTION

BASICS OF WATCHING FORAGING BEHAVIOR

Birds have to eat and eat often. A study on bird impact on arthropods (insects, spiders, and other joint-legged chitin-bearing creatures) by M. Nyffeler and colleagues concluded that birds throughout the world consume between 400 and 500 metric tons of arthropods annually, most of which are eaten by forest-dwelling species. For most species, especially small birds, eating is the dominant activity. This means that when you go birding, you often are watching birds in search of or consuming some kind of sustenance. But there is more. Birds are never really safe, at least not for long. Predation always looms. Food must be found and consumed while potential danger is ever present. And that is a delicate behavioral balancing act.

Smaller birds spend more time foraging than larger species because they have higher metabolic rates (energy required per gram of body tissue). Small species such as wood-warblers, chickadees, thrushes, sparrows, and hummingbirds continually refuel. Larger species such as gulls, geese, herons, hawks, and vultures may spend considerable idle time between foraging efforts because their metabolisms are slower and they normally take in proportionally large food items, more grams of food per meal, that must be digested over time. A Say's Phoebe and a Common Yellowthroat need to eat throughout each day. A Herring Gull and a Red-tailed Hawk do not.

Birds develop keen search images, seeking food in multiple ways among various habitats: moving through trees and shrubs or grassy fields or along tidelines and mudflats or in the air. Searching requires time and energy to locate and recognize food items, time to capture and devour them, as

A large bird, such as a Black-footed Albatross, does not require nearly the amount of food per gram of body tissue as a small bird, such as a kinglet. The albatross feeds on fish, squid, and various crustaceans.

well as risk of exposure to potential predators. As an insightful exercise in frustration, try simulating being a chickadee or some other small bird. Look carefully at tree branches and bark. Check the leaves. Look at the ground. How much food do you see? How long does it take to find a bug, a caterpillar? You likely will not see much. But birds see potential food with amazing acuity that you cannot begin to match.

A small bird such as a Ruby-crowned Kinglet seemingly forages constantly, restlessly flitting from branch to branch, tree to shrub, periodically hovering at a leaf or leaf cluster. Many birds, particularly wood-warblers, exhibit similar foraging intensity. Watching such species might make you wonder if the continual effort, the seeming hyperactivity, is worth the return. When a jay hastily loads its crop with acorns collected beneath an oak tree, it's pretty obvious that the jay is getting a good return for its energy investment. But kinglets? Well, they do find

OPPOSITE: An American Woodcock with a captured earthworm, an essential component of its diet. An old name for woodcock was "bogsucker." Woodcocks are very good at finding and extracting worms. They have to be.

The male Red-breasted Merganser with the fish did manage to keep it and swallow it even while the other male remained in hot pursuit. Once the fish (which appears to be a species of sculpin) was swallowed (head first, whole, and still alive), the duck had some digestion to accomplish and perhaps did not eat for the remainder of the day.

sufficient food, at least usually; otherwise there would be no kinglets. Consider that kinglets, as well as many other species, such as wood-warblers and chickadees, weigh precious little. Moving comes very easily to such a creature since there is little mass to move. Their seeming hyperactivity is not as energy costly as it might, at first glance, appear to be. With good search images and quick reaction times, kinglets and other small, active foragers find, capture, and swallow food so quickly that birders may miss it, even as they observe the bird performing the behavior.

Many flocking species, in particular wintering Snow Buntings, seem unduly restless. A feeding flock will frequently take to the air, the birds all calling, and make sweeping, circling flights around the feeding area, usually an open stubble field or sand dune. They then often land pretty much in the same place from which they became airborne. Why do this? Why waste energy? Merlins are why, or an accipiter. The smaller birds spend energy surveying

in an attempt to ascertain the risk of predators being around. Their vigilance is constant and they take no chances. Once one goes up, most of the rest do, and immediately. Shorebirds follow a similar pattern.

Foraging behavior is often obvious. A Belted Kingfisher or an Osprey hovers over water and dives, the kingfisher head first, the Osprey feet first. Both are attempting to capture a fish, each having seen the would-be piscine target while hovering above the water. A hummingbird pauses at a flower, its bill probing within the flower as it consumes nectar. A California Scrub-Jay comes to a mixed-seed feeder and flicks its beak among the various seeds, selecting and swallowing peanuts, rejecting other seeds, even as a Mourning Dove sits on the feeder and methodically swallows dozens of tiny millet seeds. A kingbird or phoebe sallies forth from a perch, its beak snapping audibly as it captures a flying insect. Vultures tug at carrion, extracting strands of decomposing flesh. A sandpiper or plover pulls a marine worm from the sand, just as a robin

pulls an earthworm from soil. Waxwings gulp berries whole and Pine Grosbeaks mash and swallow crabapple fruits. Birders spend many cumulative hours watching birds feed because that is much of what birds do.

Some foraging patterns are subtle. Species such as Red-eyed Vireo move methodically through shaded foliage gleaning various cryptic caterpillars or other insects, repeatedly singing as they do so. It doesn't necessarily appear that the vireos are foraging, but they are. Sanderlings scurry on beaches, scampering back and forth with the waves, quickly and adeptly capturing small invertebrates that are momentarily exposed by the retreating water. Flocks of swifts or swallows course about overhead among myriad small airborne insects (the equivalent of aerial plankton) which these birds are ingesting, one after another. They essentially fly through their food, snatching it up as they go. A Red-tailed Hawk sits and waits, idle on a utility pole along a highway, head turned toward the ground, likely searching for voles in a bordering field. What are your chances of observing that hawk make a capture? Perhaps good if you are willing to invest the time watching it. And it may require a fair amount of time. Buteo hawks are patient hunters.

Blue Jays typically swallow acorns or seeds whole, packing them into their crops before flying to cache or eat them. *(Bruce Hallett)*

OPPOSITE TOP: Look closely at the berry still visible as this Cedar Waxwing is about to swallow it. Waxwings are devoted to a diet of fruit.

OPPOSITE BOTTOM: This Say's Phoebe is about to capture an insect. It flew from its perch on a fence, hovered briefly, and then dropped into the grass to grab its prey.

THERE MUST BE 50 WAYS . . . TO CATCH A FISH

Fish are abundant, diverse, and nutritious. Collectively, the world's salt- and freshwater fish represent a diverse resource spectrum for avian predators. Among fish, size varies, habitat varies—much varies. Many families and species of birds have adapted to forage primarily on fish, but in remarkably different and distinctive ways. The high diversity of fish-consuming bird species results from fish representing such a broad spectrum of potential food resources.

Birds have evolved numerous methods to catch and devour fish. Consider how distinct from each other are the following groups of birds, all of which feed entirely or heavily on fish: penguins, pelicans, mergansers, loons, grebes, cormorants, anhingas, gannets, boobies, frigatebirds, jaegers, gulls, terns, skimmers, alcids, herons, egrets, bitterns, storks, Osprey, and kingfishers. That's a lot of species, and the list is not all-inclusive. I once rehabilitated a male Baltimore Oriole that taught himself to perch on my aquarium and catch my prized tropical fish as they came to the top to get their fish food. The oriole was a quick study.

Fish represent a driving, powerful, evolutionary selection pressure in shaping bird behavior and anatomy. Fish, over evolutionary time, have, by their

These examples represent a fraction of the diverse groups of birds whose diets are primarily made up of fish: (*top left*) Common Tern *(Garry Kessler)*, (*top right*) Double-crested Cormorant *(Jeffory A. Jones)*, (*bottom left*) Osprey *(Jeffory A. Jones)*, (*bottom right*) Great Blue Heron *(John Kricher)*.

immense diversity, directly contributed to the global diversity of birds. Insects have done admirably along those lines as well, and plants, ranging from various seeds to various fruits (and occasionally leaves), have also helped provide for a cornucopia resulting in evolution of numerous bird species, each focused on its own particular food base and with its own suite of adaptations, both anatomical and behavioral. Food acquisition is a major driver of the evolution and diversification of birds.

FINDING FOOD: SEARCH IMAGE AND FORAGING DIVERSITY

"Look for the golden arches." The twin yellow arches on a tall sign often visible at long range along an interstate highway form a search image for hungry, fast food–prone travelers. Birds are quick to develop search images both through their genetic endowments and through learning.

Feeding behavior, like most other bird behavior, is genetically based but strongly augmented by continual, often innovative learning. Newly hatched grouse and quail instinctively begin pecking at objects on the ground. Soon they learn to differentiate and recognize specific food items. Newly fledged puffins, who find their way from their nesting burrows to the sea with no help from their parents,

instinctively know to dive and pursue fish. Some birds are highly adept at exploiting novel food sources, at least novel for that particular species. Herring Gulls have been known to use bread to lure goldfish to the surface of ponds. Some Green Herons have learned the same trick, using bait of various sorts to attract fish to within striking range.

At least one Eared Grebe taught itself to forage for food without diving. Grebes normally dive in pursuit of their food, but this unique individual, documented by J. T. Wilcox, allowed itself to be beached by waves and, once the wave retreated, it stabbed at the sand, capturing amphipods, a novel behavior for an Eared Grebe. Once the waves returned, the grebe floated to another location and continued foraging, temporarily beached, its flexible neck stabbing at the sand as it captured prey. Such behavior is learned, self-taught. The ornithological literature abounds with examples of birds performing novel foraging behaviors, many first reported by keen-eyed birders.

There is a kind of "Where's Waldo" component to how birds find food. Each species is attuned to see (or in some cases feel) specific characteristics that the birds quickly differentiate as food items, to find a tasty "Waldo" among a broad spectrum of possibilities. Birds subconsciously filter out nonfood

Green Herons occasionally use some form of bait to lure fish to the surface, where the heron may capture and devour them. Most Green Herons, however, do not employ that trick. *(John Grant)*

Brant, once specialists on eelgrass, have successfully adapted to feeding largely on leafy green algae.

imagery, similar to how a person in a crowded room filters out human faces as she searches for a friend within the crowd. A spectrum of search images allows a bird to efficiently scan its environment, spotting potential food in a complex landscape. Birds such as robins, thrushes, waxwings, starlings, and others recognize fruit-laden trees and bushes by using the ripening of fruit in late summer as a visual signal. In winter woods, American Robins are quick to spot reddish orange bittersweet fruits or brightly ripened fruits of other plant species such as winterberry and sumac. Plants "advertise" their fruits to birds because the real function of a fruit is to be consumed and have the seed(s) within it dispersed.

Chickadees have search images attuned to spotting inanimate food items such as insect egg cases, but they are also skilled at finding live food. And, of course, chickadees "know" birdseed and suet when they encounter it. A Brown Creeper methodically spirals up and around tree trunks, usually starting at the base of the tree, its search image for food such as spiders, insect egg cases, and pupae triggered virtually by walking up to it. Look at bark very closely. How many food items are evident to you? Could you make it as a Brown Creeper? And, when the going gets really tough in winter, Brown Creepers, at least some of them, have learned to eat seeds and suet at bird feeders, perhaps by associating with and observing chickadees.

In a multidimensional structurally complex habitat such as a forest, various bird species are adapted to feed in distinctly different ways, something immediately obvious when comparing the foraging behavior of species such as Great Crested Flycatcher, Pileated Woodpecker, Scarlet Tanager, Wood Thrush, Ovenbird, and Ruffed Grouse, each of which may cohabit the same tract of forest but each of which perceives its food spectrum differently from the others. The tanager and flycatcher tend to feed well within the leafy canopy, but the flycatcher sallies out to snap up flying insects while the tanager moves methodically along branches and gleans food (such as caterpillars) from the leaves. The Wood Thrush hops in search of food on the ground or seeks food in the shrubs and understory trees, where it nests. The woodpecker explores tree snags from forest floor to canopy in search of bark beetles and ants. The Ruffed Grouse and Ovenbird each walk along the shaded forest floor, the Ovenbird searching for prey that would be of little or no interest to the grouse. The vertical structure of a habitat such as a forest actually provides multiple subhabitats, rather like layers of a cake, inhabited by multiple bird species.

On the other hand, less structurally complex habitats such as a grassland and meadow offer less "bird space," reducing foraging possibilities, and thus these sorts of habitats tend to have fewer bird species. It was long ago realized that what is termed *foliage height diversity*, a term encompassing the structural complexity of a habitat, correlates closely with bird species richness, the number of species found within the habitat. Forests, which provide the "high-rise buildings" of habitats, have greater foliage height diversity than grasslands or shrubby habitats. Mudflats may appear to be the least diverse habitats available to birds, utterly two-dimensional, but mudflats actually abound with diverse and different-sized invertebrates burrowing at various depths in sediment that changes with the tidal flux. Thus mudflats offer a richly diverse source of food for birds, helping to account for why so many shorebird species that feed on mudflats differ in characteristics such as body size, leg length, and bill characteristics (see "The Foraging Niche" below).

Plants provide multiple food sources for many species of birds. Many land birds feed on seeds and fruits, and hummingbirds specialize in nectar feeding. Far fewer bird species feed on leaves, though some species, in particular grouse, will devour evergreen needles, and some species clip young leaves and eat flower buds. The reason birds do not feed on leaves is because of the difficulty of digesting the cellulose and other complex structural and defense compounds contained within a leaf. Birds' digestive systems work rapidly, helping keep their weight down, sparing them of the burden of carrying heavy undigested boluses of food while flying. Hard-to-digest foods such as coarse, chemically complex leaves offer no advantage to birds.

Waterbirds feed on a wide range of food items, including submerged vegetation and multiple species of invertebrates and fish. Even algae and cyanobacteria (formerly called blue-green algae) are food for some birds. Along with feeding on various forms of shrimp, flamingos filter tiny single-celled cyanobacteria as essential components of their diet.

Eastern populations of Brant, once specialist foragers on eelgrass (*Zostera*), have adapted to consuming leafy green marine algae (*Ulva*) as well as feeding terrestrially on grass, as Canada Geese do (perhaps the Brants learned to feed on grass by watching Canada Geese). This dietary shift was necessitated by a major blight beginning in 1931 that decimated eelgrass and that, in turn, caused mass starvation among Brants. But some Brants adapted to eating green algae, and those were the ones that successfully reproduced. Gradually an overspecialized species broadened its feeding preferences, and today Brants thrive, largely by consuming green algae.

THE FORAGING NICHE

The word *niche,* when used in ecology, refers to the spectrum of needs and requirements that must be fulfilled for an organism to survive in its environment. When it's applied to food, each bird has a characteristic foraging niche that encompasses its choice of feeding habitats, locations within those habitats, search images, food choices, and methods

OPPOSITE: White-breasted Nuthatches (*top*) and Red-breasted Nuthatches (*bottom*) are each bark foragers but differ in habitat and food preferences. As is typical, the White-Breasted Nuthatch is in a deciduous tree and the Red-breasted is in a conifer.

of feeding. Niche is roughly equivalent to the bird's "profession"—how it makes its daily living.

Woodpeckers, nuthatches, creepers, and a few warblers such as Black-and-white are in the bark-foraging profession, often called the bark-foraging guild, a reference to medieval trade guilds. But within that profession the methods of foraging vary among the species, and thus the specific foraging niche of each species is in some manner distinct from those of the other species. Nuthatches typically hitch head-first down a tree trunk, probing bark for food items. Nuthatch species generally occupy different types of woodland and also have differences in range. Pygmy and Brown-headed Nuthatches are similar inhabitants of pine forests, but one species is in the West, one in the East. Red-breasted and White-breasted Nuthatches may sometimes occur within the same woodlot, but Red-breasteds are mostly inhabitants of woods abundant with conifers, while White-breasteds are inhabitants of woods with broad-leaved trees. The two nuthatch species overlap in diet but are by no means identical in food or habitat preferences.

Food diversity leads to bird diversity, not the other way around. Within a forest during summer you might observe a Red-shouldered Hawk and a Cooper's Hawk. Both are sit-and-wait predators. But the Cooper's Hawk skulks patiently within the shaded woods until it spots a potential prey item, usually a bird but sometimes a small mammal. The Cooper's Hawk then swiftly attacks. The Cooper's Hawk is an accipiter, and its rounded wings and long tail afford it high maneuverability in dense forest, helping provide the element of surprise as it swiftly navigates the branches and hones in on its target. The Red-shouldered Hawk, a buteo, cannot maneuver as skillfully as a Cooper's Hawk. It perches in a tree, sometimes in full view, often overlooking a stream or pond, where it scans for food in the form of small mammals, frogs, snakes, and occasionally birds, crayfish, or insects. Once it spots prey it will drop quickly on its intended target. Its diet is broader than the Cooper's Hawk and its foraging behavior different. The two species do not compete much with each other because their main prey choices and methods of capture are different.

Herons and egrets differ in body size, leg length, and bill length. It is possible to observe Snowy Egrets, Great Egrets, Great Blue Herons, Reddish Egrets, Tricolored Herons, Little Blue Herons, and

The Cooper's Hawk is a stealth predator. The cottontail probably never saw it coming. *(John Grant)*

Green Herons all foraging at the same time in the exact same shallow-water lagoon, estuary, or marsh. They are often joined by Roseate Spoonbills, Wood Storks, White Ibises, and Glossy Ibises. Ultimately, each of the egret and heron species will strike swiftly at the water in an attempt to make a capture, usually a fish. Each species employs various foraging behaviors. In general, Great Egrets, Great Blue Herons, Snowy Egrets, and Little Blue Herons typically stand still and stare into the water—they are stand-and-wait predators that focus intently on a target fish coming within striking distance. Then the bird leans forward with neck tightly curled, eventually striking swiftly, sometimes impaling prey with its sharp beak. Tricolored Herons often spread their wings, shading the water, attracting fish, and minimizing

glare. Reddish Egrets are known for their elaborate dances, as they scamper around, wings spread, often jumping short distances as they stir up potential prey. Snowy Egrets, as well as other heron/egret species, often stir water with their feet, presumably to expose prey. Snowy Egrets also habitually make short flights over fish-rich water, dangling their legs, their feet just touching the water surface, suddenly thrusting their neck between their feet, the beak grabbing a fish, all while still flying. The Green Heron, smallest of the group, usually forages from the bank of a pond or marsh, staring intently and using its long neck to strike prey.

Spoonbills, Wood Storks, and ibises feed differently from egrets and herons, relying on sensory organs in their bills to capture food. Spoonbills use

This Snowy Egret is skimming its feet across the water and dipping its bill as fish surface, where it captures them, one of its various kinds of foraging behaviors. *(Bruce Hallett)*

An assemblage of wading birds gathered to feast on a dense concentration of fish in a marsh. The species here are Snowy Egret, Tricolored Heron, Great Blue Heron, White Ibis, Roseate Spoonbill, and Wood Stork. *(Bruce Hallett)*

their unique bills to filter food items such as small fish and numerous invertebrates as they methodically swing their partially opened beak back and forth in the water. Wood Storks probe below the surface, beaks partially open, instantly snapping shut on fish or crayfish as they encounter them. Ibises probe deeply into the sediment for burrowing prey species. The anatomical and behavioral differences among these species reflect both the diverse array of food available as well as how each species has specialized to tap into its own specific part of the food spectrum.

Various shorebird species feed on mudflats at low tide. Most of the food they seek is within the mud. Small sandpipers and plovers extract worms and mollusks close to the surface. Ruddy Turnstones methodically probe among stones on coarse sediment or on algae-covered rocks, exposing hidden prey. Short-billed and Long-billed Dowitchers resemble avian sewing machines as they vigorously probe deeply into the soft mud, bobbing up and down as they feed. Greater Yellowlegs run short distances in shallow water, seeking surface fish but also occasionally probing for worms and other burrowing food items. Godwits walk along in shallow water, probing sediment with their long bills, sometimes with heads submerged as they search.

Shorebirds probing in sand and mud cannot see their prey but have highly sensitive beaks that signal them when they contact food. Some have kinetic upper beaks that act as flexible forceps, grabbing prey, triggered by tactile stimulation. Woodcocks, which are birds of moist forests, capture worms in this manner. Hummingbirds also have kinetic beaks, enabling them to curve their upper mandible as they snatch insects from the air.

Wood-warbler species differ from the examples above because their body sizes and proportions are similar to each other. Differences in body size and bill characteristics are not nearly as pronounced as in the case of shorebirds. The classic 1958 study by Robert MacArthur (now known among ornithologists and ecologists as "MacArthur's Warblers") showed that five structurally similar wood-warbler species that nest in boreal (spruce-fir) forest, and that often forage together in the same spruce or fir tree, are actually using different parts of the tree and employing somewhat different foraging behaviors. The five species are Cape May, Blackburnian, Black-throated Green, Bay-breasted, and Yellow-rumped (called Myrtle Warbler when MacArthur published his study). Cape May Warblers tend to feed near the top of the conifer, the spire of the tree. In contrast, Yellow-rumped Warblers are more frequent feeders in lower parts of the tree and often drop to the ground in pursuit of food, something they also do on their wintering grounds. Bay-breasted Warblers methodically probe the shaded inner branches toward the middle parts of the tree. The important point of this example is that these

This Red Knot (in basic plumage) is demonstrating its flexible upper beak tip.

These are the famous "MacArthur's Warblers" that became a model for the study of how various bird species partition the foraging niche: (*top left*) Cape May, (*top right*) Black-throated Green, (*center left*) Bay-breasted, (*center right*) Black-burnian, (*bottom right*) Yellow-rumped (Myrtle).

subtle but statistically significant separations in foraging location and behavior are not obvious upon casual observation and required hours of observation and statistical analysis to demonstrate. Unless you spend time looking and carefully monitoring the amount of time each subject bird spends in different areas of the tree as it forages, you would never know.

Some wood-warblers exhibit unique foraging niches, setting them apart from others. The Black-and-white Warbler is dedicated to bark foraging. Behaving like a combination of a chickadee and a

Black-and-white Warblers have specialized in bark foraging.

Northern Mockingbirds habitually flash their wings as they forage on open spaces such as lawns, perhaps to scare up potential insect prey.

nuthatch, the Black-and-white Warbler methodically moves around on tree trunks and branches, often inspecting the undersides, gleaning food. Yellow-throated Warblers also frequently forage on bark and tree trunks. Painted Redstarts and Hooded Warblers enhance their efficacy at capturing prey by flashing their prominent white tail spots, scaring up potential insect prey. American Redstarts do the same thing.

Wing-flashing is frequently observed in Northern Mockingbirds as they forage on the ground. A mockingbird habitually opens its wings, flashing the prominent white wing markings. This behavior is thought to aid in prey discovery, just as tail spreading does in redstarts and Hooded Warbler. The sudden "flashing" behaviors stimulate insects to move, thus allowing birds to detect them.

Foraging niche partitioning is widely represented among birds. A study performed in central Connecticut showed that Black-capped Chickadees forage differently from Tufted Titmice, at least in winter. Chickadees spend more time on upper and outer tree branches. Titmice forage more evenly throughout a tree and often forage on the ground. While this study involved careful numerical data collection, it is nonetheless possible for any birder to patiently watch chickadees and titmice forage and observe these kinds of differences.

Males and females of some species have somewhat different foraging niches. Male Downy Woodpeckers tend to forage higher in a tree than females. Like the wood-warbler example, you have to watch Downies forage for a considerable time before this subtle pattern will be evident.

Among the shorebirds, avocets represent an example of foraging specialization. With their uniquely upturned bills, avocets sweep the surface water, capturing insect larvae and small crustaceans. No other shorebirds feed in such a manner. Male American Avocets have bills that are a bit longer and straighter than those of females, suggesting foraging specialization between the sexes. Bill length also differs among godwit sexes.

Among birds of prey, females are larger in body size than males. This difference is called reverse sexual size dimorphism because in most bird species the male birds are larger than the females. Accipiter hawks, which forage heavily on birds, show the most extreme difference in size between the sexes, particularly Sharp-shinned Hawk, in which male body mass is only about 58 percent that of the female. There are multiple hypotheses attempting to explain why reverse sexual size dimorphism is advantageous, and no single explanation has satisfied the full breadth of examples. One explanation focuses on males provisioning females while the females are brooding eggs. Smaller males may capture a wider array of food items and thus be more efficient providers. It is also possible that by being larger, females can endure longer fasts while brooding a clutch (but why wouldn't this apply to many other bird species?). Smaller males also fledge sooner after hatching, which might aid in avoiding siblicide within the nest, a common occurrence among birds of prey.

Foraging niches in all of their multiple variations reflect the reality that birds must compete within

Great Horned Owls, like other species of birds of prey, have "reversed" sexual dimorphism. In this photo the female (slightly larger) is the owl on the right. (*John Grant*)

This Tennessee Warbler is doing what birds do daily by the billions—consuming animal prey, in this case what appears to be some sort of insect larva.

their own species as well as with various other bird species that seek the same array of food. Specializing, as reflected in both anatomy (including sexual size differences within species) and in kinds of foraging behavior, is to be expected in such a diverse but demanding world. Food resources represent a huge and varied spectrum of nutritional possibilities. No species could utilize anywhere near all that is available in a forest, a field, a marsh. Could a night-heron glean caterpillars from canopy leaves? Could a Herring Gull snap up small flying insects as Tree Swallows do? And why would they, since caterpillars would provide little nutrition to night-herons, and small flying insects even less to a Herring Gull. Variations in body proportions, bill proportions, and feeding behavior are driven both by competition and by resource diversity and together account for why birds are so diverse.

ANIMAL LOSERS AND PLANT WINNERS

When a bird captures and eats an earthworm, the worm loses. Whenever a bird eats an animal, whether it be a Mississippi Kite eating a dragonfly or a Peregrine Falcon devouring a Semipalmated Sandpiper, the eater wins, the eaten loses. Birds have a one-way relationship with their prey animals. Forest bird species are unwitting allies of various tree species, as it has been shown that insectivorous birds collectively reduce insect damage to leaves, damage caused primarily by caterpillars and other leaf-consuming insects. In the early part of the twentieth century, birds of many species were collected to study stomach contents, and it was soon learned that birds ingest enormous numbers as well as thousands of species of arthropods and worms, including many insect pest species, all significant parts of various avian diets. Birds eat lots of bugs. Thus birds were soon recognized to be valuable assets to maintaining healthy ecosystems.

The only animals that ever benefit from ingestion by birds do so indirectly. These are animals such as parasitic roundworm and tapeworm species that move into birds vectored by an animal, such as a fish, that the bird may happen to eat.

Many birds feed on fruits and, in doing so, may act as important seed dispersers. Plants benefit. American Robins and Cedar Waxwings vector seeds of numerous species, such as eastern red cedar, tupelo, various cherries, Virginia creeper, and wild grape, a list that barely begins to cover the full scope. Blue Jays have had a major effect in aiding the northward return of oaks and hickories after the last glacial retreat. Jays cache acorns (see "Caching Behavior," page 191) and in essence, as

California Scrub-Jays have been shown to be important species in oak dispersal, as they carry acorns long distances and bury them. *(Ed Harper)*

climate warmed when glaciers retreated, jays effectively planted new trees on an annual basis, helping move the trees northward.

Birds act as seed dispersers for invasive plant species as well as native species. It has been shown that European Starlings and American Robins feed on and likely disperse the invasive species autumn olive (*Elaeagnus umbellata*). However, native plant species remain primary to avian nutrition. Gallinat et al., in a major study, concluded that even in the presence of abundant fruit from non-native plant species, most bird species tended to feed more heavily on native species, and thus it remains essential that conservation practices preserve native plant species and reduce non-native species. One invasive plant species has been uniquely beneficial to Northern Mockingbirds when they began colonizing southern New England in the early 1980s. In winter they inhabit clumps of multiflora rose (*Rosa multiflora*) that provides both protection and energy-rich food.

Many bird species are omnivorous, devouring both plant and animal material. Seasonal omnivory occurs when a bird species switches its focus to different foods at different seasons. Thrushes typically forage heavily on fruits from late summer through much of winter but become primarily carnivorous during breeding season. Sandhill Cranes, crows,

and various blackbird species, all of which feed on animal food for much of their diet, feed heavily on corn stubble when it is available. Red-headed Woodpeckers feed on acorns during the nonbreeding season, caching many (see "Caching Behavior," page 191). But Red-headed Woodpeckers also feed on animal material extracted while bark foraging and, like Lewis's Woodpecker, they routinely take flying insects by sallying forth to capture them in midair. Their foraging method is diverse by woodpecker standards. Birds are opportunistic and innovative, availing themselves of whatever suitable food sources present themselves.

SOME UNIQUE FORAGING NICHES

Some bird species feed in unique ways or on highly specific food items, becoming foraging specialist species. Here are some examples.

The Black Skimmer is one of three skimmer species, the others being the African Skimmer and the Indian Skimmer. Nothing looks like a skimmer but another skimmer. With long slender wings, black above, white below, a stubby upper mandible and an elongate lower mandible, a skimmer flies just above the water, beak open, lower mandible skimming the surface. When it encounters a fish or crustacean, it quickly snaps its bill shut on the prey. Skimmers

OPPOSITE: Black Skimmer (*top*) and American Flamingo (*bottom*) each have unique adaptations of their mandibles for feeding. (*Top: Garry Kessler; bottom: John Kricher*)

feed during the day but, because they capture prey by tactile contact, they also forage at dawn and dusk and at night, when more potential prey is near the surface. Skimmers are a subfamily within the family Laridae, the gulls and terns, and stand out boldly within this group.

Flamingos, family Phoenicopteridae, offer an extreme example of foraging specialization. All of the world's six flamingo species feed with the head upside down. The mandibles are lined with fine comblike structures, rather like the baleen of some whales, filtering cyanobacteria (often called blue-green algae) and tiny crustaceans.

Two bird species, the Snail Kite and the Limpkin, are specialized foragers of apple snails (*Pomacea*). Apple snails occur in freshwater lakes and marshes throughout the American tropics and subtropics, including Florida. It is interesting that two species as morphologically distinct as Limpkin and Snail Kite both evolved to capture, extract, and devour apple snails. Recently it has been learned that Snail Kites in Florida are attaining a larger bill size, presumably in response to an invasive apple snail species that is larger than the native species. A fundamental question is why did Limpkins and Snail Kites evolve to specialize on apple snails?

The answer may be that these mollusks represent very rich food items. Where they occur, they are abundant, easy to find, and easy to capture, though hard to extract from their shells. Both Limpkins and Snail Kites have solved that problem, albeit with differently crafted tools. A Snail Kite typically flies low over a marsh and snatches a snail from vegetation with one of its sharp talons. The kite will then fly to a perch and extract the animal with its deeply hooked upper mandible. Limpkin beaks bear scant resemblance to those of Snail Kites but are also adept at removing the snail from its shell. Limpkins pluck snails from vegetation or probe for them in water and then carry them to land. The tips of both upper and lower mandibles of Limpkin beaks are shaped and curved in a way that both mandibles are able to gain strong purchase on the snail and deftly remove it from its shell. Though specialized to feed on apple snails, Limpkins that have been observed recently in Georgia have shown foraging flexibility, learning to feed on freshwater mussels. Watching Limpkins feed is impressive. When snails are bountiful, Limpkins capture and devour dozens in a hour or so.

Sapsuckers, of which there are four species, represent another example of feeding specialization.

The Snail Kite (*top*) and the Limpkin (*bottom*) are both specialized for feeding on large snails, but in different ways. This Limpkin is feeding a snail to its offspring.

These woodpeckers feed by excavating sap wells, small horizontal rows of holes that tap into the cambium layer of a tree and allow sugary sap to flow. The sapsucker devours the sap as well as insects attracted to it. Sapsuckers represent an example of a unique method of feeding evolving before the speciation of a group. Each of the four species is closely related to the others, all in the genus *Sphyrapicus*. They bear close resemblance, the Williamson's being perhaps the most distinct within the group. It is unlikely that they each independently evolved sapsucking. More likely there was once one sapsucker species excavating sap wells, from which evolved the current four sapsucker species.

Among the most specialized foragers are dippers and frigatebirds. The American Dipper is one of five global dipper species, each in the genus *Cinclus*. In addition to the American Dipper, found in fast-running streams throughout the American West, two dipper species occur in South America, one in Asia, and one in most of Europe and the British Isles. All feed in a similar manner, which, like the example of sapsuckers, suggests that their foraging behavior evolved before they speciated. Dippers feed along the edges of and within fast-running streams, either walking along the bottom of the stream in search of animal food or "flying" underwater in pursuit of it. Dipper metabolism is uniquely low relative to body size, and they hold more oxygen in their blood than other species, both adaptations to holding their breath while submerged. They also block their nostrils with a skin membrane, and their nictitating membrane protects against submerged debris striking their eyes. Dippers have no difficulty enduring the swift currents in streams where they nest and feed. It is difficult for humans to stand in some currents that dippers routinely inhabit.

As for frigatebirds, they are so called because of their aggressiveness toward other birds (a historical reference to piratic sailing ships called frigates), forcing tropical terns, for example, to drop their prey, rather like more northern jaegers are prone to do. But frigatebirds exhibit yet another specialization, that of capturing flying fish as the fish fly up above the water to escape predation by predatory fish such

The American Dipper exhibits a suite of unique adaptations it shares with other dippers to inhabit fast-running, often cold streams. *(Ed Harper)*

as dorados. Frigatebirds, of which there are five similar species, are outstanding fliers and amazingly adept at snatching flying fish from the air. Frigatebirds are unique in that they feed entirely at sea but cannot rest on water. If they land on water they can no longer become airborne and will perish.

Some generalist foragers have adapted uniquely specialized behaviors to capture prey. Species such as Ring-billed Gulls, which often forage for scraps in fast-food parking lots, are sometimes seen "dancing," vigorously stamping their feet on moist grassy areas or on wet sand along the shoreline. Some shorebirds such as Piping Plover will also rapidly patter their feet on wet sandy beaches, much like Snowy Egrets and other species do in marshes. What accounts for such behavior?

Stamping or pattering feet appears to stimulate worms to come close to the surface. Worms tend to do that when it rains, and thus the stamping of the birds' feet mimics the vibrations of raindrops, sending a false signal to the worms. I have watched Ring-billed Gulls foot-pattering on the grassy slopes bordering Lake Merritt in Oakland, California, and there are videos available on YouTube. It works—they catch worms. In Europe, Herring Gulls occasionally do this foot-pattering "rain dance." Piping Plovers have been photographed foot-pattering on New Jersey beaches, perhaps simulating wave fluctuations that stimulate burrowed worms to move to the surface. Ring-billed Gulls also commonly foot-patter on wet sand along the surf. The most pertinent question would seem to be this: how did the birds make a connection between rain-caused vibrations or wave fluctuations and worm behavior on beaches, and then go on to simulate such vibrations by foot stamping?

HANDLING AND CONSUMING FOOD

Some foods, like small seeds, are easy to swallow, and birds have little or no difficulty consuming such items. But the level of difficulty in capturing, swallowing, and digesting food resources—what you might refer to as handling time—varies. American Robins roam over a lawn, heads periodically cocked, staring at earthworm burrows. Once a robin grabs a worm, it tugs at it, requiring effort and not always resulting in capture. If the worm is extracted it will writhe and twist and must be subdued. Robins will beat it against the ground. Finally, the annelid is ingested, perhaps battered but still alive, more or less. Many bird species routinely consume live animal prey. When loons, grebes, cormorants, herons, and numerous other avian piscivores swallow their fish prey, the fish is usually alive, sometimes very much so.

Anhingas and cormorants frequently work for many minutes to subdue and position a large fish

A Pine Siskin sits and feasts on tiny but abundant goldenrod seeds.

for safe swallowing. It would be a potentially fatal mistake for any bird to swallow a fish tail first because fish routinely have sharp spines associated with their fins, especially the dorsal fins. These spines flatten from anterior to posterior, so if the fish is ingested head first the bird is able to move it through its esophagus with reasonable ease. But if the fish is positioned tail first, the spines will likely wedge in the esophagus, and the bird will choke to death. Loons, cormorants, and Anhingas toss fish up and catch them, often numerous times, before getting the fish positioned so that it can be safely swallowed. Kingfishers typically carry a fish to a tree and whack it over and over against a branch before swallowing it. Birds occasionally make mistakes and capture fish too large to be safely swallowed; they eventually discard those prey items.

Ospreys usually do not swallow fish whole. They typically carry a fish to a branch or nest and pluck and eat large pieces of flesh. But in taking the fish from its capture to where the bird intends to eat it, the Osprey typically positions the fish head first in its talons. The hydrodynamic shape of the fish becomes an aerodynamic aid to the flying Osprey as it carries an often-heavy fish, reducing the drag that the fish would cause if it were positioned otherwise. How can Ospreys be that smart? Easily. They literally feel the effect of holding a fish as they fly, the resistance it creates, and thus they soon learn to position the fish so that it offers the least resistance. The risk Ospreys face is to capture a fish too big to handle. If its talons lock onto a fish with too much weight to lift, an Osprey can drown, being unable to extract its talons.

Some animal food requires that the bird put considerable effort into eating it. Species such as herons and various gulls capture live crabs and usually work to remove the crab's legs, including claws, before attempting to ingest the crab. Raptor species such as falcons and accipiter hawks take many minutes to tear apart and eat a kill. These species primarily eat other birds and must pluck masses of feathers as they prepare the prey to be torn apart and eaten. It all takes time, and birds of prey remain vigilant as they are dismembering and eating prey. Raptors gulp large pieces, resulting in visibly distended crops, easily observed as the bird flies.

When a bird eats a large prey item whole, as a Barred Owl would eat a small mammal, inevitably the nondigestible parts—hair and bones in the case of a mammal—are compressed into a pellet that is regurgitated. Pellet accumulation on the ground usually indicates an owl roosting tree. However, many other species regurgitate undigested matter in pellet form. These include cormorants, grebes, herons and egrets, gulls and terns, eagles, hawks,

TOP: This female Anhinga captured a fish but still faces the work of manipulating it, often by tossing it and catching it, so that it is ingested head-first. *(Garry Kessler)*

BOTTOM: This immature Peregrine just fed on a Northern Flicker and is still carrying part of the carcass. Note the distended crop, indicative of having just fed.

vultures, various shorebirds, kingfishers, some corvids, flycatchers, Gray Catbird, kinglets, thrushes, dippers, shrikes, and various wood-warblers. There is little nutrition in the pellet contents. Attempting to digest bone, hair, chitin, and the like retards otherwise efficient digestion and adds unnecessary weight. Pellet formation requires somewhere around 9 to 16 hours from ingestion to regurgitation.

Not all birds that eat difficult-to-digest food produce pellets. Scoters and eiders feed on mussels and other shellfish species, obtained by diving. The calcareous shells are not regurgitated but are crushed and move entirely through the intestinal system of the bird. That adds weight, but scoters dive for prey and must have sufficient time to stay underwater and pry the shellfish from rocks, so added ballast is therefore not a detriment.

Birds exhibit a remarkable diversity of bill types. Beaks, like precision tools, work well but only within a spectrum of possibilities. Various bird species have specialized to focus their foraging on particular kinds of foods. In doing so, bill shape, length, depth, and strength evolve in response to the demands exerted by the bird's diet. Beak characteristics are evolutionarily labile and, if food sources change, beak dimensions may change within relatively few generations to accommodate the new range of foods.

During the annual cycle, beak color is variable among some bird species, such as Pied-billed Grebe, Cattle Egret, Atlantic Puffin, and European Starling. Usually such changes correlate with breeding, when beaks are at their brightest. But beak coloration is also an essential adaptation to successful foraging. Various flycatchers, wood-warblers, and other species have a dark upper mandible. They may or may not have a dark lower mandible. Those that have dark upper mandibles tend to forage for active insect prey. The darker pigmentation of the upper mandible reduces glare, a hypothesis that has been tested in the field by Sean Williams and Edward H. Burtt Jr. Flycatchers with artificially lightened upper mandibles (creating more glare) foraged more in shaded areas and, when foraging in bright areas, captured fewer prey. Birds with naturally light-colored beaks, species such as Northern Cardinal, American Goldfinch, Summer Tanager, and American Robin, do not commonly catch aerial insects. Some waterbirds also show the pattern of having a dark upper mandible.

FLOCK FORAGING

Some bird species feed in flocks ranging from relatively scattered individuals, such as a group of bluebirds or flickers on a lawn or grassy field, to dense flocks of blackbirds, pigeons, starlings, Horned Larks, pipits, Snow Buntings, or shorebirds. These flocks are often composed of more than one species. Brown-headed Cowbird, Rusty Blackbird, and Yellow-headed Blackbird often occur in flocks of hundreds of Red-winged Blackbirds. In winter, Snow Bunting flocks often include Horned Larks and Lapland Longspurs.

Flocking behavior offers enhanced protection from predation. Flocks typically feed in open areas where the birds could be detected by predators. Being part of a flock ensures that there are many

Shorebirds have among the most diverse bill types of any group of birds, as is obvious from this small sampling: (*top*) American Avocet, (*center*) Marbled Godwit, (*bottom*) Long-billed Curlew.

alert eyes available to detect a potential predator. Additionally, once airborne, so many individual birds swirling in a huge flock offer a difficult moving target for a predator. If one bird is captured, that is a drop in the bucket. Average survivorship for an individual bird during a predator attack on a large flock is high.

Flocks exhibit effective evasive behavior. When a Peregrine attempts an attack on a starling flock, the airborne mass of birds will quickly form a swirling dense ball of starlings. The close proximity of the birds composing the flock make it difficult for the Peregrine to single out an individual for attack (though the predator does often take a bird). Shorebird flocks fly in tight formation, abruptly changing direction and altering the shape of the flock, behaviors that require energy output but that contribute strongly to survivorship. Birds in flocks attempt to position themselves in the interior of the flock,

ABOVE AND OPPOSITE: *(Above)* Northern Gannets, as well as boobies and other birds that locate concentrations of fish, form flocks that forage together, a behavior sometimes called a feeding frenzy. *(Opposite top)* When under attack by a Peregrine, European Starlings form a tight ball of swirling starlings, greatly reducing the average risk to any particular individual. *(Opposite bottom)* Ducks forage in a variety of ways. These Northern Shovelers have found a concentrated food source and are tightly clustered as they feed. *(Above and opposite top: John Kricher; opposite bottom: Ed Harper)*

where the likelihood of capture is least. Birds on the fringe are most at risk.

Flocks often locate rich areas for foraging. There is no disadvantage to being part of the flock because often the food source is sufficiently abundant or

Northern Parula (*left*) is among the many species that have a dark upper mandible, useful in reducing glare during food capture. Yellow-billed Cuckoos (*right*) also show the pattern of a darker upper mandible. *(Left: John Kricher; right: Ed Harper)*

scattered that competition among individual birds is low. Birds ranging from Red-winged Blackbirds to Snow Buntings form large flocks, the birds densely clustered. They move across a field in a manner suggestive of a giant amoeba, the birds moving in near unison, often densely packed, but those in the rear of the flock, particularly juveniles, will "hop-scotch" by flying over the others toward the front of the flock. As this behavior continues, the flock resembles a kind of avian wave as groups of birds aerially roll over others to be at the front.

Northern Gannets as well as some booby species locate fish schools and plunge-dive almost in unison. Dozens of gannets will converge on a fish school, each diving javelin-like, head-first, sometimes from heights of 50 to 100 feet. This form of flocking is not in order to avoid predation but is instead forced by the concentrated nature of fish schooling. Birders refer to such behavior as a feeding frenzy.

Waterfowl ranging from swans and geese to many duck species also feed in flocks, likely because the depth and concentration of food causes the birds to recognize optimum feeding areas and respond en masse. Groups of teal may densely cluster at feeding patches, tipping in relative unison, or groups of goldeneyes or eiders all dive at essentially the same time. Ducks also are vigilant for aerial predators, and flocking adds a measure of protection for each bird.

Mixed foraging flocks are common components of bird communities throughout the world. In North America, forest-dwelling mixed flocks are composed of a nucleus species, usually a species of chickadee, joined by nuthatches, woodpeckers, kinglets, creepers, and perhaps various wood-warbler and other species. Birds travel together, foraging as they roam the woodland. Foraging methods differ among species, reducing potential competition. In winter, groups of chickadees defend large territories and thus know their section of woodland very well. Other birds benefit by joining the chickadees. Predator detection is enhanced, making it safer to search for food. Mixed foraging flocks result in species being clustered as they perambulate. You may enter a woodland, see and hear nothing, only to find, min-

Tree Swallows form dense flocks that seek out wax myrtles for winter food.

utes later, that you are suddenly surrounded by actively feeding birds. While it is adaptive to be part of a mixed foraging flock, birds nonetheless engage in occasional agonistic interactions.

Townsend's Solitaire, Phainopepla, American Robin, and waxwings flock in search of fruiting trees and shrubs. Fruit-rich trees and shrubs tend to be patchy resources, but once found, there is often much to be had. In fall, Tree Swallows form immense flocks that together seek out rich clumps of wax myrtles (*Myrica*). They ingest the fruits, derive energy, and in essence become winter herbivores. The same is true of flocks of eastern Yellow-rumped Warblers, once called Myrtle Warblers.

CACHING BEHAVIOR: HOARDING FOOD FOR A RAINY DAY

Many bird species cache food and then must remember exactly where the food is in order to recover it. Shrikes impale prey on thorns and barbed wire, sometimes to be eaten in short order, sometimes to be eaten later. Caching likely developed in shrikes because of their feet. Shrikes are passerines and have typical passerine feet, not equipped with powerful talons to hold and manipulate prey. So putting an insect, bird, or mouse on a sharp spine or barbed wire helps the shrike keep the prey in place as it eats. Given that the prey, once dead and impaled, goes nowhere, the shrike is free to eat what it wants and then leave it and return as it deems necessary. All it has to do is recall exactly where the prey item is stashed. In Florida, Loggerhead Shrikes feed on large and toxic lubber grasshoppers (*Romalea guttata*). Shrikes typically leave the grasshopper impaled for one or two days, enough that the toxicity has sufficiently degraded to allow the shrike to consume the insect. Shrikes evolved caching behavior long before Florida's Loggerhead Shrikes began feeding on lubber grasshoppers, but the caching behavior itself—leaving the dead insect for up to 48 hours—helped adapt shrikes for consuming a prey item that other birds normally cannot eat.

Chickadees and nuthatches are scatter hoarders, caching food in bark crevices and other hiding places. The birds must remember precisely where numerous individual food items have been cached. Obviously these birds have this ability; otherwise the behavior would be useless, indeed maladaptive. But the degree to which each bird does indeed recall where it has cached *all* of its food items is not known and likely varies among individuals. What is known is that the spatial memory as to specifically where cached seeds are placed resides in the large hippocampus of the brains of seed-caching birds of all species. Not surprisingly, among chickadees and tits, those in the more northern regions or higher elevations, where winters are more severe, cache

Canada Jays (*top*) cache seeds using sticky saliva. Blue Jays (*bottom*) are essential seed dispersers for oak species. (*Top: Bruce Hallett; bottom: Garry Kessler*)

Jaegers are aggressive kleptoparasites of terns. This Parasitic Jaeger has already made the Common Tern drop its fish. *(Garry Kessler)*

proportionally more seeds than those of lower latitudes and more mild climates.

When it comes to seed caching, corvid species set the bar high. In boreal forest, Canada Jays cache seeds using a gluelike substance in their saliva to adhere food to trees. Blue Jays feed widely on mast crops such as oaks and beech, individuals selecting various caching sites, a behavior that is easily observed in autumn when acorns and beech mast become abundant. In the western states Steller's Jays are similar in their behavior, caching acorns and pine seeds. In autumn Blue Jays diligently bury acorns but they likely do not remember where to find all of them, perhaps not even most of them, which is why Blue Jays are key species in propagating oak trees. I have watched Blue Jays repeatedly carry acorns to a location where, presumably, they bury them. One study (by Darley-Hill and Johnson) showed that over 28 days, a group of about 50 Blue Jays cached approximately 133,000 acorns, representing 54 percent of the crop among a group of 11 pin oak (*Quercus palustris*) trees. Acorns buried but not recovered may become oak trees.

Clark's Nutcrackers are locked in an ecological interdependency with whitebark pines (*Pinus albicaulis*). Individual Clark's Nutcrackers disseminate many thousands of whitebark pine seeds, digging holes and burying seeds, a behavior conducive to allowing seeds that are not recovered and eaten to subsequently germinate. A nutcracker is able to transport as many as 90 pine seeds at one time, contained in an expansive throat pouch. A single bird may bury up to 30,000 seeds in scattered caches. How do they find them again?

Recall that corvids (crows and jays) rank among the most intelligent of birds (chapter 4). One form of this intelligence is spatial memory, the ability to recall precisely where a bird has cached each of its seeds. As mentioned above, seed-caching birds have a well-developed hippocampus in the brain, the area that stores long-term memory. Besides recollection of where seeds are located, some jays attempt to steal from the caches of other individuals that they have surreptitiously observed caching. Jays exhibit behaviors that appear to reduce the likelihood of their own caches being taken by other jays. Woodhouse's (formerly Western) Scrub-Jays have remarkable spatial memories, but also pay close attention to other jays, choosing their caching sites so as to reduce the possibility of another jay discovering and essentially stealing the seeds. Jays make choices dependent on whether or not other jays are present, jays that could witness where seeds were being placed.

KLEPTOPARASITISM AND COMMENSAL FEEDING

Nature, to use a common expression, "is what it is." And what it is often does not appeal to human sensibilities. It is difficult, indeed impossible at this point, to know if birds are capable of what we humans call hate. But it is not difficult to imagine that Ospreys have no fondness for Bald Eagles or that terns of various species might well despise jaegers. Bald Eagles often attack Ospreys that are carrying a fish, forcing the Osprey, by persistent harassment, to drop the fish so that the eagle can take it. Jaegers and frigatebirds relentlessly harass and pursue terns in an often successful attempt to force them to drop fish from their beaks. Great Black-backed Gulls that reside around seabird colonies will await puffins as they return with food, attempting to force them to drop their larder before getting to their nesting burrow.

These acts represent a form of behavior called kleptoparasitism. The prefix *klepto* refers to the irresistible urge to steal. Kleptoparasitic bird species forage by spotting other birds with food and then attempting to steal that food item for themselves, an act of parasitizing the victim. It is a common form of agonistic (combative) behavior likely learned by watching others. Aggressive kleptoparasitism is not always easy but it carries little risk to the perpetrator and apparently succeeds sufficiently often to be worth the energy output.

Some birds exhibit kleptoparasitism only on occasion. American Coots are sometimes kleptoparasitic toward diving ducks such as scaup. The ducks dive more deeply than the coots, and as they surface with vegetation in their bills the coots attempt to grab some of it. But coots may also be victims of kleptoparasitism by dabbling ducks such as Gadwalls and American Wigeons, surface-feeding ducks that attempt to snatch plant food as the coots surface. It is often enlightening to watch a flock of foraging ducks when there are coots among them. All sorts of interesting behaviors ensue.

Kleptoparasitism is commonplace within bird species that forage in close proximity. Just watch a Royal Tern or a Herring Gull with a newly captured food morsel and see how other birds chase it and attempt to take the food. Kleptoparasitism is also common among shorebirds, herons, egrets, and ibis. Gulls will also harass diving birds that surface with food items, attempting to steal from them. Great Black-backed Gulls routinely attempt to steal from surfacing eiders and mergansers.

Being a victim of kleptoparasitism is a fact of life for many bird species. They have no choice but to try and avoid losing their food, but they often do lose it. Because passerine birds feed on small insects, worms, and/or plant food such as seeds, kleptoparasitism is not commonly observed among them.

A few bird species have adapted to utilize other animals as aids in their quest for food. The Cattle Egret is a prime example. As the bird's common name implies, Cattle Egrets associate with various bovines or other hoofed mammals. The large animals act as beaters as they walk through pasture grazing, causing insects and other arthropods to become more active and conspicuous, so that the

Cattle Egrets are able to find more prey than they could in the absence of the cattle. This is a commensal interaction. The cattle have no vested interest one way or another in the Cattle Egrets. Franklin's Gulls are commensal with farmers in some areas of the Midwest. As fields are mechanically plowed the farmer's actions result in stirring up numerous insects as well as exposing earthworms that the gulls devour. Other inland gulls respond similarly. Brown-headed Cowbirds (chapter 15) are thought to have historically followed huge herds of bison as the herds stirred up potential food for the cowbirds.

Common Ravens engage in a commensal interaction with packs of wolves that inhabit Yellowstone National Park. In winter, ravens follow the wolf pack as it moves through the park in search of elk to kill. Once a kill is made, the ravens remain after the wolf pack has been satiated and feed on the remaining scraps of the carcass. This is not the same sort of behavior practiced by species such as vultures that typically soar high and are constantly scanning over a wide area for food. The Yellowstone ravens are methodically following the wolves around, having learned that wolves are their best source of obtaining calories on a regular basis.

BEHAVIOR AT BIRD FEEDERS

The previous example of Franklin's Gulls being attracted to mechanized agricultural activity could be thought of as a form of bird feeding. The human farmer is making food available to the birds. By the same token, gulls often congregate around fishing boats, obtaining discarded bait or pieces of fish. But of course these examples of bird feeding are not the same as purposeful backyard bird feeding, as practiced by millions who wish to attract birds close to their homes.

Bird feeding has never been more popular. Combinations of various seeds (millet, sunflower, thistle) and peanuts, as well as suet, fruit, jelly, and sugar-water, attract an array of species, from chickadees and jays to orioles and hummingbirds. In 1987 the Cornell Laboratory of Ornithology began a program called Project FeederWatch. Thousands of people annually send records to the Cornell Lab covering the months of November through April, when bird feeding is most practiced. Project FeederWatch (feederwatch.org/) continues, and participation has expanded since its inception. It represents an example of data collection by citizen science.

Bird behavior at feeders is mostly agonistic. Typically, but by no means always, larger birds displace smaller birds. A jay will swoop in and off go the goldfinches, at least for the moment. Downy Woodpeckers may be aggressive, displacing one another and even displacing larger birds. I once watched a

Downy Woodpecker displace a group of about eight Evening Grosbeaks from a tray feeder. Every time a grosbeak would attempt to land, the woodpecker would rush at it and the grosbeak would retreat. Northern Cardinals are often agonistic toward one another. Flocks of juncos and sparrows show displacement reactions, one bird attempting to land on the spot occupied by another. Two siskins, two goldfinches, or two Evening Grosbeaks will go beak-to-beak as one tries to take the perch on the feeder from the other or even if they get too close to each other.

Data taken from Project FeederWatch has helped shed light on complex behavioral interactions of feeder visitors. Analysis of over 7,000 records on avian interactions generated by Project Feeder-Watch (feederwatch.org/blog/who-is-the-toughest-bird) confirms (unsurprisingly) that larger bird species tend to dominate feeder access over smaller species. Size does matter. For example, of the 136

American Coots (*top*) often exhibit kleptoparasitism toward one another or toward diving ducks such as Ring-necks. The Snowy Egret (*bottom*) is moving toward the foraging White Ibis, and should the ibis bring up prey, the egret is likely to attempt to snatch it.

(*Top*) This female Red-bellied Woodpecker successfully intimidated the Blue Jay, and the jay left the feeder. (*Bottom*) Hummingbirds (such as these Ruby-throats) become highly pugnacious when crowded at feeders. (*Top: John Kricher; bottom: Robert Askins*)

species recorded, the most dominant species was Wild Turkey. But there are numerous nuances. Mourning Doves are about the same size as Blue Jays but jays dominate doves. Crows and blackbirds also are disproportionally dominant, even when their size is factored in. Some dominance interactions are species specific. For example, European Starling is dominant over Red-headed Woodpecker, which is dominant over Red-bellied Woodpecker. But Red-bellied Woodpecker is dominant over European Starling, forming a triangle of dominance. Squirrels are also omnipresent at some feeder areas and clearly are dominant to birds. Among the most pugnacious feeder birds are swarms of hummingbirds, who often jostle for position on hummingbird feeders.

Birds, regardless of dominance tendencies, are not in their comfort zones at feeders. Bird feeders are typically exposed, easy for predators to find, or

American Crows, as well as some other bird species that benefit from bird feeding, are often predators of various nestling birds. *(Jeffory A. Jones)*

near a house. A bird feeding area has no rules, no orderly lines, no civility. Not only must birds contest for access to the feeders and cope with the social stress imposed by many individuals attempting to obtain the same concentrated source of food, but they also must weigh the risk of predation. Feeders attract predators ranging from feral cats to accipiter hawks. The increasing abundance of Cooper's Hawks wintering in the Northeast is thought to be linked with increased bird feeding. In the case of the hawk, it's the clientele of the feeders that provide the food, a kind of backyard food chain.

Feeders in areas with severe winters probably aid bird survival. They also contribute to the survival of species that do not normally visit feeders, those that come when the going gets particularly rough, species such as Hermit Thrush, American Robin, Varied Thrush, and Brown Creeper. Each of these species may consume seed, having presumably been drawn to the feeders by the presence of other species. At the same time, Blue Jays, Common Grackles, and American Crows thrive at feeders, and these species habitually prey on other birds' eggs and nests. Thus feeding birds may indirectly contribute to increasing nest failure among some species.

The growth of winter bird feeding over recent decades almost certainly contributed to northern range expansions of such species as Tufted Titmouse, Northern Cardinal, Carolina Wren, and (in the East) House Finch. But at the same time, the density of birds that assemble at bird feeding areas sometimes results in the spread of disease from feeders that have not been regularly cleaned. Some years ago some eastern House Finch populations began developing a disease severely affecting their eyes. The disease, a form of conjunctivitis, was caused by a bacterium, *Mycoplasma gallisepticum*. The disease slowly spread through eastern and midwestern House Finches and eventually reached the West Coast. The lesson is that people who do put out bird feeders should regularly subject their feeders to a thorough cleaning.

It seems intuitive that bird feeding creates a dependency among the birds using the feeders. However, birds are mobile and if a feeder ceases to supply food, birds will quickly go elsewhere, just as they do when fruit on a tree is fully consumed. Little data exist on the impact of stopping bird feeding in midwinter, but one study (by Brittingham and Temple) looked carefully at survival rates of wintering chickadees with access to feeders compared with those that did not visit feeders and found no difference in winter survivorship.

THE OMNIPRESENCE OF PREDATION ON BIRDS

Birds ranging from juncos to chickadees to various sandpipers would live different lives were it not for predators. Nest success would be profoundly higher in a world without predators. But predators are and have been for 500 million years a diverse, major, and essential component of ecosystems, whatever form ecosystems may take or have taken. For today's birds, as for all of their ancestors, predation is a fact of life. Birds must adapt and learn. Smaller birds, such as many of the shorebird species and almost all passerines and other small land birds, face a daily threat of predation from both ground and aerial predators. Larger birds are less apt to fall prey but are still at some risk. Detection of predators is essential and is both instinctual and learned in birds.

Birds that venture into unfamiliar areas or are not associated with a flock are particularly vulnerable. Imagine, for example, being fortunate enough to observe a Ross's Gull, a great rarity, only to witness your "big tick" being preyed upon by a Peregrine Falcon. This event, as documented by Pete Sole, actually happened on January 14, 2017, in midafternoon at Pillar Point Harbor in Half Moon Bay, California. The Ross's Gull had been feeding alone on the ground and was vulnerable. As birders watched, the gull took flight, only to find itself suddenly being pursued by two Peregrines. It lost. The gull tried to evade the falcons but was quickly captured by one of them and that was that. Had the gull been among more birds, say a large mixed flock of gulls and terns, one of them might have detected the falcons and all would have been airborne and raucously calling, taking evasive action. The Ross's Gull might have been spared.

Predatory birds are adapted to avoid detection. They are skilled at attack and capture. Obviously predators often succeed, as there are many of them. Adult Cooper's Hawks, for example, are successful in about one-third of their attempts to capture and kill a bird and presumably juveniles, with less experience, are less successful, a high learning curve. The most difficult time for predatory birds is the critical learning period that follows fledging. It is not easy to learn the trade of being a Cooper's Hawk or a Merlin. But once the skill set is attained, many predatory birds enjoy a lifetime of success. Bald and Golden Eagles are known from banding records to live over 30 years, and species such as Red-tailed

Looking skyward, a Piping Plover, though cryptic in its environment, is still ever vigilant about the potential for a predator to attack it.

Hawk, Great Horned Owl, and Peregrine Falcon are known to live beyond 20 years in the wild. Cooper's Hawks feed largely on birds and can sometimes live nearly 20 years, and a lifespan of 12 years is considered normal for them. That's a lot of kills.

Birds detect predators in a variety of ways. First, they have keen search images. Second, most take no chances. Many birders take note of how wary birds can be, taking flight at the slightest perceived risk, a clear example of "better safe than sorry." If you watch a flock of foraging sparrows, Horned Larks, or Snow Buntings, note how frequently various birds among the flock turn their heads and look skyward. The same is true for shorebirds and waterfowl. Watch a flock of blackbirds, most of which are foraging on the ground, but notice that some individuals remain in trees or on bushes, where they are

This Merlin (*left*) had just captured and killed a White-winged Crossbill. The Great Black-backed Gull (*below*) is one of the primary predators in the bird world. This one is feasting on a Black-legged Kittiwake that it just killed at a large kittiwake colony. (*Left: John Kricher; below: Bruce Hallett*)

best positioned to detect a predator. Jays and crows typically have sentinel birds that scan, presumably for predators, while other flock members feed. Birds on the outer edge of a flock scan more frequently than those in the middle of the flock. Whether airborne or on the ground, individual birds are constantly jostling to get into the middle of a flock, where it is safer. When roosting, Northern Bobwhites form a tight circle, all birds facing outward. This arrangement allows the birds to take flight instantly if a predator happens on them.

Studies with humans who approach European Starlings have indicated that the starlings pay close attention to the eyes of the humans. Starlings are more apt to fly if humans are staring directly at them. Turning your head or merely shifting your eyes is often enough to delay a starling's evasive action. Directly approaching a flock of ground-feeding birds may cause them to fly sooner than if you were to obliquely approach them and try not looking directly at them as you are shortening the distance between you. Many birders who photograph have lamented that sometimes the easiest way to make a bird fly is to raise your long lens directly at it. Off it goes.

Once birds detect a predator, they share that information. Should a Peregrine Falcon fly over a marsh where waterfowl and shorebirds are feeding, the ducks and shorebirds will simultaneously take to the air, swirling in tight flocks until the perceived threat is over. Birders know to look skyward for the would-be predator if feeding flocks suddenly become airborne.

PREDATOR HARASSMENT AND MOBBING

Many birds respond to predator detection by harassing or mobbing the predator. Examples abound. Northern Mockingbirds are well-known harassers of crows and hawks, sometimes actually striking them in the air, a habit also often seen in Red-winged Blackbirds and Eastern Kingbirds. Mockingbirds also harass ground predators such as cats and have been known to attack humans who inadvertently approach their nest. Crows harass Common Ravens. A raven is much larger than a crow, and crows seem to know this because, unlike with mockingbirds, which usually act alone to harass larger birds, crows harass ravens only in groups of at least three crows. Crows also harass owls and hawks of assorted species. Various jay species as well as kingbirds frequently harass larger birds.

Harassment sometimes involves an element of risk to the bird doing the harassing, particularly if it is harassing a bird of prey. But usually the bird being harassed just flies away until its tormentors give up on it. And that is the likely advantage of harassment to those doing the harassing: a potentially dangerous species is driven away. It is adaptive for kingbirds to harass larger birds since kingbirds habitually feed in open areas from exposed perches where they could be vulnerable to an avian predator. Better to see the predator first and make it go away.

Colonial birds appear to recognize strength in numbers in their harassment behavior. Enter a tern colony and you will immediately be pummeled by groups of noisy and aggressive adult terns. Terns will strike at the top of a would-be predator, which is why humans entering a tern colony always wear a hat and often carry a small stick above their heads to redirect the beaks of the defending terns.

Nesting birds defend their eggs and nestlings to the best of their often-limited abilities by attempting vigorous and potentially risky harassment of predators, including cats, raccoons, snakes, and birds. Only when the predator succeeds in getting to the eggs or nestlings do the parent birds move on. There is no point in further risking themselves.

Mobbing is a particular form of harassment in which multiple birds of various species converge on a predator and badger it with vocalizations and threats. Passerines routinely mob roosting owls and hawks. Mobbing also occurs with ground predators such as cats as well as snakes, some of which climb trees.

Mobbing makes other birds aware of the presence of a predator and perhaps may force the predator to leave. For multiple individual birds to know a predator is in their immediate neighborhood helps protect each of the individuals. All birds are more wary once they know of a predator's presence and thus each benefits from having more eyes alert for the predator.

Birds, depending on habitat, exhibit different behaviors in response to a predatory event. Should a predator such as a Cooper's Hawk or a Merlin swoop in but be unsuccessful in its attempt to take a bird, open-area birds such as larks, blackbirds, and sparrows will fly and relocate elsewhere. Shorebirds, which feed in concentrated mudflats, sometimes remain airborne until the predator is gone and then return to the site where they had been feeding. Forest birds, already having some cover because they are among trees, will often "freeze." They become stiff and immobile for many minutes until somehow they decide it is all clear. Freeze behavior is commonly observed around bird feeders after an attempt at predation by an accipiter or Merlin.

OPTIMAL FORAGING AND PREDATION

Birds have to balance the need to feed, often in the open and with considerable focus and concentration, with the need to guard against sudden death by predation. Beyond that, birds have to make choices about how to forage and upon what to

THIS PAGE AND OPPOSITE TOP: Blue Jays (*top*) commonly harass hawks, such as this Red-tail. Northern Harriers (*bottom*) confront Snowy Owls, perhaps to protect their hunting territory or out of concern for their own safety. (*opposite top*) A male Boat-tailed Grackle in pursuit of a Swallow-tailed Kite. (*Top and bottom, this page: John Grant; opposite top: Holly Merker*)

forage. For a bird such as a Royal Tern, is it better to stay inshore and feed on many small fish or to fly farther to sea and feed on larger fish? Which gives the most reward relative to the effort involved and risk taken?

Many years ago animal behavior scientists developed a theory of optimal foraging. The theory is an attempt to understand how natural selection would act to optimize the calories obtained by a bird relative to the energy spent obtaining them. In other words, what is the most economic means to gather the most energy with the least expense of energy?

Optimal foraging theory was tested with multiple mathematical models, some of which became complex because of the number of variables involved. Indeed, in a way it was a tribute to birds (and other animals) in that it required considerable calculus to describe what these animals routinely do on a daily basis.

In choosing where to forage, a bird's decision may be dependent on several variables, ranging from innate habitat recognition to detection of food sources such as insect swarms and fruiting shrubs. Associated with habitat choice is dietary choice, namely what to eat and how to obtain it. If the bird is nesting it must decide how far to venture to obtain food not only for itself but for its young. And finally, the bird has to make a choice as to when to abandon a foraging location and seek another one. In other words, how long should an American Robin remain on a lawn seeking earthworms before choosing to move elsewhere, because of either a lack of worms relative to effort or anxiety about potential predation? Each of these variables must include the constant potential for exposure to predators. So optimal foraging gets complicated.

Consider two bird feeders placed in different locations. One is out in the open, easily visible, a flat tray full of energy-rich, already husked sunflower seed. Birds need only land, gather some seed, and be off. The second feeder is near a dense shrubby area that offers some close protection to birds should a predator appear, plus the feeder is not nearly so obvious. The mixed seed in this feeder includes some sunflower seed that still has its husk and thus needs processing after a bird takes a seed. Birds need to exert somewhat more effort to use the seed near the shrubby area (because they must remove the husk), but they are more exposed to potential predation should they choose the feeder in the open area. What is the optimal choice for birds?

This would seem to be a classic trade-off situation where there is really no *right* choice. But birds are of necessity forced to make such decisions each day. Given that their lives depend on their behavioral decisions, a conservative approach might seem optimal, even if the energy payback is somewhat less. Choose the feeder near the shrubs. If a known predator were in the area, birds would almost certainly choose the more protected feeder.

Birders who visit sites frequented by migrating passerines and other small birds soon come to realize how important plant cover is to these birds. Indeed, a broad study (McCabe and Olsen) looking at land bird migrants at six banding stations— a total of 10,000 migrant birds—concluded that patches with a high density of plants, providing thick cover, outranked all other variables, including fruit abundance, in predicting where birds would be. Birds preferentially chose sites with thick cover and fruit availability but, absent of fruit availability, dense plant cover was the best predictor of bird abundance.

Another study (Goullaud et al.) showed how the risk of predation impacts foraging decisions, making optimal foraging that much more complex. In this case a population of Horned Larks in an alpine area of British Columbia was tested to determine how predation risks influenced provisioning behavior, the parents' efforts to collect food for offspring still in the nest. Using decoys of a fox and a raven (both nest predators) plus a Savannah Sparrow decoy (not a predator or competitor and used as a control), the researchers learned that perceived predation risk diminished visits by parents to the nest by about 66 percent. The parent birds brought back fewer food items to the young when they perceived that a predator was nearby. However, once the predator decoy was removed, the female birds began increasing visits to the nest, compensating for the time lost when they perceived the predation threat. The Horned Larks had taken the reality of potential predators into consideration, varying their provisioning behavior accordingly.

THE AUDITORY BIRD

HOW BIRDS MAKE SOUNDS

Sound is an essential component in avian communication and social behavior, taking many forms and serving numerous functions. Birds offer an impressive, indeed inspiring, spectrum of sounds and do so in a variety of ways. Avian sounds are mostly vocal, but nonvocal sounds also abound. Mechanical sounds, often referred to as *sonations*, include bill clattering by albatrosses and storks, bill snapping in owls, "pumping" by bitterns, drumming by woodpeckers, and various sounds made by wing and tail feathers, such as wings whistling in flight, as occurs with goldeneyes; wings slapping together, as with pigeons and doves; wings whirring loudly on takeoff, as in pheasants; wing twittering from modified primary feathers, as in American Woodcock; winnowing by modified tail feathers, as in Wilson's Snipe; various booming-type sounds made by courting prairie chickens, Pectoral Sandpiper, and Common Nighthawk; wing drumming by Ruffed Grouse; and wing twittering sounds made by some hummingbird species such as Broad-tailed and Rufous. Anna's Hummingbird is unique in that courting males emit a highly audible sharp *chirp* note using the fifth rectrices of their tail. As the male bird makes his steep courtship dive, he spreads his tail and produces the uniquely loud sound.

Vocal sounds are produced as muscles contract when air passes through a bird's syrinx, an anatomical feature unique to birds. Air passes through the syrinx during breathing, just as it does in a human's larynx. And just as no sound happens in humans unless the vocal cords are intentionally engaged, so it is with birds, but in a different manner. Birds use various muscles to control the membranes of the syrinx and thus to generate specific sounds.

Birds have no vocal cords in their larynx. The syrinx in almost all bird species is located at the base of the trachea, where it fuses with the two main bron-

It is common for birds to open their beaks widely when vocalizing, as is the case with this singing male American Robin. But because birds make sounds in their syrinx, not their larynx, they can vocalize with beaks closed.

chial tubes that go to the lungs. Syrinx structure varies among major groups of birds, as does the complexity of musculature surrounding the syrinx. Most birds have only one or two muscle pairs governing the syrinx, accounting for why the sounds of many birds are not very complex. But oscine songbirds have eight or nine muscle pairs and thus are able to generate far more complex, highly melodic, songs. When a bird sings, the syrinx muscles change the positions of various membranes within the syrinx and that, in turn, produces the sound, putting the specific notes to the music, so to speak.

OPPOSITE: This male Rose-breasted Grosbeak is typical of oscine passerines, the true songbirds, emitting a rich, melodious song.

Sonations are common in many bird species. Species such as American Woodcock (*top*) and Wilson's Snipe have modified feathers that make distinctive sounds when in flight, used in courtship. Ruffed Grouse (*bottom*) select drumming logs and rapidly beat their wings against their bodies to attract females by producing a low-frequency drumming sound. *(Top: John Kricher; bottom: Garry Kessler)*

Most bird species make sounds that are largely if not entirely innate. Nonetheless, within each species there may be a considerable repertoire of sounds. Large gulls, such as Great Black-backed, Western, Herring, Glaucous, and Glaucous-winged Gulls, make close to a dozen distinct and easily recognizable calls. For example, these species all make a protracted and loud "long call," in which they throw back their heads as they call, usually ending by dropping their heads low. The long call signifies that the gull is behaving aggressively. It is a challenge call, something that often becomes obvious if you watch the subsequent behavior of the gull performing the long call. It may also serve to attract a potential mate. Long calls differ in tone among gull species, but it is not surprising that so many gull species make long calls. Gulls occur commonly in mixed-species flocks. A gull doing a long call may wish to express its agitation and dominance beyond its own species. Also, large gull species are evolutionarily closely related, their genetics quite similar, and the long call may have evolved before the gulls speciated, becoming modified somewhat in each of the various species.

The guttural squawks of herons, the hooting of owls, the quacking of ducks, the whistled calls of shorebirds, and the hissing of vultures are part of the genetic makeup of each species. The moanlike utterances of male Common Eiders and the low guttural quacking sounds of female Common Eiders are as much a part of the birds as their plumages. Eiders, like most other bird species, are hatched already knowing sounds that characterize their species and their sex as well as how and when to make them.

Even though nonsongbird species essentially inherit their pattern of vocalization, there is still variation among individuals. That variation is the key component to understanding much of why birds vocalize during dominance interactions, courtship, and other activities. Common Loons, for example, make a diverse range of distinctive yodeling sounds, part of their vigorous territorial behavior during breeding season (chapter 14). Vocalizations can also have an emotional component. A gull doing a long call may be very agitated, "feeling" the need to make the call. Birds make daily decisions about each other based in part on the information they derive from the type of vocalization, its intensity, and the demeanor of the vocalizing bird.

Oscine songbirds, described below, as well as parrots and hummingbirds are unique in that they have the capacity to learn and modify their songs, setting them apart from other bird species. But birds capable of learning songs also make calls that are largely innate. The basic song of each species, though much influenced and polished by learning, is also genetically influenced. Blue Jays will never hatch from an egg and then *caw* as crows do.

Many species of gulls, including the Glaucous Gull (*top*), make the long call sound, an innate component of their vocalization repertoire. Species such as Sandhill Cranes (*bottom*) use vocalizations, in this case a deep bugling sound, for flock cohesion while migrating. *(Top: Ed Harper; bottom: John Kricher)*

Birds within the order Passeriformes, globally over 5,000 species, are called perching birds because of their toe arrangement in which their feet are uniquely adapted to grasp branches. Passeriformes include two major groups, which substantially differ regarding vocalizations. The suboscine group in North America includes only the tyrant flycatchers (Tyrannidae). But in Central and South America suboscines include not only more than 300 species of tyrant flycatchers, but also such groups as antbirds, antpittas, woodcreepers, tropical ovenbirds, manakins, and cotingas, which, taken together, represent hundreds of species. All suboscines have simple and generally nonmusical songs that are essentially innate. Eastern Phoebes are hatched with the full knowledge that their song is now and always will be *vee-be-OH,* just as Least Flycatchers are hatched knowing *che-bek!* They do not learn their song.

But the true oscines, birds such as corvids, shrikes, vireos, wrens, chickadees, mimic thrushes, thrushes, starlings, waxwings, finches, sparrows, wood-warblers, and blackbirds and orioles, rely on a basic blueprint of innate knowledge significantly augmented by learning their songs from their father and other males of their species who are in their immediate neighborhood. Oscine songbirds isolated after hatching without having heard their species' song will sing, but the song is garbled and not typical of their species. Learning is essential, and it continues throughout their lives. Song learning requires the commitment of neural energy, which is likely why songbirds manufacture new neurons in parts of the brain dealing with song (chapter 5) as part of the annual cycle. Oscine songbirds can enrich and alter their song repertoires throughout their lives.

The oscine syrinx is more complex than that of any other bird group. Complex musculature is present around and within the syrinx. The birds are able to generate elaborate songs by compressing and moving air through the syrinx in various ways, independently changing the orientation of the membranes on both sides of the syrinx, actions that determine the song. It is possible for the muscles on either side of the syrinx to contract differently, blending the sound made by the right half with that of the left, making for a richly detailed, complex song. Essentially the bird is singing two songs at once, integrating both into a single harmony. This amazing reality is revealed when a songbird's recorded song is vastly slowed down. The sonogram, a graphic illustration of the pitch and frequency of notes, reveals two songs that seamlessly combine as one. The bird's brain, of course, determines the actual song pattern. Young birds must practice their songs, as described on page 213.

CALLS COMPARED WITH SONGS

Birders often use the terms *call* and *song* interchangeably, and for good reason. For example, Northern Bobwhites whistle, *oh Bob White!* and Mourning Doves plaintively sound off with *oo-ooohh—hooo, hooo, hooo.* Are these species calling or singing? Examined functionally, the sounds are

Gray Catbirds make a wide variety of call notes, including catlike mewing, as well as a varied and complex song. They have a broad sound repertoire.

The Common Loon (*top*) is well known for its extensive repertoire of haunting yodels, a familiar sound on lakes in its summer breeding areas. Black Phoebe (*bottom*), along with all other tyrant flycatchers, is a suboscine member of the Passeriformes and thus its song is innate, not learned.

American Crows have a rich series of vocalizations, but none of them are particularly musical, at least not like those of thrushes and orioles. Nonetheless, all corvids are oscine passerines, true "song-birds." *(Craig Gibson)*

songs, uttered to declare territory and attract females. Jays make a variety of mostly nonmusical sounds, nuthatches make simple repetitive vocalizations, grackles and cowbirds make gurgling, nonmusical sounds, and waxwings make simple thin buzzy sounds, all of which are often termed calls, though in fact they function as songs, which can be understood with some context.

A Tufted Titmouse and a Baltimore Oriole may be loudly vocalizing in nearby trees. Many birders would say the titmouse, with its simple and persistent loud *Peter, Peter, Peter* whistle, is calling and the oriole, with its rich and melodious whistled notes, is singing. That is because the titmouse whistle sounds simple, repetitive, and relatively nonmusical to the human ear while the oriole is producing just the opposite, a richly textured song. But each bird is loudly asserting its territoriality, so why say one is calling and the other singing? Titmice have harsh, buzzy call notes, as do orioles. In the example described here, both of them are functionally singing. They are declaring territory or advertising for mates.

Ornithologists consider singing, no matter how the bird sounds, to be any vocalization asserting and reinforcing territoriality or intended to attract a mate or maintain a pair bond. It does not matter if the song is innate or learned, simple or complex, harsh or musical. What matters is how it functions.

Call notes serve to communicate many important things: begging for food (in nestlings); soliciting food (by females forming pair bonds); alarm associated with detection of a predator; distress upon being attacked; location of food sources; dominance within a flock; agitation; and to communicate flock direction. This list is not all-inclusive. Call notes are the daily language of a bird species. Numerous distinct call notes each signify specific messages. When a mixed flock of birds converge to harass a roosting owl, most will be vigorously making call notes distinctive to their species, each of which signify alarm and agitation. Call notes serve to communicate among pairs, or within a family, or among flock members. Red-winged Blackbirds, typical of many species, make an impressive variety of short, harsh, and variable *chek* notes whose function, depending on the note, ranges from flock cohesion to aggression to bird-to-bird contact.

Long ago, bird songs ranging from those made by caged canaries to the ethereal flutelike notes of thrushes, to the rich bubbling warble of a soaring Skylark ("Hail to thee, blithe Spirit," as Percy Bysshe

Shelley wrote in "To a Skylark" [1820]), were once naïvely thought to serve for human pleasure, the joy of listening to birds. That outrageous belief changed for good in 1920 when Henry Eliot Howard's classic *Territory in Bird Life* was published. Howard recognized that song functions as an aggressive declaration and defense of territory as well as an attractant to females. A singing male Bullock's Oriole, or Lazuli Bunting, or American Goldfinch, is either signaling his territorial boundaries, using song to aggressively advertise that he is in residence and other males should keep out; or letting would-be mates know that he is available for nesting; or telling his mate that he is close by. Song is often augmented by position—the bird may sing from a high and exposed perch where the sound carries well but where the singing bird is exposed, creating some risk. An individual may repeatedly use a specific song perch as it daily reasserts its dominance.

Many bird species use specific songs for communicating territoriality different from those used for communicating with their mates. For example, the Black-throated Green Warbler, as well as many other wood-warblers, typically sings two easily distinguished buzzy songs, one aimed at other males, a territorial song, the other directed toward females. Among songbirds, each singer typically has several songs that it uses for different functions or at different times of day. Many species have dawn songs that differ from songs given later in the day. Most songbird species develop impressively large song repertoires (see page 214).

HOW TO LEARN CALLS AND SONGS

For those of us with a good ear for music it is rather easy to pick up on bird songs and calls and soon commit dozens, indeed hundreds, to memory. For others, it is more difficult to learn bird vocalizations, but it is worth trying as there is much to be gained. There is no substitute for field time spent listening. The trick is not to leave the vocalizing bird too soon. Singing birds offer much in the way of vocal repertoires, and spending a half hour or so listening to one singer offers a great learning opportunity. Birders should not undervalue the pleasure of just listening to birds. It is widely known that during nesting season birds are most active in the early hours of the morning, and singing is at its peak (see "Dawn Chorus," page 215). But birds will call and sing all day, and some, such as the Northern Mockingbird, will sing into the night. Of course, night is also the time for owls and nightjars to vocalize.

Mnemonic devices help describe and then recall some bird calls and songs. Consider how many birds have names that are variously descriptive of their vocalization: Willet, Laughing Gull, Killdeer, chickadee, Whip-poor-will, Chuck-will's-widow, Bobolink, bobwhite, phoebe, pewee, Clapper Rail, Dickcissel, Warbling Vireo, Gray Catbird, Chipping Sparrow. Describing a vocalization in words is often helpful. I was once told that I would always know a Song Sparrow's song because it sings *Madge, Madge, Madge, put on your teakettle-ettle-ettle*. That little phrase more or less works, but Song Sparrow songs vary considerably from bird to bird as well as within repertoires of individual birds. Madge is not always putting on her teakettle quite the same way. Perhaps the most well-known description of a sparrow song is *old Sam Peabody, Peabody, Peabody,* the phrase depicting the plaintive song of a White-throated Sparrow. Many birders successfully develop word phrases that correlate closely with the cadence of a bird call or song.

For the musicians among birders, it is possible to put musical notes on bird song and express it as one would write a musical score using standard music notes, structure, and script. In 1921 F. Schuyler Mathews published *Field Book of Wild Birds and Their Music,* a book that uses the scales and notes of sheet music to describe the songs of various wild birds. Truth be told, applying human constructs to try and depict birdsong is not very effective. Playing the song of a Wood Thrush on the piano does not sound like a Wood Thrush. It lacks the rich and melodious flutelike tonal qualities. For species such as Winter Wren, it is doubtful that its notes could be played rapidly enough by human fingers on a piano.

Bird songs and calls on CDs and various phone apps, used properly, are good tools for both learning and field identification, but be aware of birding ethics with regard to playing songs in the field (chapter 2). There are now some apps available that attempt to identify the bird species after you record it on your phone. An additional approach is to study the sound graphically using either a spectrogram or an oscillogram. Spectrograms, also called sonograms, show the frequency range of the sound on the Y-axis of a graph and show the duration of various components of the song on the X-axis. Oscillograms show sound wave pressure on the Y-axis and time on the X-axis. Both types of graphs are best studied in concert (so to speak) with listening to the actual song or call notes. See Pieplow (2017, 2019) for a guide to using spectrograms in identifying North American bird sounds.

It is possible to make your own sonograms and spectrograms with programs that are easy to download on a personal computer. Finally, you may access thousands of bird calls and songs by using the Macaulay Library at the Cornell Laboratory of Ornithology (macaulaylibrary.org) and, in turn, if so inclined, you may wish to add some of your recorded songs to their collection. The Macaulay

Library is an accessible compendium of bird sounds as well as sounds of other forms of wildlife. Xeno-Canto is another useful online resource.

CALL NOTES AND THEIR FUNCTIONS

Most birders probably hear more call notes than songs because birds are calling all year round, not only during breeding season, when song is peaking. Keeping in mind the various functions of call notes, it is sometimes possible to "translate" what the call means. A flock of Canada Geese honking as they fly overhead are making contact calls. If you hear two Carolina Wrens alternately calling in the understory of a woodland, you are also hearing contact calls. If you surprise a Green Heron along a stream and it abruptly takes flight uttering a demonstrative and loud *skee-yowl!* you have heard an alarm call. When a flock of Snow Buntings flies up in front of you from an open field you'll hear quite an array of high *cheew* and *pipp-pipp* calls interspersed with a flat, buzzy *zhipp* call as the flock circles around, often to eventually land at about the same spot from where they took off. You heard a combination of alarm and flock cohesion calls.

Black-capped Chickadees have diverse call notes with multiple meanings. The begging *dee* is given by young birds about to fledge and just after fledging. A female chickadee gives a similar call, a broken *dee,* when breeding, usually upon completion of the intended nesting cavity. The chickadee call complex is the most common of the call notes, an array of calls that are variations of the staccato *chick-a-dee-dee-dee* and are frequently given by both males and females, especially during the nonbreeding season. Black-capped Chickadees also make hiss and snarl calls, twitter calls, and *tseet* calls. The different calls have multiple functions, ranging from flock cohesion to warnings of potential predators.

Chicka and high *zee* calls transmit specific information about the size and location of a potential predator as well as the threat level it represents. Male chickadees increase the frequency of *chick-a-dee* calls and add more *dees* when they detect a small raptor. Black-capped Chickadees are much more responsive to calls of Sharp-shinned Hawk compared with Northern Goshawk, the latter being so large that it is unlikely to pose much of a threat to chickadees. Northern Pygmy-Owl calls also elicit

Black-bellied Plovers have a distinct high-pitched and plaintive call note emitted when taking flight and while in flight.

The (Slate-colored) Dark-eyed Junco is known for its sharp and distinctive "smacking" call note emitted frequently in winter flocks. Its song is quite distinct from the call, a pleasant trill somewhat similar but more melodious than that of a Chipping Sparrow.

vigorous warnings in the form of *chick-a-dee* and *chicka* calls. These small owls represent a serious threat to chickadees.

Red-breasted Nuthatches, a species that often associates with flocks of Black-capped Chickadees, have developed an understanding of the meanings of chickadee calls and react accordingly, a case of a bird species learning the language of another bird species to its advantage. Playing the high *seet* call of the Tufted Titmouse, which is an alarm call, will cause a Veery singing at dusk to react to the titmouse's alarm call and to make alarm calls typical of its species, essentially spreading the warning. Blackbird flocks will take flight when they hear the alarm call of a Killdeer, and feeder birds will quickly fly to cover or freeze if a Blue Jay emits an alarm call.

Chickadees are small and vulnerable birds. One would not expect Golden Eagles, Great Blue Herons, Double-crested Cormorants, or numerous other species to exhibit such language complexity in their range of call notes. Colonial birds are quite vocal as they are forced into close proximity, a stress that stimulates vocalization.

Many species give flight calls, which serve various functions. Goldfinches and other finches give flight calls even for short flights. Larks and other open-country birds use flight calls as well as elaborate courtship songs given from aloft. Pileated Woodpeckers often emit their repetitive *kuk, kuk, kuk* calls while in flight, helping pairs stay together when one bird moves away from the other. Black-bellied Plovers make a plaintive, high, whistled flight call.

During migratory periods it is possible to go out at night and hear the flight calls of migrants as they fly overhead. Flight calls are common both in diurnal and nocturnal migrants and obviously function in flock cohesion and perhaps in other less obvious ways. Most migratory birds make flight calls. The challenge of successfully identifying the various calls has been met, and birders now may consult www.oldbird.org and various other websites. Being able to identify flight calls of nocturnal migratory birds allows you to know just what species are migrating.

SONG FUNCTION

As Eliot Howard correctly surmised in 1920, birdsong functions as a form of aggressive behavior used in territorial establishment and defense as well as for mate attraction and pair cohesion. Song is part of the bird's extended phenotype, helping determine its degree of competitiveness as an individual within its species. It is as important as any other signal, such as plumage brightness or aggressive vigor, that a bird might use in attaining a territory and acquiring a highly fit mate. Song is evident throughout breeding season, but some species sing to some degree all year. Adult birds often briefly intensify singing after young are fledged. This is because juvenile birds are actively learning their songs soon after having been fledged. In many species females sing, so song is not, as was once largely assumed, confined to males (see "Female Birdsong," page 218).

Song begins in earnest before breeding and during spring migration, before the singer has reached its breeding grounds. Groups of wood-warblers foraging at stopover sites far from where they will breed sing to some degree, though often not with the vigor or volume that will be the case once they reach their ultimate destination. Most other migrants also do this. Singing during migration is likely a result of enhanced testosterone in the migrants as they come into breeding physiology. Some species, such as Northern Cardinal, Song Sparrow, American Tree Sparrow, Golden-crowned Sparrow, Loggerhead Shrike, Western Meadowlark, Northern Mockingbird, Carolina Wren, American Dipper, Townsend's Solitaire, and American Robin, are known to sing to some degree all year, including on their wintering grounds. Long-distance migrant bird species sing briefly when they reach the wintering ground in the Caribbean or Central or South America as each bird establishes its winter territory. Once a bird secures its winter territory it will typically defend it with call notes. Wood Thrushes on winter territory in Central America make sharp *whip, whip, whip* calls at dawn and dusk. In some species, such as Northern Cardinal,

singing diminishes during winter in more northern parts of the birds' range but continues in the southern parts of the range.

Some species have elaborate songs termed *emotional release* songs. This term is purposefully vague. The nocturnal song of the Ovenbird has been suggested to fit this category. An Ovenbird will fly at dusk or during the night above the woodlot or forest where it is nesting, singing a rich warbling song, different from the well-known *tea-cher, tea-cher* song it makes diurnally. The function of the dusk/night song is unclear.

INNATE AND LEARNED VOCALIZATIONS

Most species of birds do not need to learn to vocalize—they are hatched knowing how. Ducks know how to quack from the get-go, though the vocalizations of ducks often vary between males and females within a species. In some species, including some ducks, quail, and murres, birds begin to vocalize prior to hatching.

Innate vocalizations are the norm for most species that are not oscine songbirds. But for oscine songbirds, a significant amount of learning is required. It is usually the male bird, the father, that

Ovenbirds sing often during the day during breeding season but also have a less often heard and very melodious dusk and night song. *(Garry Kessler)*

Ducks, such as these Common Eiders (*top*), have innate vocalizations that differ between the sexes. Male Common Eiders emit a low-pitched moanlike sound while females make a quite distinct guttural croaking sound. Brown Thrashers (*bottom*), in stark contrast to species with innate vocalizations, have been known to learn up to 2,000 songs.

is the model, the tutor, though hearing other males of the same species augments the learning process. With the recent realization that in many species females also sing, it is probable that learning also occurs via female song. Many bird species have a sensitive period which begins about two weeks from hatching and lasts just short of two months, during which they effectively learn their song. The learning process may represent a full curriculum of over a dozen song variations typical of the species.

Young birds begin softly singing what is called a subsong, a low-volume rendition of their tutor's song. As they gain confidence they increase the volume of their singing. It should be obvious that birds are receptive only to their own species' song and not to the songs of other bird species. An offspring of a Western Bluebird may hear a Swainson's Thrush and a Western Tanager as well as other species during its sensitive period, but it learns only its own species' song.

Bobolink males are unusual in that they appear to learn the song of their species when they return in the spring following fledging because the males have stopped singing by the time the young males fledge. A Bobolink male in his first breeding season returns to his colony and learns the song type of that colony.

Some species, such as Northern Mockingbird and Brown Thrasher, mimic other birds' songs and calls and incorporate them into their own repertoires. These species are called open-ended learners. They continue throughout their lives to learn new songs, usually self-taught. The Northern Mockingbird is a species whose very name reflects its ability to mimic. Brown Thrashers are known to acquire repertoires that reach up to 2,000 songs.

SONG AND ITS VARIOUS MANIFESTATIONS

The sharp and loud *wheep-wheep-wheep* of a treetop Great Crested Flycatcher is attention getting and far from musical, but it represents the basic song of this suboscine species. The song is somewhat variable, with two principal *wheep*-like phrases and gradations between them, so if you listen to a Great Crested Flycatcher you will soon hear some of these variations. They serve for declaration and defense of territory and for contact between mated pairs. The song is often accompanied by individual repeated or isolated *wheeps* and *whips* given throughout the day. It is a simple repertoire but does consist of more than one kind of vocalization.

An American Robin, a true oscine, has a considerably more complex song, with more notes to it, and far more musical to human ears. It is basically described as a repeated *cheerio cheerilee* or *cheerilip, cheerilee,* a rolling, repeated, melodious phrasing. But phrasing varies, often including an added component called a *hissely* phrase. Sometimes a robin will do repeated *hiselys*. Because the song is learned, two male robins may sound decidedly different. The song of an American Robin is more complex than that of a Great Crested Flycatcher.

Once a bird species learns its song, it may modify it. Most individual songbirds develop their own repertoires of songs, which are variations of the basic song pattern. A male Song Sparrow usually has between 5 and 15 song types. In her classic study of the Song Sparrow, Margaret Morse Nice wrote that when listening to the varied songs of two rival male Song Sparrows, "as far as variety of singing went, it was as if fifteen birds of different species had settled on our grounds" (*The Watcher at the Nest*, 1939, page 34). Individual Marsh Wrens are known to sing over 200 song patterns. Why be so versatile? Evidence from field studies suggests that a large repertoire is effective in dominance relationships among males as well as impressing females. Think of it as a musical demonstration of the bird's résumé, its qualifications for holding a territory and being a worthy mate. It is what evolutionary biologists call an honest signal. The bird had to learn and develop its vocal versatility that, along with other qualities such as plumage, leads to reproductive success.

Watch a territorial male singer and note the frequency with which he switches from one song type to another. Song-type switching is common to bird species and correlates with level of aggression and dominance.

Song matching, also called matched countersinging, is sort of the avian version of dueling banjos. Two territorial males match one another's song. As one of the birds offers a new rendition the other will match it and then offer his own new rendition, only to be matched by his rival. It's like betting in a game of poker: "I'll see your song and raise you one melody." It is possible to hear this behavior in the field if you listen closely to two males near their territorial boundaries (chapter 14). In many habitats it is possible for a male bird to hear multiple other males and females of his species singing. New males typically assimilate into their neighborhoods by quickly learning and then matching the songs of others of their species. Song matching is a form of nonviolent competition among the males, a means of reinforcing their territoriality.

Variation in song output is related to time of day. Some birds sing mostly in the early morning beginning with the dawn chorus, described on the next page. Some sing throughout the day but with less frequency and intensity than they do in the early hours. Thrushes, including American Robins, are well known for singing at dusk as well as at dawn. When darkness approaches, a Wood Thrush will sing vespers, so to speak, its song gradually trailing off, ending with simple *whip* call notes before the bird goes quiet. Weather affects song in that birds sing less on windy days and rainy days.

Bird song types vary with habitat. Many bird species occurring in open habitats usually have high-pitched or buzzy songs that carry well in such open terrain. Think of a Field Sparrow and Grasshopper Sparrow, for example. Bird species inhabiting closed forests or marshes have low-pitched songs that carry well in such habitats. Think of a Yellow-billed Cuckoo or a Sora. Many birds sing from high perches, so some forest-dwelling bird species emit high-pitched songs. Some wood-warbler species such as Blackpoll and Cape May Warblers are examples.

Time and rate of singing vary among species. Vireos stand out as perhaps the most persistent singers, and such persistence, while recognized for many years, defies a clear explanation. One classic study, cited in virtually all ornithology texts, was per-

The Red-eyed Vireo has long had a well-earned reputation as a persistent singer.

formed by Louise de Kiriline, who spent a very long day counting exactly how many songs were sung by a single male Red-eyed Vireo. That bird sang 22,197 songs in 10 hours. But why? Vireos sing as they forage, so at least we know that the bird was not fasting for the day. But that begs the question. Why must a Red-eyed Vireo be such a persistent singer when a Scarlet Tanager, a Black-and-white Warbler, and a Rose-breasted Grosbeak—each of which may be nesting in the same woodlot—are far less persistent in that regard?

There is no simple explanation as to why songs vary so broadly in complexity. Wrens have more complex songs than waxwings. Woodpeckers substitute simple but loud drumming for song when establishing territories. While the overall functions of birdsong are clear, the numerous variations and nuances, from individual to individual and from species to species, add to the great richness of bird vocalizations and pose many questions.

DAWN CHORUS

The dawn chorus is a well-known phenomenon of birdsong, particularly during breeding season. At first light, and even a bit before, birds begin singing, one species after another. As light increases, more singers join in a cacophony of birdsong. At the height of the temperate breeding season, the dawn chorus is remarkable and it is a challenge to birders

to pick out all of the singers, kind of like picking out conversations at a loud, crowded party. But it is not a challenge to the birds. They are quite adept at recognizing others of their species among the throngs of singers. As it becomes full light the dawn chorus begins to decline. Species slow their rate of singing and drop out, beginning their foraging. Dawn chorus is also evident in the equatorial tropics, occurring to varying degrees throughout the year.

Why a dawn chorus? During breeding season, when dawn chorus peaks, most singers are defending territory, and dawn singing may reinforce territorial boundaries. Typically, skulking nonterritorial "floater" males (chapter 14) are present, seeking territories. Informing floaters that a territory remains occupied is advantageous to the territory holder. A singer also lets conspecific neighbors know that it is still out and about. Further, the singing could attract females that are not yet mated or are seeking additional copulations (chapter 15).

Studies of Black-capped Chickadees during dawn chorus support the hypothesis that the dawn chorus is a vital component of the social dynamics of the chickadee community. By extension, it may and likely does have the same function for other bird species. Breeding chickadees that were part of the same winter flock were studied (by Foote et al.) and compared with chickadees from different flocks using a technique that employed multiple microphones called acoustic location systems. It was learned that in spite of the noise created by other singing birds, dominant males easily detected the songs of other chickadee males and were vigorously engaged in song matching with them. Males whose territories were close together were the most frequent birds to song match, a clear indication of aggressive territoriality. Birds who had been in the same winter flock, who were familiar to each other, song matched less than birds that had been in different flocks.

Dawn chorus appears to be a collective effort to maintain territories. Winter Wrens, which maintain year-round territories, sing at dawn during the nonbreeding season, which makes the case that dawn chorus does function to reinforce territorial defense and boundaries. Dawn chorus is a daily reset button of the dominance relationships of a bird community.

One reason for the dawn chorus may be a pragmatic one related to foraging. During breeding season most bird species feed on arthropods, particularly insects, and the cool dawn hours are a

The Northern Mockingbird sings both night and day and is representative of a species that has a unique ability to memorize and repeat numerous songs, each in great detail.

time when most insects are not yet fully active, so the birds are not losing foraging time by singing. Dawn is also a time when the air is calmest and song will carry farther and be clearer, with less degradation, helping individuals recognize the exact identity of conspecifics.

Birds also have a dusk chorus, less dramatic than dawn chorus but noteworthy. Thrushes in particular sing at dusk, but other species become more active singers as the sun begins to set.

REGIONAL DIALECTS

Songs of the same species can sound different from one place to another. Having spent time in both Massachusetts and Georgia, it was obvious to me that Northern Cardinal and Tufted Titmouse, among others, sang distinctly differently in the two places. The songs were easy to recognize as being those of cardinals and titmice, but the cadence and patterns were not the same between the two locations. What I noticed were differences in regional dialects. Just as human speech patterns differ from one region of the country to another, so do song patterns of various wide-ranging bird species. This results from birds' abilities to vary their song along with cumulative song matching through generations. In other words, it is akin to cultural differences. Regional differences in birdsong occur in numerous species.

WHY MIMICRY?

One of the most watched YouTube videos (and there are multiple examples available) is that of the Australian Superb Lyrebird, part of David Attenborough's *The Life of Birds* series (1998). This lyrebird inhabited a forest near a car park and learned to mimic car alarms and motor drives of cameras, among many other things. Lyrebirds have long been admired for their ability to reproduce many bird sounds in addition to their own rich repertoire. In North America, the birds that routinely practice mimicry are the Mimidae—mockingbirds, thrashers, Gray Catbird—and the Sturnidae, namely the European Starling. In addition, Blue Jays and crows frequently mimic the calls of hawks. Blue Jays are particularly drawn to mimicking Red-shouldered Hawks. Some other bird species such as White-eyed Vireos are also occasional mimics.

Mimicry involves skillful incorporation of the calls and songs of other species into the repertoire of the bird doing the mimicking. Many birders might think they hear, for instance, a Belted Kingfisher, until they also and immediately hear a Northern Flicker, Summer Tanager, Chuck-Will's-Widow, Tufted Titmouse, and Orchard Oriole, along with the kingfisher, all emanating from exactly the same place, and interspersed with a variety of chortling

This is a male Painted Bunting doing his soft song while quivering his wings. This bird is behaving very aggressively toward the nearby intruder.

and call notes. The singer is, of course, a Northern Mockingbird doing what mockingbirds do.

It remains unknown exactly why mimics actually mimic. There is a deep genetic history among some mimics, since all of the birds in the families Mimidae and Sturnidae have some measure of mimicry skills. The Sturnidae include not only the starlings but also the famed myna birds. One species, the Hill Myna of tropical Asia, is a particularly renowned mimic. But in some families one species that mimics, such as the Marsh Warbler of Europe, is the only member of the family to exhibit complex mimicry.

There is as yet no widely agreed upon explanation for why mimicry evolved in various bird species. Considering just the Northern Mockingbird, it has been suggested that it may wish to discourage rival mockingbirds by creating a soundscape through its territory, a useful ploy if food were to be limited (this hypothesis suggests rather deep thought on the part of the mockingbird and its ancestors). More to the point, a singing male mockingbird may accumulate a long repertoire to impress would-be mates and discourage competitors, a demonstration of his fitness for territorial defense and fatherhood. Male mockingbirds that lack territories sing at night, sometimes throughout the night. This behavior might prevent direct confrontation with a territory holder and yet advertise the presence of the singing bird.

Birds such as mynas, parrots, and crows have the ability to mimic human speech, forming various words and phrases. Other bird species also have such abilities but not to the same degree. Parrots differ from mynas and crows in syrinx structure, as mynas and crows are true oscine songbirds and parrots are not. Parrots are now considered to be most closely related to falcons, and falcons are not mimics. Regardless, this human speech mimicry is a byproduct, not a fundamental adaptation showing fitness. Neither parrots nor mynas secure a territory by reciting phrases from the Gettysburg Address or emphatically saying "Hi there!" Parrots, recall, have proportionally the largest brains of all bird groups, accompanied by personality distinctions and a very high overall intelligence. Only they know why they mimic our words. We'll leave it at that.

THE SOFT (OR WHISPER) SONG

Many birds, including some that are not songbirds, will, on occasion, sing softly, the volume turned way down. This is the soft song, also called the whisper song. Perhaps surprisingly, the soft song can be a signal of aggression on the singer's part. A male bird singing a soft song is ready to physically attack a perceived rival. The function of low volume may be to force a potential rival into close proximity, eyeball to eyeball, so to speak, a clear threat.

I observed a male Painted Bunting do a soft song directed at a visible, close potential rival. The territorial male approached to within a few feet of its potential rival. The entire demeanor of the territorial male indicated intense aggression but it did not attack. Instead, as he sang very softly he quivered his wings and thrust his head forward, his posture an obvious warning to the other male. The interloper broke off and flew away, and the territorial male chased him to what I presumed to be the boundary of his territory. Some minutes later the territorial male was back on his usual song perch singing at normal volume.

FEMALE BIRDSONG

Suppose you are watching a singing Loggerhead Shrike, or Gray Catbird, or House Wren. In each of those species males and females look alike. But the bird you see must be a male because it is singing. Or look at that Baltimore Oriole singing atop the tree. Even though it is early May it "must" be a juvenile male because it looks like a female and it's singing. But what about that Northern Cardinal? It is obviously a female, and she is singing. In recent years it has become increasingly apparent that female songbirds of many species do sing and, further, that incidence of female singing among birds has been woefully under studied.

As early as 1943, famed ornithologist Margaret Morse Nice documented female singing in Song Sparrows as well as in some other songbird species. David Lack, in *The Life of the Robin,* describes female singing of European Robins and traces the observation back to Darwin. But the widespread occurrence of singing female birds remained relatively unnoticed. Only recently has interest grown in female song and what might be its functions.

A survey compiled by Lauryn Benedict (femalebirdsong.org/) provides documentation of female singing in 25 families of North American passerine species. Female singing is not a rare occurrence. Tropics-focused ornithologists have learned that in many if not most species of tropical passerines—species that are on the same territory for most if not all of the year—females routinely sing.

As suggested above, it does not matter if males and females look distinctly different. Female House Finches, Red-winged Blackbirds, Purple Martins, Black-throated Blue Warblers, and Scarlet Tanagers are all known to sing. Female singing occurs in the majority of passerines, based on a global survey of 1,000 species in which 64 percent of the females were found to sing. This has been an amazingly overlooked area of potential research to which dedicated birders may make contributions (see Odom and Benedict 2018).

It is perhaps not surprising that female singing occurs in species such as Song Sparrow, Northern Mockingbird, Northern Cardinal, Pygmy Nuthatch, and Black-capped Chickadee, each of which are known to occupy the same general area all year. But long-distance migrant species such as wood-warblers and tyrant flycatchers also have many species in which females sing.

Perhaps it should not be surprising that females sing. Female song is often not as robust or persistent as male song, but it seems to function similarly, namely in defense of territory and for maintaining social contact with mates. Some females are known to sing while on the nest. Vibrant, loud, frequent song, such as is typical in male birds, both incurs an energy cost and puts the singing bird at some risk of discovery by a potential predator. Females usually incur a greater energy cost than males do during the building of nests, and the act of both nesting and singing would impose an additional burden. But it is also in the female's best interest to defend the territory, especially while the male might be out provisioning for the brooding female or newly hatched nestlings. Female singing during breeding season may also act to signal other males of potential mating opportunities even if the female doing the singing is already mated (chapter 15).

For females to sing requires nothing unique in physiology or anatomy. Research on birds in the Icteridae family (Price et al.), the orioles and blackbirds, suggests that females lost or reduced singing while evolving into long-distance migrants. Since males return before females and establish territories, there is less incentive for females to sing. In icterid species confined to the tropics, female blackbirds and orioles all sing. Perhaps loss of singing represented a trade-off in terms of the rigors imposed by both migration and a compressed breeding season.

DUETTING

In the tropics it is sometimes the case that males and females join in singing, producing what sounds like a single song. Typically, a male will begin the cadences and the female will take over and complete the song. Males and females singing in close proximity make it difficult to distinguish between the singers as they seamlessly work in concert with each other. The function of the duet is likely to both allow separated birds to stay in contact and to serve as a declaration, by both sexes, that they hold the territory. It may also function to strengthen the pair bond, as tropical birds tend to live for a long time,

Carolina Wrens duet and are active singers throughout the year. This bird is obviously undergoing molt but is still singing.

longer than birds in the temperate zone, and may stay together for years.

Duetting is far less prevalent in the temperate zone than in the tropics. In North America, 21 species of Passeriformes are reported to duet, a group that includes Eastern Wood-Pewee, Pygmy Nuthatch, Wrentit, Painted Redstart, Western Tanager, California Towhee, Red-winged Blackbird, Brown-headed Cowbird, Baltimore Oriole, and House Sparrow. One species commonly observed duetting is the Carolina Wren, but the duet is modest. A female may add a note or two as the male concludes his normal repeated *teakettle* song. The American Crow is also known to duet, but only in association with aggressive behavior. Nonpasserines also duet.

Duetting is well known among owls, though it typically takes the form of a male and female calling back and forth to each other. Sandhill Cranes have two kinds of loud duet calls (which are termed the *unison call* in cranes), which serve as essential components of their pair bonding and territorial declaration. In all, about 400 species of birds globally duet, covering 40 percent of bird families.

BIRDSONG AND URBAN NOISE

There is growing interest among ornithologists in how the increasing level of suburban and urban noise might affect birds' singing patterns. Many studies (see Luther and Baptista) have shown that birds are responding to urban noise by altering their song. When confronted with low-frequency urban noise, birds alter their songs toward a higher frequency, and they also (unsurprisingly) sing louder. In laboratory-controlled conditions, House Finch singing changed when the birds were confronted with increasing noise. House Finch song rose in minimal frequency when low-frequency noise was played, and lowered in frequency in the absence of noise. In other words, the birds sang at a higher pitch with a background of low-pitched noise, presumably because that would help overcome the background noise. The same result has been reported for White-crowned Sparrows compared over a 30-year time span. Urban noise is typically loudest at low frequencies; thus, by raising pitch and volume, the birds' songs would be more effectively heard.

Eastern Wood-Pewees, in a study done in Washington, DC (Gentry et al.), were found to alter their song in keeping with whether there was high traffic noise or low. This result demonstrates the sensitivity and flexibility birds have for coping with their auditory environment.

The fact that some bird species are altering their singing frequencies and volume in response to urban stimuli is an example of avian cultural evolution. These birds are adapting to the reality of loud background noise, but this trait is likely not yet incorporated in their genes. Each generation of birds learns to sing at the best pitch for its environment.

REAL ESTATE, MATE ATTRACTION, AND PAIR BONDING

WHY COURTSHIP?

Why don't birds just mate randomly? How seemingly easy that would be, saving immense stress, energy, and time. A female and male bird encounter one another, identify each other as being of the same species (innate recognition, of course), and mate. Mate, nest, and that's that.

But, for good reason, nature does not work so simplistically. Nesting season is complicated by competition to obtain and hold territories, to attract and hold a mate, to copulate and produce an egg clutch (and often more than one clutch), and to fledge offspring, all of which impose high energy and risk demands on both males and females. Birds have no choice but to devote an inordinate, indeed taxing, effort to reproduction, and they must be able to detect meaningful differences in each other. That matters mightily. Males size up other males in their efforts to establish and defend a territory and obtain access to females. Females size up males in making decisions (and yes, it is normally a decision made by the female) on mating partners. Females also compete among themselves for access to prime males and prime territories. Competition is unavoidable. Territories, nesting space, food, and number of potential mates all stimulate competition.

A fundamental axiom of nature is that resources are not infinite and the more demand on resources, the higher the level of competition. The essence of natural selection is that limited resource availability, in whatever form, forces intraspecific competition that selects for survival of the fittest. Simply put, not every bird gets to breed. The fit are those whose genes successfully move into the next generation. That is not a random event. It is based on how birds, as individuals, succeed or fail. It is in the interest of

Two male Purple Martins contest for access to court a female early in the breeding season. The males may occasionally engage in physical combat, though it's usually brief. *(Jeffory A. Jones)*

any individual bird to try to successfully mate and produce young.

I have heard birders remark that various adaptations in birds occur "for the good of the species." Birds flock "for the good of the species." Birds mate "for the good of the species." There is no such thing. The species, from an individual bird's viewpoint, is a meaningless abstraction, a word we humans use to describe a group of animals that mate among themselves (chapter 6). The individual—and, by implication, its genes—is fundamental to evolution. It is individuals that matter to birds attempting to breed. Each individual is egocentric. No male Bewick's Wren will voluntarily opt out of nesting so that another male will have the opportunity to nest. No female Black-headed Grosbeak will yield her nesting site to another female. There is no "good of the species." There is only individual fitness.

Because birds are able to mate more than once, fitness in birds is measured by lifetime reproductive

---

OPPOSITE: These Red-headed Woodpeckers are copulating, the male on the back of the female. They look alike, but they nonetheless "know who is male and who is female." *(Jeffory A. Jones)*

The female Northern Cardinal is soliciting copulation from the male, culminating the courtship efforts of both individuals in the pair.

success (LRS). Achieving a high measure of LRS is not easy. The birds you watch are the current winners in this annual ongoing generational contest that defines nature. As one noteworthy example, consider Wisdom, a female Laysan Albatross that nests on Midway Atoll. As of 2018, Wisdom was at least 68 years old. She was first banded in 1956 by Chandler Robbins, who rose to legendary status in ornithology. Wisdom is estimated to have produced at least 30 chicks during her life, most of them fledged and grown to adulthood. Her most recent chick (at this writing) hatched on February 6, 2019. She is the oldest known albatross and she is still reproducing. To put it mildly, Wisdom has enjoyed significant lifetime reproductive success, a clear winner in natural selection. There are numerous websites and videos devoted to Wisdom, all readily available on the internet. Just google "Wisdom the Albatross." Wisdom may even be "friended" on Facebook.

HOME RANGE AND TERRITORY

Many bird species inhabit what is called a home range for some part of the year. This is an area where the bird is highly familiar with its habitat. Bald Eagles, as might be expected, typically occupy large home ranges. One study (Garrett et al.) done in the Columbia River estuary in Oregon and Washington looked at nine breeding pairs of eagles that inhabited the area year-round. The average home range area was 8.5 square miles (22 km²) but ranged from 2.3 to 18 square miles (6 to 47 km²). Some ducks on their wintering ranges occupy a home range confined to an estuary or even a large pond, but if the water should freeze, the ducks move on to seek open water. Home ranges are generally flexible, changing with conditions. Swallows and swifts fly from their colonies to share feeding areas, and gannets, puffins, and murres fly from their nesting colonies to acquire food for themselves and their young, so these species have fluid home ranges. Albatrosses range so widely that for some species much of the world's ocean area is fundamentally their home range. Home range takes many forms among bird species. Winter flocks of redpolls, waxwings, Snow Buntings, and Horned Larks sometimes briefly settle in a particular area but otherwise move nomadically throughout the nonbreeding season, eschewing a winter home range.

Loyalty to a temporary home range is why birders

frequently find a rare species such as a Northern Shrike or a Snowy Owl to be present throughout much of a given winter in the same area where it was initially discovered. The bird remains, presumably because there is adequate food for it and it has learned the area, reducing risks associated with constantly moving to a new area. One key point is that birds that inhabit home ranges in whatever form normally do not defend them from others of their species. Actual defense of real estate is termed *territoriality*.

In 1966 the playwright Robert Ardrey published a bestselling book titled *The Territorial Imperative*. This volume followed his previous bestseller, *African Genesis* (1961). Both books were devoted to showing the evolutionary connectivity between animals and humans. In *The Territorial Imperative* Ardrey made much of the fact that so many animal species, birds prominent among them, rigorously defend areas that we term territories. As mentioned in the previous chapter, Eliot Howard established the reality of avian territoriality in his classic book *Territory in Bird Life* (1920). Territoriality takes multiple forms but is generally a fundamental characteristic of any bird species' annual cycle.

A territory is defined as a defended area used for

Bald Eagle pairs, shown here copulating, occupy a large home range but often tolerate other pairs of eagles nesting in close proximity. *(John Grant)*

breeding. The bird advertises its claim on its territory and attempts to prevent others of its species, and on occasion other species, from crossing the invisible lines that, to the territory holder, define boundaries. Passerine birds sing from an exposed perch, often while displaying, declaring their territories. If you visit a habitat on a daily basis during breeding season and mark the specific locations of singing males on a map, those dots, taken together, will roughly reveal territorial boundaries. This method, called spot-mapping, has been frequently used in ornithology. Singing, displaying, and chasing characterize territorial establishment, including occasional fights.

Territories vary in value, so there is intense competition for prime real estate. This competition is all the more intense in the temperate zone, and particularly the northernmost latitudes, because breeding season is compressed into just a few months or even a few weeks for most bird species. In essence, all birds are seeking territories at the same time. If food is abundant, territories tend to be smaller because a bird need not defend a greater area. If food is limited, territorial boundaries become larger, requiring more intense territorial competition and defense. Many bird species return to the same territory year after year (see "Philopatry" below). A male bird holding a territory has difficulty keeping out 100 percent of other males all of the time, a reality that leads to extra-pair copulations (see "Copulation," page 232). Territorial boundaries in passerines may change during the breeding cycle. It is not uncommon for birds to gradually reduce their territory size over the course of nesting.

Territory structure varies. One kind of territory frequently seen in many passerines, such as Song Sparrow, Baltimore Oriole, Black-headed Grosbeak, Ovenbird, and Least Flycatcher, is one in which the pair—particularly the male—claims the territory for all of their feeding and nesting needs, keeping out others of their species. In species such as Eastern Bluebird and American Robin, overlap in territorial boundaries is tolerated and birds share feeding areas. Robins often feed in open areas, and the presence of other robins may improve the birds' abilities to detect potential predators. Many bird species ranging from Barn Swallows to Black-legged Kittiwakes defend only the tiny territory in and around their nest. Colonial seabirds are obvious examples of shared feeding areas since individuals are unable to defend a patch of ocean from other foraging seabirds. Beyond that, fish and various other marine prey move around, and thus, as more birds are searching, they more easily discover schools of fish. Gannet colonies are neatly structured, the nests all about equidistant from one another because the territory is defined merely as the distance from its nest that the gannet can reach to stab at a neighboring gannet, which they do.

What happens to males that fail to obtain territories? Where do they go? Perhaps nowhere. They stay in the neighborhood, surreptitious, keeping a low profile within other males' territories. These birds are floaters. It has been shown many times that if a male bird is removed from his territory, the territory will almost immediately be taken over by a floater. Suppose you see a male Indigo Bunting on a song perch. A Sharp-shinned Hawk makes a quick strike and takes the bunting. The next morning the Indigo Bunting is right back singing on his song perch. But of course it's not the same Indigo Bunting—a floater male has claimed the territory. Floaters typically skulk, not attempting to challenge a territory holder, instead waiting for an opportunity to take over the territory or to perhaps sneak a copulation with a female. The very existence of floater males demonstrates that available territories do act as limiting factors in bird populations. There is no reason not to think that there may be populations with floater females as well. Females are just as driven to mate as are males and thus in areas with dense female populations there may be nonterritorial, nonbreeding floater females.

PHILOPATRY

Many bird species exhibit natal philopatry, individual loyalty to their breeding site, returning each year. In 1804 John James Audubon put what he called "light threads" on nestling Eastern Phoebes, and was able to confirm that some returned the following spring to their same nesting area at Audubon's home in Mill Grove, Pennsylvania, outside of Philadelphia. Breeding philopatry has been confirmed for a large number of bird species of many families. If you habitually visit the same "patch" to see birds year after year, you will surely observe some of the same individuals you have seen in previous years, whether or not you realize it. The Gray Catbird you see in the shrubs may have wintered in Belize, but it has returned to the same shrubby patch in which it nested the previous summer. The Piping Plover on a familiar stretch of beach is likely to be the same bird you saw last summer and perhaps the summer before that. The same holds for the nearby Osprey at its nest.

Breeding-site philopatry is basically forced with regard to colonial nesting birds. Those species annually return to the same colony since potential colony locations are limited. This means that the same individual egrets and herons return to their colonies and that Purple Martins annually return to their same colonies of gourds or martin boxes. Chimney Swifts return to their same chimneys.

Natal philopatry is the basis for conservation initiatives to re-establish seabird colonies. Stephen W. Kress of the National Audubon Society successfully re-established Atlantic Puffins, which used to breed on several islands off of Maine, including Eastern Egg Rock Island. He did this by bringing chicks from a Canadian puffin colony and successfully fledging them on Eastern Egg Rock (with no help from any resident puffins, since there were none). The chicks, imprinted on Eastern Egg Rock, fledged into adults, lived at sea for some years (as is typical of puffins), and then some of the birds returned to Eastern Egg Rock and bred, which they are still doing. They were stimulated to mate and reproduce with the use of decoy puffins and mirrors, so that the birds could see animated puffins (themselves!). There are now over 1,000 birds on the island. The model Kress developed to re-establish puffin colonies has served very well in re-establishment of other endangered or threatened seabird species throughout the world.

Many bird species show philopatry to their non-breeding range, sometimes called winter-site fidelity. Waterfowl have long provided examples of winter philopatry. Nonbreeding-season philopatry sometimes takes the form of winter territoriality. Both male and female Townsend's Solitaires defend winter territories where juniper trees are rich with berries, their food source. They use vocalizations as well as direct chasing to expel other solitaires from the territory. Both Red-headed and Red-bellied Woodpeckers defend winter territories. Male and female Wood Thrushes are among long-distance migrant species that defend individual territories on their wintering range, in this case in Central America. Hooded Warblers and American Red-starts are examples in which males and females independently select and defend winter territories and that females often inhabit areas that have less resource value, such as scrubby fields, while males and dominant females establish territories in more

Project Puffin, directed by Dr. Stephen Kress, used decoys and mirrors to entice Atlantic Puffins to nest at Eastern Egg Rock Island in Maine. *(Stephen Kress)*

Townsend's Solitaire males and females defend winter feeding territories. *(Ed Harper)*

food-rich forests. Multiple banding studies over many years have confirmed that many, indeed most, species do return to the exact or nearly exact site where they spent the previous nonbreeding season.

The advantage of winter philopatry and territoriality is that it provides the bird with access to resources within a familiar area. Risk is reduced. However, some long-distance migrant species, such as Acadian Flycatcher, do not defend a winter territory because they join mixed-species flocks of tropical bird species. Some Neotropical migrant species from North America join resident birds that follow army ant swarms. The nomadic ants flush up prey species of insects that the birds feed upon.

AGGRESSION

Physical aggression often occurs in territorial defense or in struggles to obtain food. Birds have evolved ritualized behaviors, songs, and other signals, each of which acts to diminish outright aggression. But aggression underlies territoriality and is sometimes outwardly manifested. It is not uncommon to see birds ranging from Song Sparrows to Northern Mockingbirds engaged in fights. One gull fighting another gull for food might grab the other gull's wing, sometimes causing injury. House Wrens are combative not only with their own species but with others. Wild Turkey males often fight for access to females. Recall from chapter 5 that many bird species attack their own image in mirrors or windows.

Common Loons compete for access to lakes that offer rich food and suitable nesting places. Lakes are limited resources that vary widely in value to loons. Pairs vigorously defend their territories, but intruders are frequent because there are many more loons than there are suitable breeding lakes. Not surprisingly, Common Loons are philopatric, successful breeders returning annually to the same lake. Male Common Loons, particularly older ones, will on occasion fight to the death to remain on their breeding lake. The haunting yodels of Common Loon pairs act as a complex signal to other loons. Male

Common Loons vigorously defend their territorial lakes during breeding season. *(Garry Kessler)*

As other coots ignore the battle, two coots have at it. High levels of aggressive fighting are common in coot society.

loons of the largest body size have lower frequency yodels, thought to warn potential rivals of the strength of the territory holder. Young loons between two and four years old generally search for unoccupied lakes, but older loons aged five or six aggressively attempt takeovers. Sometimes younger, more robust birds kill older males (that may exceed ten years of age) in taking their territory because the older birds will fight to the death. They have no other options.

American Coots routinely engage in physical combat. Coot pairs physically defend their nesting areas against intrusion by other coots and sometimes also by moorhens. A territorial coot will rush at an intruder and attempt to grab it with one foot while kicking at it with the other. Coots that remain on their breeding areas all year continue to defend territories and thus remain aggressive with other coots even when not breeding. It is common to witness coot aggression. Coots are nonaggressive toward one another only when not on territory, sometimes forming large flocks with little in the way of agonistic interactions.

Mute Swan pairs are aggressive in defense of their territories and particularly the immediate area around their nest. A male Mute Swan will rush at any intruding swan as well as other waterfowl, including geese and ducks, to drive them away. As the other bird approaches, a territorial Mute Swan assumes an aggressive display posture with wings held up and spread, head drawn in deeply, neck curved. Once close the swan will lunge at the intruder and beat it with his wings. Mute Swans have been known to attack humans in this manner.

COURTSHIP AND PAIR BONDING

Volumes have been written about the diversity of avian courtship behavior, which often includes elaborate displays. Display enhances visual stimulation, the "presentation" of one potential mate to another, often with strong auditory components. Displays are stereotyped rituals that, from a human perspective, exaggerate the appearance of plumage, showing off its most striking characteristics. Some of the most iconic images of courtship among birds include the elaborate displays of birds-of-paradise, the lekking of species such as various grouse and shorebirds (see "Lekking," page 236), the synchronized dances of some grebe species as they scamper together across the water, the locked talons of eagle pairs as they fall through the air, and the enlarged throat sac of a male frigatebird as he points his head skyward and vocalizes. Add to those the elaborately displayed nuptial plumes of various egret species, synchronized and exaggerated displays in some duck species, the presentation of fish to mates by various tern species, the energetic looping courtship flight displays of various hummingbird species, and, of course, the peacock's remarkable tail, a list that still barely scratches the surface.

The outcome of courtship is pairing, which may endure throughout the nesting cycle or be brief,

The Mute Swan in the rear (*top*) is intimidating the one on the left, attempting to drive it away. Eventually that bird will move away or (*below*) a fight will ensue. Such altercations are almost always won by the bird holding the territory.

involving just copulation. A male and female accept each other as mates and proceed with copulation. Pair bonding takes many forms, ranging from lifetime commitment to no commitment whatsoever (which, of course, essentially means no pair bond). Lifetime commitment is more meaningful in some species than others. Song Sparrows, which generally occupy the same area year-round, sometimes mate for life and sometimes don't. "Divorces" are commonplace, as females switch mates. Beyond that, Song Sparrows are normally not very long-lived so it is not uncommon for one member of a pair to perish and the remaining bird to then seek another partner. On the other hand, in long-lived species such as eagles, pairs may remain together for many years. Swans, geese, cranes, albatrosses, vultures,

This pair of Cedar Waxwings (*top*) is engaged in courtship feeding, a behavior common among numerous bird species and one often observed in the field. (*Bottom*) The Yellow-billed Cuckoo male (the bird on the top) is using a dragonfly to court the female, and is already preparing to copulate. *(Left: Susan Scott; right: Garry Kessler)*

condors, eagles, ospreys, gulls and terns, various alcids, parrots, and other birds, including some passerines (such as jays, crows, and ravens), are known to pair for life (which, of course, means until one of the pair perishes).

Long-term pair bonds and philopatry are tightly associated. Pairs may separate during migration and occupy different wintering ranges outside of breeding season, but they come together again to breed. In a study (by van Leeuwen and Jamieson) performed on Dunlins on their Alaska breeding range, 74 percent of males and 54 percent of females returned. Pairs from the previous breeding season were quickly re-established. Only when one member of the pair failed to return or returned late did individuals form new pairs.

Pairs of many species, such as Ospreys, various other raptors, shorebirds, gulls and terns, and seabirds, separate to occupy different winter ranges. As is the case for Dunlin, when the pairs come together at the breeding site the following year the pair bond is re-established, and in some cases they use the same nest as in previous years. Numerous studies have confirmed that the individuals are able to recognize each other as they reunite. A study (by Davis et al.) done using geolocators, tiny devices attached to birds that allow precise determination of their location over time, showed that a mated pair of Sabine's Gulls that nested in the Canadian Arctic took very separate "winter vacations." The male of the pair wintered off the coast of South Africa and the female off the coast of Peru. Geolocators are discussed in more detail in chapter 16.

Avian courtship is easily observed in the field. Male birds sing, often from an open perch, distinct courtship songs that function to attract potential mates, and often chase females, as well as each other. The degree to which female singing acts to help form or strengthen pair bonding is still not fully understood.

Males of species ranging from Horned Larks to American Woodcock perform aerial displays. Bobolinks, which routinely mate with more than one female, typically pursue females, often several males doing so at a time. Males also chase each other in their efforts to secure territories.

Courtship feeding is evident in many species, including herons, gulls, terns, doves, cuckoos, and numerous passerine species. In courtship feeding a female solicits an item of food from a male. In some species, such as Northern Cardinals and Tufted Titmouse, the female assumes the posture of a begging juvenile, squatting, mouth open, wings vibrating. The begging posture triggers the male's instinctive reaction to feed. This has led to some confusion by human observers who mistakenly believe the female is being submissive. Instead, she is being dominant, using the male bird to her advantage. Demanding food is the female's way of sizing up a male's ability to continually provide for her and her offspring. Courtship feeding continues well after the formation of the pair bond, and in many species males bring food to brooding females on the nest.

Sandhill Cranes, a species in which pairs mate for life, engage (as do other crane species) in an elaborate series of highly stereotypic dancing behaviors that precede the formation of a pair bond and are then continually augmented by other equally stereotypic dance steps after pair bond formation. Pre-pair bonding displays represent an effort for a crane

These Florida Sandhill Cranes use various dance steps to both establish and reinforce their pair bond.

Ducks have long been observed exhibiting species-specific stereotyped courtship displays. The posture of this drake Red-breasted Merganser is an example. Males often display in groups. *(Garry Kessler)*

to sell itself to a potential mate, to overcome any reluctance to mate, and, as well, to show prowess to potential rivals. Post–pair bonding display dances act to continually reinforce the pair bond. Many species, including waterfowl, loons, grebes, gannets, cormorants, herons and egrets, storks, gulls, and terns, form and reinforce pair bonds with a series of stereotypic postures. These displays are commonly observed in the field.

Courtship does not always result in pair bonding. Species such as hummingbirds, some shorebirds such as American Woodcock and Buff-breasted Sandpiper, various grouse species, and Wild Turkey separate immediately after copulation, and females alone assume the responsibilities of nest building, incubation, and fledging the young. In some species, this type of courtship has taken the form of elaborate lekking (see "Lekking," page 236).

While courtship in most species occurs on the breeding grounds, most waterfowl species form pair bonds on the wintering range. This is not surprising because ducks cluster together on bodies of water, ideal settings for males to compete for females and for females to select mates. Male ducks display to females and attempt to intimidate other males. Pairs form and reveal themselves within flocks of wintering ducks as the male and female of a pair remain in close proximity within the flock. Among passerines, Cedar Waxwings and American Goldfinches form their pair bonds well before they nest. Waxwings are semicolonial nesters and travel in flocks, so when a flock settles on a breeding area the pair bonds are formed and the birds are ready to go directly to nesting. American Goldfinches delay nesting until seeds are available in midsummer.

In some species, traditional sex roles are reversed and the female assumes the primary role, courting the male or various males instead of the other way around. This is not common but is seen in species such as phalaropes, Spotted Sandpiper, and jacanas. Phalaropes provide a clear example. Females are more colorful and larger in body size than males. Females display to males and compete among themselves for males. Once a male accepts a female and copulates, she will see to the nest and lay the clutch, but then it is upon the male to incubate eggs and raise the chicks. Females often continue to solicit males, form another brief pair bond, and lay a second clutch, again leaving it up to the male to provision for and fledge the chicks.

The White-throated Sparrow exhibits one of the more unusual mating systems known in ornithology. Though there is no sexual dichromatism in this species, it has long been recognized that there are two morphological forms (morphs). One morph, called the white-striped morph, has a bright white superciliary crown stripe. The other morph, called the tan-striped morph, has a tan superciliary crown stripe. It is common to observe both morphs in the same flock. The difference in morph type is caused by a genetic character called a supergene, which is a chromosomal inversion on chromosome 2, involving a total of about 1,000 genes, approximately 10 percent of the total genes present in the species. Tan-striped morphs overwhelmingly pair with white-striped morphs, and vice versa. There seems no obvious reason why this should occur but it does. Various hypotheses have been offered. There are strong behavioral differences between the morph types in such things as song frequency, aggression, and parenting skills that perhaps more or less balance out within the pair. Regardless, the preference of one morph to mate with its opposite dominates throughout the species. This is an example of what is termed a negative assortative mating, very unusual among birds.

The female Red-necked Phalarope (*left*) is larger and more brightly colored than the male (*right*).

The tan-striped morph of the White-throated Sparrow (*top*) looks quite distinct from the White-striped morph (*bottom*).

COPULATION

The obvious outcome of courtship is copulation, and it is relatively common to observe copulation among birds during breeding season. In many species, once a pair bond is formed, copulation is frequent. Males usually instigate, and there are various precopulatory behaviors common to each species. A female, once she has accepted a male as her mate, often assumes a solicitation posture, inviting the male to copulate. Fertilization, as in all land animals, is internal, though the vast majority of male birds have no penis. Instead a male typically lands on the back of a female, maneuvers into position, balancing with his wings, and turns so as to extend his cloaca to make brief contact with the cloaca of the female, at which point the bird ejaculates. At breeding season there is normally an enlargement of the male cloacal area, termed a *cloacal protuberance* (chapter 3). This serves a similar function to a penis. Ejaculation takes mere seconds and thus the copulation event is normally of short duration, likely a benefit to birds with regard to predators. Given that birds weigh so little, it is not burdensome for a male to mount a female by perching on her back, but it does require considerable balance, and birds use their wings to steady themselves. Both birds must twist their bodies to align cloacas.

Waterfowl do have penises, but copulation in waterfowl otherwise appears similar to that in all other birds, with one exception. Groups of male Mallard ducks have on occasion been observed harassing a female and eventually forcing copulation. Normally when Mallards pair, the male and female perform a mutual head-bobbing display that precedes copulation, the male mounting the female in the water and holding her by the neck during copulation. But unpaired males sometimes join together, drive away a mated male, and aggressively force themselves on the female.

Copulation in most birds is normally followed by preening. Swifts, perhaps not surprisingly, copulate in the air.

SEXUAL SELECTION

Charles Darwin and Alfred Russel Wallace independently put forward the theory of natural selection, a powerful and well-tested theory that accounts for how species actually evolve. Darwin's monumental work, *On the Origin of Species*, was published in November of 1859. The two men were in full agreement about the essence of natural selection, but Darwin took it one step further when he wondered why, in some groups of animals, such as birds, many species occur in which males and females look different, sometimes very different.

The most annoying bird in the world for Charles Darwin was the male Indian Peafowl, commonly known as a peacock. (Darwin wrote that he lost sleep just thinking about peacocks.) This is the bird with the outrageously large fan-shaped tail adorned with multiple iridescent eyelike spots that are put on glorious display when the male spreads and vigorously shakes his tail as he courts females. Darwin found it extremely vexing to try and account for how a male peafowl may have evolved such elaborate and clearly cumbersome plumage. There was

The bright, deeply blue plumage of the male Indigo Bunting is one of many examples of sexual selection among birds. The blue plumage is at its most intense during breeding season.

nothing in the bird's tropical forest environment that would suggest him to be more fit by dragging around a heavy and oversized tail, to say nothing of the rest of his bright plumage. Ah, but there was. There were females, peahens. Darwin hypothesized that it was females who were the actual agents of selection, doing so by choosing among males that, to them, were the most impressive in appearance, the most beautiful. Those winners got to mate and mate often. Darwin thus concluded that in some cases it was the female that acted to provide the essential selection pressure in shaping the appearance and courtship behavior of the male.

Further, female choice, in Darwin's view, could be capricious and whimsical as the females selected for traits that pleased them but failed to add to the male's probability of survival and might actually reduce it. But males would have no choice. They must mate. Darwin thus concluded that female choice over generations determined the often outlandish appearance as well as the elaborate courtship behavior of the males of some species.

Darwin called this concept *sexual selection* and wrote a book about it, *The Descent of Man, and Selection in Relation to Sex*, published in 1871. Four chapters were devoted to birds. And he took sexual selection one step further. Darwin was also keenly aware that in many animals, including numerous bird species, males must compete, sometimes physically, for access to females (such as when rams clash horns). Thus another essential element of sexual selection is male-to-male competition. Such competition, as Darwin visualized it, tended to result in males being larger than females and, in the case of birds, more brightly adorned. Male-to-male competition works hand in glove with female choice.

There has been considerable effort to understand female choice of males. How do females decide? It has been shown that in European Barn Swallows, tail length and symmetry of the long tail feathers in males appear to be the basis for female choice, while in North American Barn Swallows, females select males with the richest orange on their undersides. Experiments with House Finches show strong female preference for the reddest males. The red pigmentation is obtained from what the bird eats—carotenoid pigments in the environment—and could signal health in the male, having obtained sufficient nutrition to be colorful. Thus the intensity of red on a male House Finch may serve as an "honest signal" to females that the bird would make a good mate.

Dark-eyed Juncos offer another example. The degree of white in the outer tail feathers of male juncos has been shown to correlate with the male's body size and degree of dominance over other males, thus

The male Ring-necked Pheasant is a somewhat extreme example of sexual selection, with his vibrant coloration and long tail feathers. But many bird species exhibit sexual selection in various forms.

From the intensity of his red coloration, this male House Finch appears to be a dominant bird.

signaling the male's overall fitness. Males with a greater amount of white in the tail are preferred by females as mates. Blue Grosbeak males with the most vivid blue plumage have been shown to be healthy with regard to fat deposition and low parasite load, as well as able to defend food-rich territories and provide good parenting. The blue in a Blue Grosbeak is a structural color, not a pigment, and females see it not only in blue wavelengths but also in ultraviolet wavelengths (chapter 5). Thus female Blue Grosbeaks see males that may appear more different to them than they do to a human observer. And females make their choices accordingly.

House Sparrow males have black bibs (called a badge) on their throats, and the bib size signals degree of dominance. The larger the bib, the more dominant the male. But does having a large black bib actually mean that a male House Sparrow is more fit than one with a smaller bib? Will he mate more frequently? Does a large bib mean good health? Bib size does correlate with dominance and aggressive behavior in males. Aggressive behavior correlates with testosterone levels—the higher the testosterone, the more aggressive the bird. But high testosterone levels actually act to suppress the bird's immune system. House Sparrows with large black badges are therefore not necessarily healthier than those with smaller bibs. Investigations (see refer-

ences) with House Sparrows have not confirmed that females select males on the basis of bib size, but females may mate more often with dominant males simply because these males are, indeed, dominant. Which means they have larger bibs. Hmm.

Maybe in some species females select males with cumbersome or very bright plumage because these males are capable of handling the handicap imposed by such plumage. Perhaps males with disproportionally long and elaborate feathers, such as various pheasants, peafowl, the birds-of-paradise, or the long-tailed African widowbirds, are showing off their prowess to females by literally bearing the burden of their own plumage. This has been called the handicap hypothesis. If true, this would also suggest that Darwin was mistaken in claiming that capricious females were driving males down cumbersome evolutionary alleys, so to speak. Instead, they are forcing extreme demonstrations of fitness potential. Well, maybe—and maybe not.

Richard Prum, an ornithologist at Yale University who has worked extensively with sexually selected lekking species such as Neotropical manakins, has resurrected and championed Darwin's view that "beauty happens" by a female-driven genetic "runaway process." Both males and females inherit the trait that initially inspired female preference for a particular male characteristic, and the result is to continually exaggerate that characteristic through generations. Female choice might lead to males that bear plumage and patterns that are not adaptive with regard to enhancing male survivorship potential. Prum argues that Darwin was fundamentally correct. If Prum and Darwin are correct, the elaborate breeding plumes of egrets are mere enticements, no more, no less. This explanation sounds unsatisfying, but that alone does not make it false.

Sexual selection is not "one size fits all." Ring-necked Pheasants and Bobolinks are strongly sexually selected. Males are conspicuously distinct in plumage from females, larger in body size, and tend to mate with more than one female, a characteristic typical in strongly sexually selected species. Some males sire many more offspring than others. Northern Cardinals are less sexually selected. Males have markedly different plumage from females, but they are not openly polygynous and do not differ from females dramatically in body size, nor is there a huge disparity in reproductive success among cardinals that secure territories. Cardinals form traditional pair bonds and are socially monogamous (chapter 15)—pairs stay together. But studies have shown that the males with the deepest red pigmentation are disproportionately chosen by females, which is an example of sexual selection. Many duck species are sexually selected, the males more colorful and often larger than females, but ducks nonetheless form strong pair bonds, many species pairing on their winter range and migrating together. Once nesting ensues, however, males do not participate in raising young.

Hairy Woodpecker males, as well as other woodpecker species, differ from females by having bright red on the back of the neck, but this is likely not an example of sexual selection but rather of sexual recognition. Likewise, Pileated Woodpeckers differ in the amount of red on the head, males having more than females, as well as a red malar stripe. Pairs of Pileated Woodpeckers stay together, forage together, and do not exhibit behaviors at all typical of a sexually selected species. Flicker sexes differ only by the presence or absence of a malar stripe (present only on males), a characteristic that has been experimentally shown to function for sex recognition. And finally, species such as numerous tyrant flycatchers, wrens, corvids, sparrows, waxwings, shrikes, and mimic thrushes typically are not dichromatic, the males and females indistinguishable in the field.

LEKKING: REAL SEXUAL SELECTION

Lekking represents a clear example of sexual selection. A lek is an arena where male birds periodically gather for no other reason than to court females. Females, in turn, visit the lek for only one reason: mate choice. North American birds that exhibit lekking include Sharp-tailed Grouse, sage-grouse and

Male and female Northern Flickers differ in plumage by the presence of a red malar stripe on males. This is the Red-shafted subspecies.

Many grouse species, such as the Greater Sage-Grouse, shown here, engage in lekking, an essential aspect of their courtship behavior. *(Frank Caruso)*

prairie-chicken species, and Buff-breasted Sandpiper. Though not a North American breeder, the Ruff, a shorebird, presents an outstanding and perplexing example of a complex lekking species (see "The Case of the Ruff," below).

Most leks, like those of prairie grouse species, are traditional, the birds using them year after year. On a typical lek birds concentrate, obvious to one another, in close proximity. Various species of grassland grouse and prairie-chickens assemble in traditional lekking areas in spring, where the males begin their booming, dancing, and courtship posturing at just before dawn, continuing for a few hours. Leks contain variable numbers of males, and squabbles among the males are commonplace.

The essential thing to understand about leks (and about true sexual selection) is that males vary widely in reproductive success. Only a few males, perhaps merely one in a few cases, will actually succeed and mate with the majority of the females that visit the lek. Female choice governs which males mate. Virtually any female may succeed in mating, but that is not the case with males, where success at fatherhood is variable. Once the mating is completed the female attends to all of the nesting duties.

American Woodcocks are not considered to be a lekking species, but they come close. Though males and females look alike, woodcocks represent a behaviorally sexually selected species. Males gather in open fields at dusk and perform their flight displays, complete with both vocalizations and unique

sounds generated by modified flight feathers, all in efforts to attract a female with which to mate. Males have nothing to do with nest building or provisioning. Their reproductive lives are spent in display competition for females. This is sexual selection.

It is not clear why leks form. Leks, at least in the temperate zone, seem confined to grassland and tundra species. Grouse inhabiting forests, if they form leks at all, usually lek in fields outside of the forest interior. A European species, Black Grouse, provides an example. But another European species, the Western Capercaillie, leks within forests. It is noteworthy that many kinds of tropical bird species, including birds-of-paradise, manakins, cotingas, and some hummingbirds, form leks within forests.

Various models have been suggested for how and why leks have evolved, but it could be as fundamental as males historically forced together because females found it easier to compare among males that way. Lekking could be a form of female-driven sexual selection.

THE CASE OF THE RUFF

Many questions remain in order to fully explain sexual selection and lekking. By way of example, consider the Ruff. In this European shorebird species (which occurs regularly as a rarity in North America), males are considerably larger than females and, like other sexually selected bird species, wear an elaborate plumage (well, most of them do). The name Ruff derives from the long, colorful, and

The "independent" morph of the Ruff is the most common morph and is dominant to others. The white "satellite morph," shown here assuming a subordinate position, is less numerous but still successfully breeds.

thick neck feathers used during display. Females, called reeves, are relatively cryptic brown with variable amounts of black wing edging. Males gather on a lek in a marsh or field to display, often up to a dozen or more males, all in close proximity. Females visit the lek, and once a female selects a male, they mate, and she then goes off to attend to nesting and raising the chicks. No pair bond is formed. Thus far, this sounds like typical sexual selection. But it becomes complicated.

There are two obvious male morphs. The most common morph, called an independent for reasons that will become clear, is remarkably variable. Body color ranges from black to russet and the ruff plumes range from grayish to orangey to black. The other morph, called a satellite, has an all-white head and white ruff. The independent males are dominant in both numbers and display behavior, representing up to 85 percent of male Ruffs. Aggressive

independents display at one another and frequently engage in actual fights on the lek. The white satellite males are permitted on the lek but are submissive to the dominant independents. If attacked by an independent, a satellite male Ruff will quickly yield, head down, and the fight breaks off. Although a submissive minority, satellite male Ruffs do successfully mate on the lek and father offspring. That is interesting.

Why two male morphs? Good question, but the situation is even more complicated because there are actually three male morphs. A third and rare male morph was only recently discovered, one that looks very much like a female (hence it was overlooked) but is larger than a female, though still smaller than a typical male Ruff. This male morph has been termed a *faeder*, an Old English word for "father." Faeders, which represent only about 1 percent of male Ruffs, enter the lek, and these cryptic male Ruffs also successfully mate with females. The average testes size of a faeder is 2.5 times as large as those of males of the other morphs. One would think that such testes size would mean lots of testosterone and faeders would be pugnacious, but they are not.

Why would a female that is supposedly descended from generations of females that have sexually selected males for elaborate plumage and robustness ever mate with a morph that, for all intents and purposes, looks like her? Testes size? How does she know? Perhaps more to the point, all three male morphs are genetically determined (by a series of genes representing an inversion on chromosome 2), all are fertile, and all get mated, at least to some degree. Ornithologists are attempting to sort out just what is happening. Why do dominant independent Ruffs permit subdominant birds into their leks? In the case of the satellite males, it has been suggested that they are permitted because they add to the number of displaying birds and thus help attract females. Faeders are considered to be sufficiently cryptic to just slip into the lek without fanfare. But satellites and faeders do successfully breed with females that supposedly, in the deep time of Ruff evolutionary history, sexually selected the black and orangey independent Ruffs that continue to strongly dominate the Ruff population. If female choice is such a powerful driver of sexual selection, why do some choose outside the box, so to speak? If Darwin was kept up late thinking about peacocks, I can only wonder what he would think when confronted with Ruff sex.

THE RED-WINGED BLACKBIRD: A SUMMARY EXAMPLE

The Red-winged Blackbird provides an example of territoriality, sexual selection, male-to-male competition, female choice, and female-to-female competition. Red-winged Blackbirds are perhaps the most heavily studied of any North American bird species, at least as measured by the immense number of scientific papers written about them. This is because they are widespread and abundant, easy to observe, and easy to manipulate. Male Red-wings are larger than females and look different, being uniformly shiny black with conspicuous red wing epaulets. Females are smaller, with brown streaking, making them somewhat resemble a large sparrow. Females have only the faintest suggestion of orangey epaulets. Immature males more closely resemble females, likely reducing aggression directed toward them by dominant males. Immature males take a year to fully molt into male plumage.

Male Red-wings return early in spring, weeks before females. Flocks of migrating males gradually begin thinning as individuals drop out to select

Male Red-wings (*top*) are larger than females and look different, being uniformly shiny black with conspicuous red wing epaulets. Females (*bottom*) are smaller, with brown streaking, making them somewhat resemble a large sparrow.

OPPOSITE: Male Red-winged Blackbirds expose their red epaulets (top) as part of their display, which involves singing from an exposed perch with plumage exaggerated. When not displaying, male Red-wings typically cover their epaulets (bottom) to reduce aggression from other male Red-wings.

breeding territories in various marshes. Wings spread and red epaulets exposed to maximum size, the males sing their characteristic *konk-kah-reee* song from tall stems of marsh grasses or from trees as they establish territories. Their song is accompanied by exaggerated posturing, called the song-spread. Red-wing males also perform a territorial song flight. This is meant to emphasize the size of the male as he flies slowly, epaulets open, tail spread, wings fluttering. Chases among males are commonplace as competition for prime territories ensues, and fights between males occasionally occur.

Red-winged Blackbirds exhibit strong philopatry, the same dominant males returning annually to their previous territories. Territories have been shown to differ considerably in value to the birds, some much richer in food and shelter for nests than others. When many males reside in the same marsh, territories have been shown to compress, as it is not possible for a male to defend a really large territory from so many other contenders.

Males expose their red epaulets to various degrees. When in full territorial display they are maximally exposed. When males are behaving less aggressively, such as when they might be feeding together, the epaulets are partially to fully covered by scapular feathers. Epaulet exposure is a clear signal of aggressiveness in males. If a territorial male has his red epaulets painted over, he inevitably loses his territory. Reduced epaulet size lowers a male's competitiveness. Males on territory respond instantly and aggressively to male mounts with prominent red epaulets. Not every male will secure a territory, and floater males, all of which keep their epaulets covered, are commonly found. Some of these males may never mate because the dominant males use their territories for several years. This is a clear case of male-male competition as it affects sexual selection.

When females return, they compete among themselves for access to prime territories, paying little attention to male epaulet size or to the displays of particular males, instead choosing the best territories in which to nest. There is some evidence suggesting that females make choices among males with regard to song repertoire size and how intense the male is in courtship. But fundamentally, females select real estate and whatever male goes with it.

Red-winged Blackbirds are polygamous, one male defending a territory with anywhere from one or two to up to ten females nesting. Females construct their nests but males assist to various degrees in feeding young. Females in large harems (as these assemblages are quaintly called) occasionally mate with males other than the territory owner. Dominant experienced males have the largest harems, and the females have likely chosen wisely as these males are also good provisioners of food for the young.

For Red-wing females, access to males who help feed nestlings is essential, and they compete with each other for the best prospects. Experiments in which stuffed mounts of females were placed in a marsh in which the mounted birds were in positions indicative of precopulatory solicitation resulted in attacks on the stuffed mounts by other females. Both males and females will defend a nest against a potential predator, vocalizing and diving at the predator, sometimes striking it.

Male Red-wings have more variable lifetime reproductive success than females. Some males are very successful, some not. Females usually find mates, so there is less variability in their reproductive success. And that is some of what Red-winged Blackbirds do.

Male Baltimore Orioles are sexually selected such that they presumably appeal to female orioles . . . as well as to birders.

NESTING BEHAVIOR

OBSERVING NESTING BEHAVIOR: USE CAUTION

All birds lay eggs and the vast majority of species construct nests. You may notice a Song Sparrow carrying nesting materials, always flying into the same bush or tree, or you may see a Gray Catbird or Northern Cardinal repeatedly flying into the same dense shrub, a clue that these birds have nests there. An Eastern Phoebe with a beak full of insects will inevitably fly to its nest and feed young. Watch where it goes. An oriole flying from a tree with a white fecal sac in its beak is obviously leaving its nest. You may daily watch House Wrens, Tree Swallows, or bluebirds coming to and from a bird box.

In general, though, aside from those that nest colonially, birds are furtive regarding nesting behavior, and in many cases it is not easy to find their nests. That is because most nests are not meant to be found. Nesting is high risk for birds, and nests are generally placed to be cryptic. Only colonial birds—herons and egrets, cormorants, gulls and terns, alcids, and swallows—typically nest in colonies where nesting behavior is obvious.

Birders who have participated in atlasing projects know how difficult it is to confirm breeding in many bird species. Finding nests is not an easy task. A problem with observing the nesting behavior of passerines and other small birds is the risk that you may impose on the birds by just being there, unknowingly revealing the location of a nest to a nest predator, such as an American Crow or a red squirrel. Should you come upon the nest of, say, an Eastern Towhee and decide to observe it on a daily basis, it is very possible that nest predators such as Common Grackles, Blue Jays, crows, or a feral cat could discover the nest by observing your behavior. Another risk is that nesting birds may abandon the nest, dismayed by your proximity.

If a bird brooding its eggs detects you, it will

A Tufted Titmouse with nesting material. *(Jeffory A. Jones)*

sometimes freeze, remaining stock still until it perceives it is safe. That means you are way too close. This is similar to the freeze behavior birds exhibit after a predator has made an attack. Sometimes a pair of nesting birds will become agitated at your presence. As mentioned earlier, the risk to brooding birds is that you, the observer, will unknowingly reveal the location of the nest to a predator. More than one human nest observer has subsequently reported returning to observe the nest only to find it empty.

It is important to rigorously follow birding ethics at all times with regard to nest observation and photography. A description of the code of conduct to follow is at nestwatch.org/learn/how-to-nestwatch/code-of-conduct/.

WHY MAKE NESTS?

Of necessity, because all birds lay eggs, birds must care for the eggs as the embryos within them complete their development, hatch, grow, and fledge. Birds have little choice but to parent the eggs through complete fledging of young. The nearest

---

OPPOSITE: A male Pileated Woodpecker dutifully feeding his newly fledged and nearly grown daughter.

(*Top*) The male Swallow-tailed Kite (in flight) has just presented his mate with a fledgling House Sparrow. Such a gesture is essential in keeping a strong pair bond between the kites. The male Cattle Egret (*bottom*) ceremonially presents a stick to his mate, who will place it in the nest.

living evolutionary relatives of birds, the crocodilians, make nests, lay eggs, and care for them, as well as care for the newly hatched babies. Theropod dinosaurs, from which birds originally evolved, also laid eggs and made nests. Dinosaur nests with fossilized eggs and/or hatchlings have been found in many places, and one species of dinosaur, an oviraptor, was found brooding the eggs positioned identically to a modern incubating bird. Nests have a long history.

A nest allows birds to protect and incubate their clutch of eggs and care for the developing young birds until they fledge. Most bird species, including passerines, doves, woodpeckers, cuckoos, hummingbirds, parrots, loons, grebes, cormorants, waterfowl, gallinaceous birds, and shorebirds, build nests on an annual basis, a considerable task in many cases. Some passerine species such as Song Sparrow make several nests (for different clutches of eggs) during the course of a normal breeding season. Large species, such as Bald Eagle, Red-tailed Hawk, and Osprey, reuse nests, annually repairing and augmenting them with additional material. Nest construction is often labor intensive.

In many bird species nest building falls entirely to the female. In other species both sexes are involved in nest construction but the female does most of the work. In still others males and females are both equally involved in the building process. In many species males bring nesting material to the female during nest assembly. In some species, particularly in larger birds such as loons and herons, males begin the nest-building process and females take over.

In some species, such as Osprey, Red-tailed Hawk, and various heron and stork species, males continue bringing sticks to the female throughout the nesting cycle. The female then places each stick into the nest. The reason for this behavior is that nest construction is part of pair bonding, which continues not only during the initial nest construction but while the female is incubating the eggs. This behavior is commonplace for colonial water-

birds, such as herons, egrets, Roseate Spoonbill, and Wood Stork, but is also part of the nesting behavior of birds of prey such as eagles and Ospreys. The presentation of sticks by the male and their incorporation into the nest by the female is easy to observe when watching these birds during nesting season.

TYPES OF NESTS

Birds' nests vary from the most minimal, essentially no nest whatsoever, typical of vultures and caprimulgids (such as Whip-Poor-Will), to elaborate and intricately constructed shelters such as the woven baglike nests of orioles and the elegant tiny cup nests of hummingbirds. Nest form is part of a bird's extended phenotype, determined by innate behavior but also augmented by learning. Birds are hatched knowing how to construct their nests, at least mostly so, but their ability improves with experience. Nest location, shape, materials, and construction vary little within a species, which is why you can identify it as that of, say, an American Goldfinch or a Gray Catbird. A Great Crested Flycatcher could observe a Blue Jay pair constructing a nest and it would never occur to the tyrant flycatcher to make a similar structure, any more than it would occur to the jays to nest in a tree cavity and

Hummingbird nests are tightly and intricately constructed. This is the lichen-covered nest of a Ruby-throated Hummingbird. Hummingbird nests hold only two eggs, the clutch size for all hummingbirds. *(Garry Kessler)*

Great Horned Owls sometimes usurp other nests. This owl appears to have taken ownership of an Osprey nest. *(Garry Kessler)*

weave shed snake skin into the fabric of the nest. Vireos always choose a forked branch upon which to locate their hanging cup nests. Swifts break off sticks to form the nest and then use their saliva for glue as they construct their nest, attaching it to a vertical structure. Crows also break off sticks for their nests. Ospreys sometimes break off sticks, but they also collect them from the ground. Just as the feather structure of swifts, crows, and Ospreys is encoded in their genes, so are the nest blueprints and instruction manual. The same is true for all birds.

Very few bird species make no nest. As mentioned above, caprimulgids do nothing other than lay eggs on the ground and then incubate them. But female caprimulgids incubate during the day and feed at night, so the cryptic appearance of the brooding bird is essential in making the nest very difficult to find. Vultures make little in the way of a nest, but instead find a protected cave, ledge, or tree trunk and then deposit their egg clutch within. Great Horned Owls, which occur in many different kinds of habitats, will, in the East, often use an old Red-tailed Hawk nest, but in the West they are known to nest on rock ledges and in the fork of giant cactuses. Throughout their range Great Horned Owls will also nest in tree cavities and snags.

Some birds make ground nests called scrapes. The endangered Piping Plover and widely distributed Killdeer are examples. Scrapes, as the name implies, represent minimal structure, just a cleared area surrounded by pebbles moved and arranged by the bird as it clears the scrape. In scrape nesters, eggs are cryptically colored because the brooding birds are often actively foraging during daylight hours, leaving eggs exposed. Incubating birds on scrape nests are cryptically marked and inconspicuous when brooding. If disturbed by the often accidental proximity of a potential predator (including humans and their dogs), birds may exhibit some form of distraction display. In the case of plovers, this usually takes the form of what has been called a broken-wing act. The adult bird vocalizes as it shuffles slowly along the ground with a wing exposed and its tail spread. The bird obviously moves away from the nest. Once at a distance deemed sufficient (how do they know?) the adult bird flies off, but not directly back to the nest. Presumably the cryptic nature of the nest and eggs has now made it far less likely that the potential predator will return and discover the eggs. Soon the parent bird is back brooding its eggs. The distraction displays of shorebirds (and a few other bird species) all appear to be innate behaviors.

Many kinds of birds, particularly gallinaceous birds, waterfowl, grebes, loons, cormorants, rails, shorebirds, gulls and terns, alcids, some wood-warblers, and some sparrows, make ground nests of various sorts. Swan nests are large mounds, somewhat conical, raised up from the water level, with a depression in the middle where the eggs are located. Many other birds, including pelicans, loons, and Limpkin, make mound-type nests either on the ground or floating in water. Grebes and some ducks make floating nests of various marsh grasses and debris. Eider nests are generously lined with thick down feathers from the mother bird, but other bird species typically feather their nests as well. Ground nesting, while common among many larger bird species, is not required merely because of the birds' size. Herons and egrets, storks, spoonbills, and large birds of prey make large stick nests in trees. Double-crested Cormorants, which are colonial nesters, may nest either on the ground, as they do on some islands, or in trees.

Many passerines, including pipits, larks, meadowlarks, many wood-warblers, Dickcissel, and numerous sparrow species, make ground nests in fields and meadows. Some birds, such as Song Sparrows, make ground nests and, if they fail, make nests in shrubs, off the ground. These are typically open cup nests (see page 250) that are well concealed.

Puffins, storm-petrels, and shearwaters make subterranean nests among rocks on their nesting islands. Some passerines, such as Snow Bunting, Canyon and Rock Wrens, Northern Wheatear, rosy finches, and Townsend's Solitaire, nest in rock crevices.

Approximately 85 species of North American birds are known to use tree cavities as nest sites. Examples are some duck species, such as whistling ducks, Wood Duck, goldeneyes, Bufflehead, and Common and Hooded Mergansers; some trogons; some tyrant flycatchers; some falcons; many owls; some swifts; all woodpeckers, nuthatches, chickadees, and titmice; some wrens; some swallows; some thrushes (i.e., bluebirds); European Starling; House Sparrow; and Prothonotary and Lucy's Warblers. Tree cavities offer good protection but are often in short supply and in heavy demand. Only a few cavity-nesting bird species, most particularly woodpeckers, actually excavate the cavity; others

Many species of ground-nesting birds have distraction displays. This Piping Plover is an example.

must find existing ones. In addition to finding or excavating cavities, birds must defend them from other birds of both their own and other species, as well as from all kinds of other animals, including mammals such as squirrels. This means that availability of nest cavities factors heavily into the population dynamics of cavity-nesting birds.

Tree Swallows and Eastern Bluebirds have greatly increased in some areas thanks to humans providing nest boxes, just as colonial Purple Martins have been aided by clusters of hollow gourds and multi-chambered nest boxes set out for them. However, even as Tree Swallows and Eastern Bluebirds have increased, there is still competition for the boxes, including with species such as House Sparrows. European Starlings may aggressively take over a tree cavity from species even larger than themselves, such as Northern Flicker, and starlings have been suspected of contributing to the widespread decline of Eastern Bluebirds. Bluebird conservation now is focused on providing an abundance of carefully constructed bluebird houses whose narrow openings exclude starlings. Chimney Swifts are unique in that most of their population has for many years utilized chimneys rather than their traditional nesting site, hollow trees and tree cavities.

Chickadees, such as this Black-capped (*top*), here excavating its cavity nest, and all woodpeckers, such as these Pileateds (*bottom*), are cavity nesters. (*Top: Garry Kessler; bottom: John Kricher*)

The need for so many bird and other animal species to find tree cavities has conservation implications for forestry practices. Ample numbers of dead trees or live trees infected with various forms of fungus rot (which facilitates excavation) should be permitted to stand in any woodland or forest as these trees are the kinds that woodpeckers excavate for nesting and wintering cavities as well as for food such as insect larvae within the decomposing wood. Most woodpeckers do not excavate in healthy trees. The good news for other bird species dependent on finding cavities is that woodpeckers rarely reuse their previous excavations. Instead they annually excavate a new cavity in which to nest. Some woodpeckers, such as Yellow-bellied Sapsucker, may excavate a new nest in the same tree several seasons in a row.

Inside, cavity nests range from well lined to lined with nothing whatsoever. Cavity-nesting owls add no form of nesting material, though the cavity may already have material accumulated in it from previous occupants. Woodpeckers, once they excavate a cavity, immediately lay the clutch without adding any sort of nesting material. Interestingly, baby woodpeckers have padding on the scales of their ankles, enabling them to better cope with the hard cavity floor. In contrast, other cavity nesters, such as House Wrens, House Sparrows, Tree Swallows, and bluebirds, construct a cup nest primarily of plant material within the cavity. In the case of Tree Swallows almost all nests contain white feathers woven within the grassy nest structure. Bluebirds also occasionally use feathers.

Embankment nesting is a variation of cavity nesting seen in species such as Bank Swallows, Northern Rough-winged Swallows, and kingfishers. The two swallow species, after excavating burrows, line their nests with grassy material and usually add feathers. In contrast, kingfishers merely excavate their nest tunnel and add nothing in the way of nest material. Belted Kingfishers are known to make nest tunnels up to eight feet in length. Burrowing owls also excavate nests but often use existing burrows in prairie dog colonies when available.

Cliff Swallows make mud nests representing hundreds of separate trips to bring mud in their beaks and form it into the nest. Barn Swallows, which sometimes nest in close proximity with Cliff Swallows, also use mud as well as sticks and saliva to form their nests, which are then augmented with straw, paper, and lots of feathers, often from chickens (common around barns). Much effort goes into the construction of both kinds of nests, requiring about one to two weeks of work by the birds. Almost all Barn Swallows nest in barns and under bridges, but the species is obviously new to barn nesting. Historically, before European settlement,

Cliff Swallow nest building is labor intensive (*top*). Cliff Swallows gather beaks of mud to transport to and incorporate into their nests (*bottom*). *(Top: Bruce Hallett; bottom: John Kricher)*

they nested in caves, and some occasionally still do, as in the Channel Islands of California. Barns, in combination with forest clearance for farming, undoubtedly augmented the North American Barn Swallow population.

The House Sparrow, a species in the family Passeridae (Old World Sparrows), is closely related to Old World weaver finches, which make huge communal nests, mostly in Africa. House Sparrows place their nests in almost any kind of cavity, including around buildings, and in bird boxes such as those meant for bluebirds or Tree Swallows. House Sparrow nests are not communal but are typically large and bulky affairs that require both sexes to make multiple trips with straw, grass, and various other nesting materials. If the House Sparrow pair is nesting in a bird box, nest sticks often protrude out of the entrance hole. House Sparrows are aggressive and will try to evict other birds from cavities.

Species such as egrets, herons, storks, spoonbills, and cormorants construct platform nests made of large sticks. The birds often compete not only for

House Sparrows frequently compete, often successfully, with Tree Swallows for access to nesting boxes.

favorable nest sites but also for sticks and twigs, sometimes robbing each other's nests of sticks. A typical platform nest is generally flat and may appear fairly crude. Most platform nests have some finer plant material in the center, where the eggs are placed. Both sexes contribute to nest building but male birds routinely bring sticks to females even when they are brooding. Typically, there is a form of ritual between the pair as the male presents a stick to the female, who then incorporates it into the nest.

Hundreds of bird species make open cup nests constructed of a variety of materials with a depressed cuplike area within, usually lined with soft material. Mud is sometimes used to strengthen cup nests. These nests are either hidden on the ground or located in trees or shrubs. Shrubs offer the protection of dense branches, and trees add to that the protection of height. Small birds often locate their nest toward the end of a branch, making it more difficult for large predators such as squirrels or large birds to reach. Brooding birds typically remain quiet and still, an adaptation to protect against discovery by a potential predator. Some small birds are thought to place their nests in proximity to the nest of a predator such as a Great Horned Owl. The large owl will not be interested in small passerines but would act

as a potential deterrent to nest predators, such as screech-owls or various squirrels.

NEST CONSTRUCTION

The Rose-throated Becard, seen in the United States only in southern Arizona and south Texas, makes a large, complexly woven nest of bark strips and grasses, augmented with multiple other plant materials, including pine needles and lichens, a dense structure with thick walls and an entrance near the bottom of the nest, always to one side. The nest requires a huge effort, generally around two weeks of dedicated work by both male and female, the female doing the bulk of it. Even as the female is incubating, her mate may continue adding to the nest.

Why build such a huge edifice? Birders who seek Rose-throated Becards typically look for the big nest and then wait for the residents to put in an appearance. If you seek becards, enjoy watching the birds, but also take time to admire their amazing nest. Then wonder why it is that some bird species, such as Rose-throated Becards, apparently are genetically hardwired to put so much more effort into nest construction than most other species.

Other bird species make elaborate nests, including Verdin, a small bird that constructs a large stick nest with a side entrance. Bushtits construct a large pendent nest that abounds with various forms of plant material, feathers, and fur, all carefully woven with spider web material; and Ovenbird is named for its domed nest located cryptically on the ground of a forest.

It is difficult to make a bird nest. Try it. The simplest nests are scrape-type nests but what about those hundreds of bird species that spend so many hours to weave open cup nests or construct mud nests? It is beyond the scope of this book to review all types of nest construction, but I have provided references. When examined species by species, the diversity of forms that nests take as well as the materials used and the precision of nest construction are impressively broad.

Some cup nests, such as those constructed by Mourning Doves, are minimalist. The simple stick nest of a Mourning Dove is so casually assembled that it is usually possible to see openings in it by looking at it from the bottom. The male dove brings sticks to the female dove and she constructs (such as it is) the actual nest, which appears fragile but basically holds together. Contrast this with the tightly woven basketlike nest of a Baltimore or Bullock's Oriole, or even that of an American Goldfinch, nests that often endure the wind and snow of win-ter. Orioles put much more effort into nest building than doves. Cerulean Warblers, a canopy-nesting species, make a tight and compact nest in the forest canopy that includes intricately woven plant material, ranging from grasses to mosses, combined with hair and spider webs.

Compared with a Mourning Dove, oriole nests and Cerulean Warbler nests appear far more substantial. But look at the abundance of Mourning Doves in North America compared with Baltimore Orioles and Cerulean Warblers. And consider that for Mourning Doves (1) the nest is very basic, (2) only two eggs are in a clutch, and (3) the eggs are white, not at all cryptic. There appears to be no correlation between the kind of nest, the intricacy of construction, the time and effort that goes into nest building, and the abundance and survivorship of the bird species. Some bird species exert far less effort in nest construction than others, and some of them, like Mourning Doves, are nonetheless abundant. Although doves make flimsy nests they have a very long breeding season, lasting from April to October in most areas, and sometimes longer. They keep trying.

Birds sometimes nest in atypical locations and often incorporate unusual objects into their nests. House Wrens have been documented to add rodent

Construction of the immense nest of the Rose-throated Becard (male shown) requires a great deal of effort on the part of the birds. *(David Larson)*

skulls and bones, nails, and wire to their nests. Common Ravens have nested in numerous odd locations and have been most creative in the materials incorporated into their nests, including barbed wire. I have seen a House Sparrow nest with a deceased Tree Swallow woven in among the fibers of plant material. Perhaps the House Sparrow killed the swallow and usurped its nest. House Sparrows also will nest among the dense sticks of Osprey and Bald Eagle nests.

I once watched a Northern Rough-winged Swallow pair successfully nest in an outflow pipe near the waterline on a docked and abandoned cabin cruiser. House Wrens are known to nest in many unusual places, including inside a bovine skull and on an automobile axle. Carolina Wrens are equally creative in selecting nest locations, including flower pots, mailboxes, and, on one occasion, atop the radiator of a functioning automobile. One of my students watched a Mourning Dove pair nesting in the ledge of her dormitory window.

House Finches and House Sparrows sometimes add cigarette butts to their nests. This odd behavior might be adaptive. Perhaps the butts, with their various chemicals, reduce the parasite load within the nests. Incorporation of cigarette butts into nests appears to be a learned behavior, and the question becomes whether or not the birds recognize that it is actually beneficial to use them. It is unlikely that birds foresaw the potential adaptiveness of nicotine and its associated compounds in the nest.

Caspian Terns, which are scrape nesters, sometimes decorate the fringe of their nests with various mollusk shells, but in other cases a nest of this species may be little more than a scrape with some plant material or flotsam around it. Other bird species, including some shorebirds, also decorate the edges of their nests.

Male Marsh Wrens typically construct a series of "dummy" nests, before a female wren, and sometimes several females, select one in which to nest. House Wrens also make several nests and sometimes, as is the case with Marsh Wrens, attract more than one female to make use of these nests.

NEST FAILURE IS COMMON

Nests are potentially vulnerable, and annual nest failure often exceeds 50 percent, sometimes much higher. Nest failure rates, particularly for open cup nests, are surprisingly high. But then again, there are many nest predators ranging from chipmunks, red squirrels, raccoons, skunks, opossums, and various snakes to black bears. Many bird species, including crows, jays, and grackles, prey on other birds' eggs and nests. Birds compensate, so to speak, by repeated attempts to nest over their lifetime.

Piping Plovers commonly surround their nest with fragments of seashells, as do other species.

Nest failure is common, indeed routine. These Wood Thrush nestlings could be taken by predators at any time during the nesting cycle. *(Jim Zipp)*

In the fundamental calculus of biology, for a population of sexually reproducing animals to remain unchanged, each pair must produce two offspring that, in turn, both reproduce (one to replace the male, one to replace the female). For a species such as a Mallard with an annual clutch of 5 to 14 eggs and a lifetime of perhaps over ten or more reproductive years, this means that an overall annual egg and duckling mortality rate as high as 90 percent would nonetheless sustain the population. Southern populations of Northern Cardinals may have up to four broods per year, each brood with a clutch size of three or four eggs. Thus a pair of cardinals could conceivably fledge 16 young over a summer, and cardinals may survive for three to five years, occasionally considerably longer. It is therefore clear that mortality of eggs and young must routinely be high. The world is not awash with Mallards or cardinals.

Egg and nestling vulnerability is evident throughout the nesting process. For a typical passerine, it requires about five to six weeks to lay eggs, incubate, hatch the eggs, and feed and fledge the young. For an American Robin nest with a clutch size of four eggs it is likely that three of the four young, presuming they all hatch, will have died within a few months of fledging, often within a few weeks. The one survivor is still at daily risk because of inexperience.

Gulls and terns are at risk as fledglings, some killed by others of their own species. Herring Gulls have been commonly observed cannibalizing chicks in their colonies as the chicks inadvertently wander from their nest. One European study (Parsons) of a Herring Gull colony showed that of 329 chicks eaten by adult gulls, a mere four gulls were responsible for 167 of the dead and devoured chicks.

In order to fledge, a Common Murre chick must jump from its nest high on a narrow cliff ledge to the ocean below, its father encouraging it to take the long plunge. Once in the sea, the father of the inexperienced bird remains nearby while its chick must quickly learn to dive for food as well as dive to avoid predaceous gulls. Many murres perish during this difficult learning curve. Female Common Eiders must try to face down Great Black-backed Gulls as the predaceous gulls attempt to grab downy chicks in the sea. Similarly, passerines will attempt to defend their nests against predators such as jays and crows, often to no avail.

In areas where forests have been fragmented into scattered woodlots, species such as Wood Thrush have experienced elevated rates of nest failure, to the point that they fail to fledge any young. Fragmented forests greatly increase edge area, making it much easier for nest predators such as crows, jays, grackles, raccoons, and squirrels to locate nests. The curious thing is that each spring Wood Thrushes reappear, sing, establish territories, and nest in these small and fragmented woodlots, only to again fail to raise any young. Why don't they go extinct from such areas? The answer is that relatively nearby large tracts of intact forest where Wood Thrushes are successful nesters produce a surplus of young birds most years. Some of these birds subsequently colonize the fragmented woodlots, thus maintaining the appearance of a healthy population when, in fact, there is no successful reproduction occurring in such fragments.

This phenomenon is called a *source-sink* model. The source of surplus birds, the large forested area, replenishes the sink, the fragmented woodlots where reproduction inevitably fails. This makes it difficult to rely on techniques such as breeding bird surveys to be accurate indicators of nesting success (though the birds are, indeed, breeding, just not with success). Birders might hear no distinction between Wood Thrush singing in areas of source populations compared with sink populations, but the outcome in nesting success is starkly different. Many species of forest birds are sensitive to source-sink dynamics.

Long-distance migrant birds run a high risk of not returning after their initial migration to their wintering area (chapter 16). These fully fledged but inexperienced birds, ranging from Blackpoll Warblers to Ospreys, suffer considerably greater mortality than adults. They must make a long and often arduous migration to an area where they have never been, find a territory or home range in which to spend the nonbreeding months, and then find their way back to their breeding area. They must somehow avoid many unfamiliar hazards. Once a long-distance migrant has completed a journey and returned to its nesting region, it tends to have greater survivorship on subsequent migrations.

EGGS, EGG LAYING, CLUTCH SIZE, HATCHING

Eggs vary in size, shape, and coloration. Variation in size is not surprising, since bird species have a considerable body size range. Turkey eggs ought to be larger than sparrow eggs.

Chicken eggs are often thought of as typical of birds' eggs, and in some ways they are, but the shape of a chicken egg is only one of several shapes taken by birds' eggs. Duck eggs are oval, as are those of hawks and hummingbirds. Shorebird eggs are more sharply pointed. Grouse eggs, not surprisingly, resemble chicken eggs in shape. Eggs may be elliptical, pyriform, or oval, with degrees of variation among each of those categories.

Ornithologists have wondered if variation in egg shape is indicative of adaptation. Why might birds have differently shaped eggs? One widely cited example is that of the Common Murre, whose eggs are highly pyriform, sharply pointed at one end. Since Common Murres nest on narrow cliff ledges it seemed logical to suggest that the egg shape allows the egg to roll in a tight circle and not drop off the cliff. But it has become clear that such is not the case with murre eggs. Studies by renowned murre researcher Tim Birkhead have concluded that Common Murre egg shape provides stability on the narrow nesting cliff, reducing the chance that the egg will roll. What is adaptive about murre egg shape is that it resists rolling.

One study of egg shape was based on a unique statistical analysis employing a computer program called Eggxtractor. The study compared the dimensions of about 50,000 eggs of 1,400 bird species. The analysis measured the degree to which eggs varied from being perfectly spherical in shape. In other words, how pointed or how round? Once that was determined, it was soon learned that egg shape among bird groups did not correlate with nesting considerations such as nest type, clutch size (though see below), or nest location. What it did correlate with was wing dimensions. But why? Wing dimensions are indicative of flying ability. Eggs that were most pointed at one end, such as the egg of the Common Murre, were found to be typical of birds that were frequent, good, and often long-distance fliers.

The egg dimensions suggested that eggs were principally adapted to fit the pelvis dimensions of the various bird species. Long-distance migrants such as various shorebirds, or birds such as murres that spend much time in the air, have the most asymmetrical eggs because the skeletal characteristics of these birds, particularly pelvis width, represent an adaptation to flight that in turn selects for eggs that are easily accommodated into the corpus of a flying bird. If this study is supported by further work, egg shape is in large part a byproduct of flight demands as they vary among species.

Eggshell color and patterning (chapter 3) ranges widely. Birds that nest on the ground in the open or in open cup nests tend to have eggs that are at least somewhat if not highly cryptic. But even in open cup nests, eggs do not necessarily need to be cryptic because one of the parents will be brooding the eggs throughout the day. Mourning Doves, as noted above, lay white eggs in flimsy nests—but the doves stay on the eggs. Many cavity-nesting birds, such as woodpeckers, and some swallows, such as Purple Martins and Bank Swallows, generally lay white eggs, but other cavity nesters, such as nuthatches, chickadees, and Great Crested Flycatcher, have patterned eggs. Common Murres lay the eggs with the most variable coloring and patterning. No

Clutch size is variable among bird species but usually within a narrow range. Four eggs is a typical clutch size for Gray Catbird. *(Jonathan D. Moulding)*

one knows why. Many colonial nesting waterbirds have basically plain, usually white eggs. The eggs are obvious because the colonies, with multiple nests, are obvious. One member of the pair will usually stay with the eggs while the other forages. Many duck species lay white or whitish eggs, but the nests of ducks tend to be well hidden.

Once the nest is constructed, the female bird begins laying her egg clutch. It obviously requires effort and energy to produce and lay eggs, so it is typical for a female bird to lay one egg every day or every other day until she has laid the full clutch. Many bird species delay incubation until the clutch is complete, which results in the young birds hatching at about the same time. This means that no single individual is favored by having hatched early and grown before its siblings hatched, but it also means that the parent birds must feed the entire clutch as the young birds grow together.

In some species, particularly birds of prey, incubation begins as soon as the first egg is laid. This, of course, means that the young birds hatch asynchronously, and should there be three or four nestlings,

they will be of various sizes. The smallest will not hatch until the first to hatch has been fed for perhaps a week and is growing quickly. This reality obviously puts the smaller nestlings at a disadvantage regarding parental care since they must compete with larger siblings for access to food brought by parents. Siblicide may occur.

Depending upon species, clutch size varies widely. Waterfowl and gallinaceous birds have large clutches with up to a dozen eggs in a clutch, but usually fewer. Among passerines, cavity nesters—particularly chickadees, titmice, nuthatches, and wrens—have large clutches, sometimes up to nine eggs. Kinglets, which are not cavity nesters, also have clutches of up to nine eggs. Open cup nesting passerines (in the temperate zone) have clutch sizes that vary but are usually between three and six eggs.

Several factors affect clutch size. First, the bird must be able to incubate the clutch. For an Eastern Phoebe to have, say, a dozen eggs in its nest is problematic in the extreme because should they all hatch, the nest would not accommodate them all as they grow. The bird would waste immense energy

Peregrine Falcons are among the bird species that begin incubation soon after the first egg is laid. This means that the chicks will hatch on different days, be different ages, and be different sizes, oftentimes to the detriment of the smallest chick. *(John Grant)*

attempting to feed all the baby birds, some of which would be doomed by falling from the nest before fledging. The usual clutch size for an Eastern Phoebe is five eggs, but the range is three to seven eggs.

Why a range? Why not always five eggs? Many factors likely contribute, but certainly the health and robustness of the female is one variable, along with the degree to which food is available to feed both the incubating bird and the nestlings. Clutch sizes may be annually adjusted to environmental realities. In years when lemmings, an important food source, are abundant in the Arctic, a female Snowy Owl may lay a clutch of up to 15 eggs, though the usual clutch varies from 4 to 10. In years when lemmings are scarce, Snowy Owls may suspend nesting.

Ornithologists, by experimenting with artificially increased or decreased clutch sizes, have concluded that clutch size is based on the optimal rather than maximum number of young that can be raised in a given year. This is because what matters is really lifetime reproductive success of the parent birds (chapter 14). Birds forced to work much harder when more eggs are added to their nest (by researchers) experienced diminished health and thus their lifespan was potentially shortened, ultimately reducing their lifetime reproductive success. Natural selection

has acted to select for optimal clutch size, allowing the bird more opportunities in future years.

Clutch sizes of birds nesting in the temperate zone are larger than those of similar species nesting in equatorial regions. Many tropical passerines that make open cup nests have clutches of only two eggs. Why? Nest failure rates are uniquely high in the tropics, typically over 90 percent. Having only two eggs may help in avoiding discovery by predators and in speeding the time to fledging. Food is also less concentrated in the tropics, where there is usually not an annual short seasonal flush of abundant insects, such as occurs in spring in the temperate zone, when birds begin nesting. And finally, tropical birds in general live considerably longer than temperate species and enjoy a more relaxed annual breeding season. They are able to keep trying, so to speak. It is doubtful that the lifetime reproductive success of tropical passerines is any less than that of comparable species in the temperate zone.

Clutch size is genetically determined by natural selection and adjusted by local conditions, as described above. A bird thus knows when its clutch is complete, presumably not by counting the eggs but instead by a physiological feedback system that stops ovulation. The species in which this occurs, comprising most of the world's birds, are called *determinate* layers. If an egg or two is removed from the clutch, the bird will not compensate by laying more eggs to replace the lost ones, though it will likely re-nest if the entire nest is predated.

Some, but by no means most, bird species are *indeterminate* layers. Chickens are an example. Once chickens are in physiological egg-laying mode, eggs may be removed from the nest daily and the hen will replace them. There is a famous example (see Lanyon) of an experiment done on a Northern Flicker, also an indeterminate layer (as are other woodpeckers). An egg was removed daily from the flicker's clutch and it continued to lay, replacing the lost egg. The bird produced a total of 71 eggs in 73 days.

Birds, both determinate and indeterminate layers, are quick to respond to failures in clutches. Should a clutch fall to a predator, parent birds usually re-nest and produce a new clutch. Thus a determinate layer is not restricted to a fixed number of eggs per season. Birds may try several times after nest failures.

Many bird species are normally double or in some cases triple brooded. Typically, while a female is incubating a second brood, the male parent will be feeding and attending the newly fledged birds from the first clutch.

Eggs require incubation, and it is often the case that parent birds take turns incubating. Females generally do the majority of incubation, but males augment the process. The function of incubation is to not only protect but, more importantly, warm the eggs, speeding the biochemical reactions that allow the growth and development of the embryo. Both sexes develop what are called *brood patches* in the belly area. These are areas of bare skin, heavily vascularized to bring heat directly to the eggs. Brood patches vary from one to three depending on species. The parent bird positions itself over its clutch to allow the brood patches to be in contact with the eggs. Most birds have brood patches but some do not, including cormorants and Anhingas, gannets and boobies, and pelicans and frigatebirds.

Hatching is a routine, obviously necessary, but difficult process and it is accomplished by the baby bird with no assistance from its parents. Birds within eggs have what is termed an egg tooth on their upper beak, a hard protuberance that helps the baby bird to break free of the eggshell. The egg tooth is lost after hatching. Some birds begin to vocalize before hatching, and parent birds respond. Sometimes other birds of the clutch, still in eggs, also begin to vocalize, and this may aid in synchronizing hatching within the clutch. Birds do not move very much when hatching because there is literally no room in the shell to allow for much movement. When the chick has grown so large that it fills the shell and is sufficiently developed to exert relatively continuous pressure on it, the eventual result is *pipping* the shell, breaking it, and hatching. Once the bird

The egg tooth is visible on each of these Mallard chicks.

breaches the shell it continues to push for anywhere from hours to a day or so (depending on the kind of bird) to make its full escape.

Once a bird is hatched, the parent birds remove the eggshells, flying off and dropping them when distant from the nest. This behavior aids in clearing the nest of shells that would take up room needed for the clutch of growing birds, affect nest sanitation, and, perhaps most important, potentially expose the nest to a predator. Colonial nesters often merely drop the eggshells beneath the nest.

PRECOCIAL, ALTRICIAL, AND IN BETWEEN

Newly hatched chickens, ducks, and Killdeers are cute. Newly hatched parrots, robins, and crows are not cute. Baby chickens are downy, eyes open, alert, peeping, ready to follow the mother bird. Baby crows are naked, reptilian, eyes shut, fairly hideous looking, barely able to move. The reason for this disparity in appearance is the time spent in the egg before hatching. Waterfowl, gallinaceous birds, and shorebirds spend more time in the egg and thus they are significantly further developed by the time they hatch. An American Black Duck incubates eggs for almost a month, a duration typical of most ducks. Grouse require just over three weeks of incubation, and Wild Turkey eggs must be incubated for about 28 days. A Killdeer incubates between 24 and 26 days. In contrast, an American Crow incubates its eggs for about 18 days before hatching, and most passerines require much less time than that. For a Blue-gray Gnatcatcher, the incubation period is only 15 days. For most wood-warblers incubation requires only 12 days.

Gallinaceous birds, waterfowl, and shorebird chicks are *precocial;* another word for it is *nidifugous.* Having had more time to develop in the egg, the chicks hatch alert, eyes open wide, mobile, endothermic, and ready to abandon the nest. Passerines and many other birds, such as pelicans, cormorants, gannets and boobies, parrots, doves and pigeons, cuckoos, hummingbirds, swifts, and woodpeckers, are *altricial,* or *nidicolous.* The chicks are hatched featherless, eyes still shut, not yet endothermic, and completely helpless. The one thing that altricial birds are able to do upon hatching is to stretch their necks upward, open their beaks, and demand food, usually vocalizing as they do. They exhibit this instinctive behavior as soon as a parent comes to the nest. As altricial birds grow they must acquire all of their feathers and develop endothermy (warm-bloodedness), which accounts for why these birds require frequent feedings and continued incubation.

A fully precocial bird compared to a fully altricial bird represents quite a contrast, but there are birds that are somewhere between those extremes. There are semi-precocial birds, including gulls, jaegers, terns, alcids, and caprimulgids, in which the hatchlings are covered by down, their eyes are open at hatching, they are alert, endothermic, and able to walk, but they remain in or near the nest until they further mature. And there are also semi-altricial birds, such as storm-petrels, herons, egrets, ibises, spoonbills, and hawks, in which the hatchlings are down covered, generally endothermic, with eyes open, but are not physically able to leave the nest until more fully developed. Owls are considered to be semi-altricial but their eyes remain closed until a day or so after hatching. The difference between semi-precocial and semi-altricial is rather minimal.

SOCIALLY MONOGAMOUS BUT GENETICALLY POLYGAMOUS: EPCS AND EPFS

It has long been taught in ornithology classes that the class Aves, the birds, represents the most monogamous of vertebrates. Strong pair bonds form between the sexes during breeding season and both sexes are (in most cases) active in nest construction, incubation, and caring for nestlings and fledglings. There is commitment by both males and females. To account for such devoted parenting it seemed obvious that the eggs could not develop and hatch, nor could the young survive, without significant parental care, more than could be provided by the female acting alone. Fathers were necessary to enhance the likelihood of nesting success. Thus, natural selection selected for females and males that formed strong loyalties to each other and to the nest and its young contents. It was genetically in both of their best interest to do so. Though a male Verdin, in theory, has the option to just fly away after copulation, leaving the female with the eggs, it is not in that male's ultimate genetic interest to do so. Better that he should construct multiple nests from which his mate will choose one in which to lay the clutch, diligently bring food to the female while she is brooding, and aid in feeding young both in the nest and after fledging. This is the best route for the male Verdin's genes to successfully find their way into the next Verdin generation.

Anyone who studies life cycles of birds understands the reality of social monogamy. A few species are clearly polygynous, such as some grouse and prairie-chickens, as well as Red-winged Blackbirds and Bobolinks; and phalaropes are among the few polyandrous bird species. But the vast majority of species appear to form strong monogamous pair bonds, both sexes helping with the full nesting cycle.

Yet there is another reality, more subtle and very important. In the temperate zone, nesting season is typically compressed into a few months, and in high-latitude areas, a few weeks. All of the birds are in full physiological reproductive mode and in close

Purple Martins (*top*) represent fully altricial hatchlings. Roseate Spoonbills (*bottom*) are semi-altricial. (*Top: Mary Keleher; bottom: John Kricher*)

It is now easily possible to examine human DNA and match it to a specific person, providing that a sample of that person's DNA is available. All individuals have multiple tiny DNA segments called minisatellites that do not code (thus are inactive in the genome) and are unique to each person, rather like fingerprints are unique. DNA fingerprinting looks at these unique minisatellite segments of DNA and compares them with samples from other individuals. If Martha and John have a child, Sarah, many of the minisatellite DNA markers from both Martha and John will be part of Sarah's DNA and thus confirm Martha and John as being her real parents. The technique, now fully developed and refined, is reliable and widely applied in forensic analysis and criminal investigations as well as to establish paternity. It is now also part of ornithology.

Many bird species that form pair bonds actively seek and accept extra-pair copulations that result in fertile eggs. EPFs are common though variable among species and even variable within species, depending on factors such as numbers of territories in close proximity. Leach's Storm-Petrel and Northern Fulmar are examples of species that are both socially and genetically monogamous. Albatross pairs are also genetically monogamous. But genetic monogamy is, in reality, not common.

Purple Martins and Indigo Buntings, to take but two examples, represent species in which infidelity to mates, albeit furtive, is commonplace. In a Purple Martin nest that may contain five to seven eggs, it is likely that not all were fathered by the female's mate. By the same token, that female's mate may have fathered birds that are in eggs in several other martin nests. A male has no way of knowing if he fathered all the eggs in a clutch. In some species, typically more than 50 percent of the eggs in a nest are not from the female's mate. In some cases, none may be from her social mate. Dozens of studies have been done to examine frequency levels of EPFs. For example, a study focused on Grasshopper Sparrows in Maryland showed that over a period of five years, the percentage of extra-pair young ranged from 26 percent to 60 percent. The factor that was most important was the density of territorial males. Neighboring males and females frequently engaged in EPCs. This does not seem surprising.

Perhaps surprisingly, areas where there are many territories in close proximity, several males may, in reality, father no offspring though each will dutifully attend its nest and raise nestlings it did not father. This means, of course, that most of the young have been fathered by relatively few males, the big winners. That is not the case for the females because a clutch of eggs in a nest is normally laid entirely by the brooding female (except in the case of egg

proximity. They are in synchrony regarding the urge to have sex. Once a pair bond is formed there is nothing physiological that prevents a male or a female of a pair from copulating with a bird that is not its partner. That is why, for example, should another male blatantly intrude on a territory, the male holding the territory will instantly attack.

Ornithologists, by watching the behavior of birds in and around nests, have suspected that birds might seek and readily accept what have come to be termed extra-pair copulations, or EPCs. If that occurs, it should result in extra-pair fertilizations, or EPFs. There are many potential opportunities for EPCs to happen and the actual act of copulation is quickly accomplished in birds, so dalliances outside of the pair bond should not prove difficult. But eggs do look alike, as do offspring, so it was not possible to test this idea until the advent, in 1984, of DNA fingerprinting.

dumping). This reality of infidelity among pairs of breeding birds is not something that is at all obvious in watching bird behavior in the field. That is because EPCs normally occur away from the nest and immediate nest area.

In some bird species, including martins, Tree and Barn Swallows, Bobolinks, Red-winged Blackbirds, Northern Cardinals, and Common Grackles, you may observe a male bird closely accompanying a female as she moves away from the nest or moves within the territory. The male bird is likely engaging in a behavior termed *mate guarding*. Male birds must be aware of the reality of EPCs because each is seeking them for himself. Thus it should be of no surprise that a male would attempt to prevent other males from copulating with his mate or prevent his mate from accepting EPCs. This behavior concludes when the clutch is complete because the female has laid all her eggs and there is no point in further mating.

The proclivity among many bird species to seek copulations beyond their pair bond is why ornithologists no longer describe birds as monogamous, but rather as *socially monogamous*, a term I have been using throughout this discussion. Indeed they are. Genetic polygamy is largely accounted for, as mentioned above, by the compressed nature of breeding season combined with the reality that birds have specific habitat needs and thus a pair of, say, Scarlet Tanagers can expect other pairs of Scarlet Tanagers to be nearby. Opportunities for EPCs abound. In the tropics, where breeding seasons are often much longer than in the temperate zone, EPCs appear to be less common among pairs. That is because some birds within any particular species are in physiological breeding mode when others of their species are not.

It has long been held that males, including male birds, produce the "cheap" gamete, sperm, and are thus easily able to seek numerous copulations with

Blue-gray Gnatcatchers form a socially monogamous pair, and both sexes contribute much to the nesting process and fledging of the young birds. But that does not mean that each does not attempt to solicit copulations outside of the pair bond.

little cost. In other words, one might expect males to seek additional copulations beyond their mate. A male attending his own nest has nothing to lose by attempting to mate with a female that is not his mate. It is now clear that females, who produce the egg, the "expensive" gamete, also actively seek and certainly accept EPCs. Females do not act coy. Both sexes are fully on board for the complex sex lives that seem to typify many bird species. Females continue to seek males and mate with them after their pair bond has formed with another male. Sexual pursuit is relentless. Female promiscuity is likely due in large part to the possibility that females are physiologically able to regulate just whose sperm actually fertilizes her eggs. That produces what ornithologists call *sperm competition*.

SPERM COMPETITION

Males ejaculate millions of sperm in one copulation. Copulations are frequent at pair bonding and during nest building. Most sperm do not thrive in the uterus and oviduct of a female bird and quickly perish, excreted into the cloaca with fecal matter. Only about 1 percent of the sperm survive and find their way to storage tubules in the uterus, where they may remain alive and viable for up to a month. Multiple copulations add considerable sperm to the storage tubules. And that is where it gets interesting. It appears that the last sperm to enter the female are the ones most likely to fertilize her egg. When released from the storage tubules they are first to move up the oviduct to meet the ovum. Sperm from extra-pair copulations appear to have the advantage of displacing sperm from earlier copulations, often those of the female bird's actual mate.

This is more than a little interesting because it suggests that females actively seek males that they perceive to be more fit than their own mate as they proceed through the egg-laying cycle. Numbers count. Males that are "rested" ejaculate the most sperm. In Purple Martins and some other species females continue to seek sex partners well beyond pair-bonding and, further, they tend to copulate with males more fit than their mates. Females that have copulated with several males may be able to selectively choose which male's sperm will access their eggs.

An obvious question is, why is it that females choose a less fit mate, only to then seek EPFs with another male? Why not just choose that fitter male to begin with? The reason is likely that such a male is already pair-bonded. Soliciting EPCs then becomes the only avenue to obtaining a fitter male's sperm. There are thus numerous opportunities for females to mate with the fittest males. The first is to make a pair bond, but that implies strong competition with other females when territories are formed and pair bonds cemented. The second is to go out furtively and solicit EPCs with the fittest males. Neither male nor female birds have anything to lose by doing that. And apparently, that's what they do.

THE STRANGE CASE OF DELAYED PLUMAGE MATURATION

Birders know that many species of birds, most particularly hawks and gulls, require several years to acquire adult plumage. This reality of maturation applies to both males and females. If there is an adaptive reason for such a delay in acquiring adult plumage, it is likely based on potential intraspecific competition for mates between inexperienced immature birds and mature, established adults. Normally hawks and gulls do not breed until they acquire adult plumage and even then there may be an additional delay in the onset of breeding.

In some passerines, in contrast with hawks and gulls, there is a distinctive immature plumage worn only by males during their first spring, although they are capable of breeding and some do. Purple Martin; Bullock's, Baltimore, and Orchard Orioles; Red-winged Blackbird; Summer Tanager; American Redstart; Rose-breasted, Black-headed, and Blue Grosbeaks; Painted and Indigo Buntings; and Purple Finch are examples of species that exhibit some form of delayed plumage maturation (DPM). Birders learn to recognize males of these species in their first-spring plumages, though it is sometimes possible to confuse them with females. First-year male Purple Finches, for example, closely resemble females. A first-spring Summer Tanager, in contrast, sports a combination of yellow feathers, typical of the female, mixed with red feathers, typical of the male. First-year Orchard Orioles have black bibs, making them distinct from females.

Delayed plumage maturation is relatively uncommon. Most male passerines wear full adult male plumage in their first spring. Many explanations have been offered for why DPM exists. Some are based on adaptations that would aid the bird on its wintering ground, such as looking cryptic. Other ideas deal with how appearing different from an adult male might aid an inexperienced male in obtaining access to females by not being as easily detected by older males, thus avoiding direct confrontations. Other hypotheses are physiological, based on the cost of manufacturing brightly colored feathers. There is even a hypothesis relating DPM to potentially enhanced lifespan.

What is clear is that males of species that exhibit DPM are physiologically capable of mating and fathering offspring. DPM males will sing, sometimes over longer spans of time than adult males on territory. Why females would bother to mate with males that look so different, more muted than the

A few species, including Orchard Oriole, have delayed plumage maturation. This bird is a male in his first spring.

more brightly colored males, remains an open question, but they do—at least some do. First-spring males could presumably secure extra-pair copulations by quietly staying within the territory of another male. Perhaps males showing DPM father fewer offspring than brightly colored males, as might be expected from inexperienced birds. But the fact that first-spring males exhibiting delayed plumage maturation do reproduce, at least to some degree, is in itself interesting in relation to theories about the prevailing role of sexual selection in avian evolution (chapter 14).

PROVISIONING YOUNG AND FLEDGING

Once the egg clutch is complete, a female bird ceases to solicit copulations unless something happens to the clutch. Incubation follows. Both sexes share incubation in many species, including loons, gannets, egrets and herons, ibises, storks, pelicans, cormorants, some shorebirds, gulls, eagles, Osprey, many hawks, swifts, doves, woodpeckers, and vireos. For example, in the case of Killdeer and Pileated Woodpecker, to name but two species, the male incubates at night, the female by day. In many other species the female does either all or most of the incubation. This is particularly true in passerines but also includes quail and grouse, owls, and most waterfowl. Males provision the incubating females with food.

Once the eggs hatch it is necessary to bring food to the nestlings. In the case of fully precocial (nidifugous) species, the young soon learn to fend for themselves, though they are instructed to a degree by the mother bird. But with other species the newly hatched birds remain in or near the nest as the parents return to feed them. Incubation also occurs post-hatching in species that are fully altricial and need to continue to develop endothermy as well as grow feathers. Once the feathers develop and the birds attain full endothermy, incubation ceases. The nestlings may be left alone in the nest while the parent birds forage.

Do parent birds recognize their own young? The answer appears to depend on circumstances of nesting, but normally they don't because they are not mixed with the young of others of the same species. Birds ranging from precocial Wood Ducks to altricial Eastern Phoebes do not have the ability to identify their own young because they have never been under selection pressure to do so. The common denominator is that in bird species in which the young are not mixed with young of others of the same species, parents have no need to discriminate.

Among colonial birds, such as herons and cormorants, young birds are confined to the nest for some time and thus parents return to their own

OPPOSITE: Both male and female American Redstarts (top) feed their nestlings and attend to nest sanitation, such as removing fecal sacs (center). Once nestlings fledge, parent birds often remain with them as the fledglings continue to demand attention, as is evident in these Mountain Bluebirds (bottom). (Top and center: Jim Zipp; bottom: Van Remsen)

nests and, in effect, there are their kids. They recognize the nest location, but one could replace a chick with one from another nest and the parent birds would not know. But in some colonial birds, such as some gulls, terns, and alcids, the nests, eggs, and young are packed so close together that parents do know their own offspring. It has been suggested that the multiple patterning and coloring of Common Murre eggs, a very unusual characteristic among birds' eggs, aids parents in knowing their own egg (only one egg is laid in a Common Murre clutch). Murre chicks communicate with parents such that parents recognize their own chick's voice. Herring Gulls, which nest in large colonies in marshes, know their own young while the chicks are still at the nest, a period of about five days' duration. Black-legged Kittiwakes nest in close proximity, but their nests are typically on narrow ledges where the young never wander from the nest without fatal results. Kittiwakes need only recognize their nest site. If there is no chick, it fell off. So there is no selection pressure for kittiwakes to actually recognize their own offspring. Laughing Gulls use calls between adults and young for recognition, but in Ring-billed Gulls, recognition is based on sight, not sound.

Basic nest sanitation is required for all but fully precocial species, and there are various ways in which it is accomplished. Parents of all birds usually take eggshells and discard them away from the nest, so as not to attract potential predators. Colonial birds just drop the shells since there is no advantage in taking them away from the nesting area. Sometimes eggshells may be eaten as they are a source of calcium (though the birds obviously have no idea what calcium is). Nestling birds constantly generate feces, and parent birds may eat the feces (which, surprisingly, can be somewhat nutritious because nestlings are relatively poor at digestion immediately after hatching). But more often, with passerines, nestlings produce fecal sacs that consist of a thin mucous membrane containing the feces, which parents then carry away in their beaks and drop them well away from the nest. Colonial birds such as cormorants, herons, and the like often projectile expel their feces over the edge of the platform nest. Woodpeckers eat their nestlings' feces early in the nest cycle but ignore them later, and their cavity nests often become quite foul. As the nesting cycle draws to a close, young birds are crowded in the nest and feces begins to accumulate around the edge of the

The colorful gape shown in this clutch of Barn Swallows and evident in many newly hatched birds serves as a stimulus to the parent birds to deposit food items. Even birds that eat mostly seeds typically offer protein-rich animal food to nestlings.

The Common Gallinule has been observed to attack its own fledglings, sometimes killing the young birds. *(Holly Merker)*

nest cup regardless of parental diligence. Nests may then become abundantly infested with various parasites, particularly mites, that may weaken chicks.

Nestlings have an immense need for food, and parents have to supply it. Even in the case of precocial species, parents feed the chicks after hatching until the chicks become independent feeders. Parental feeding of nestlings is thought to be strongly innate in that the parent must see some visual cue, a *releaser,* and thereupon it will deliver the food item. In typical altricial birds, the baby birds use a powerful neck muscle to stretch upward and open their mouths wide, exposing a colorful mouth and gape flange to the parent. That causes the parent to place a food item into the nestling's mouth. Nestlings vocalize as they gape, and parents respond to the sound as well as the gape. Precocial bird species, especially passerines, sometimes require feeding an average of every 30 minutes or so, making for hundreds of nest visits daily by the parent birds. Larger precocial species do not need such frequent feeding.

Parents respond to the most vigorous nestlings, so nestlings are forced to compete for access to food. In many cases this leads to the demise of some nestlings, usually the youngest or weakest, before fledging. Those birds are unable to keep up with their stronger siblings. But when food is abundant, it is not unusual in most bird species for all nestlings to fledge, though they might not be equally strong. As might be expected, in species that have staggered hatching because incubation begins when the first egg is laid, the death of the youngest nestling from starvation is common, as is siblicide by older siblings directed at the youngest. This behavior obviously results in brood reduction. It occurs in various

hawks, eagles, cormorants, herons, egrets, and boobies. It is particularly expressed in boobies, when the second chick in a clutch of two survives only if the older chick perishes.

In a study (Hunt and Evans) that looked at chick survival of Double-crested Cormorants it was learned that in clutches of four eggs, the youngest chick (the fourth egg) survived only when one of the older three chicks did not. Otherwise the fourth chick perished 100 percent of the time. Ornithologists speculate that the fourth egg of the clutch is an "insurance egg," not meant to survive unless needed. This is almost certainly the case with the second egg of booby species.

Parents will sometimes stop feeding weaker chicks, presumably because the chicks are not sufficiently vigorous to stimulate the parents' instinctual urge to feed them. There are also cases in which parent birds deliberately kill their own young. Parent coots and gallinules will sometimes attack what is presumably deemed by them to be an overly demanding fledgling, repeatedly grabbing its head and violently shaking it, a behavior that sometimes but not always leads to the death of the young bird. A clear case of infanticide of a Dickcissel chick has been documented (Coon et al.) in which the female bird evicted a live chick from the nest and left it to die. This behavior is suspected to occur in other bird species on occasion.

Because feeding nestlings is an innately based behavior by parents, it is occasionally possible for a parent to begin feeding a fledgling of a different species. There are accounts of many species that have been observed doing cross-species feeding. The reason for doing so is a combination of the fact that

This clutch of Barn Swallows is about ready to leave the nest.

parents do not recognize their own young and because birds observe fledglings of other species begging food. This behavior is sometimes mistakenly referred to as adoption, but it is not really that. It is more a misdirected manifestation of innate behavior that stimulates adult birds to respond to releasers from baby birds. Cross-species feeding of young characterizes brood parasitism and is discussed more in the account of Brown-headed Cowbird on page 268.

Food is delivered in a variety of ways depending upon the kind of bird doing the feeding. Gulls regurgitate food onto the ground for their chicks. Adult pelicans, Anhingas, cormorants, gannets, and fulmars open their mouths and young birds aggressively insert their heads and get at the food in the adults' crops. Herons and egrets deliver food items from their beaks directly to the beaks of the nestlings. Puffins bring food to their burrows and merely place it on the ground for their fledgling to pick up and eat. Doves and pigeons are unique in secreting a nutritious milklike substance of their own making to their altricial nestlings. Passerines with beaks full of food push it into the throats of the gaping young.

As would be expected, protein-based food is most commonly fed to nestlings as it contributes to balanced growth. Grouse and quail fledglings eat various arthropods they encounter. Ducklings feed on small animal matter. Cormorants and herons bring fish and crustaceans to their young, hawks bring various animal carcasses (which they tear up for the nestlings), and robins bring worms (among other things). The spring flush of insects that accompanies leaf-out in the temperate zone is a strong selection pressure in the relatively synchronized migration and nesting cycles of many passerines and other species. A few birds feed their young a vegetable diet. Highly frugivorous species such as waxwings generally feed fruit to their nestlings. American Goldfinches, which nest late in the season when food plants such as wild thistles become abundant, feed their nestlings a mash of crushed seeds.

One of the most challenging duties of parent birds is to actually force the nestlings to abandon the nest, to fledge. Many parent birds use persistent vocal cues in what are clear attempts to coax their young into leaving the nest. This behavior is sometimes easy to observe. Withholding food is also a common inducement to nest abandonment. Actual nest abandonment by fledglings is probably based strongly on innate cues and brain development. Fledglings routinely exercise their wings before leaving the nest, though fledglings of some species, such as some alcids and cavity-nesting ducks, must literally take a leap of faith and jump from a cliff-side or from a tree cavity well before they are able to fly.

Many species of birds, particularly passerines, are double or triple brooded. In such cases it is common for the male of the pair to stay with the newly fledged young of the first brood while the female incubates eggs of the second brood. Once the second brood hatches, the young of the first brood should be independent, and the male returns to aid with feeding young of the second brood. In areas with a long growing season, some species, such as Song Sparrow and Northern Cardinal, are triple brooded. Sometimes birds construct an additional nest for a second brood rather than use the first nest. Such a labor-intensive behavior may reduce parasite infestation of nests.

Newly fledged birds continue to be at risk. They lack experience, many do not yet fly with strength and precision, and they must learn to feed independently. There are some cases wherein fledging is an all-or-nothing event, such as with swifts. Once a juvenile swift takes its leave from its nest, it must fly and it must feed in flight. It grows up very fast.

Some bird species form what are termed crèches for baby birds. A crèche is a dense grouping of nestlings usually gathered together and attended by one or several adult females. Flamingos have crèches of young, as do ducks such as Common Eider. There is a documented, filmed, and widely watched account (www.livescience.com›63221-mama-merganser-76-ducklings) of one female Common Merganser with 76 chicks accompanying her. The chicks were obviously from multiple nests, and it is not unusual for chicks to follow any female, as appears to have occurred in this case. Mergansers frequently accept chicks from different nests, and only the large number in this case was unusual.

As discussed in chapter 11, a few species with high degrees of sociality routinely have nest helpers, usually the grown offspring of the nesting pair.

BROOD PARASITISM

Brood parasites are species that lay their eggs in nests of other birds. The fate of the bird within the egg is left to the owners of the nest being parasitized. It should hardly be surprising that in dense aggregations of nests, such as in colonial species, it

A juvenile Anhinga (brown head) is aggressively prompting its parent to feed it. The juvenile has its wings and tail spread, the left wing literally over the back of the parent bird.

is possible for a female to lay an egg in a nest other than her own, and some birds commonly do. Two species of ducks, Redhead and Ruddy Duck, are well known for this kind of egg dumping. So are Cliff Swallows. Other duck and swallow species also occasionally drop an egg into another bird's nest. Common gallinules on occasion lay eggs in nests of other gallinules. In years when caterpillars are abundant, Black-billed and Yellow-billed Cuckoos occasionally lay an egg in another cuckoo nest, either of their own or of the other cuckoo species.

These types of *facultative brood parasitism* are opportunistic in nature. In the case of the swallows as well as other species, when a bird drops an egg in the nest of the same species, how is the nest owner to know? The eggs won't differ in appearance.

There is another form of brood parasitism that is much more evolved and complex, exhibited by Old World cuckoo species. It is termed *obligate brood parasitism*. In these cases, the brood parasite species never makes a nest of its own but instead always lays an egg in the nest of another species such that the parasitized species raises the young of the parasitic

species. The best known and most highly evolved examples of brood parasites are most of the Old World cuckoo species as well as African honeyguides and African whydahs and indigo birds. In North America, cowbirds are obligate brood parasites. The Brown-headed Cowbird is widespread in North America, but two other species, the Shiny and the Bronzed Cowbirds, are also fully obligate brood parasites, so the behavior evolved prior to speciation. A few other obligate brood parasitic cowbirds are found in South America, plus one cowbird species that is not a brood parasite.

Brown-headed Cowbirds are documented to parasitize about 220 North American bird species. They are often but not always successful. If a cowbird should, for example, lay its egg in a duck nest, not so good. More to the point, some bird species, including Gray Catbird, American Robin, Blue Jay, Brown Thrasher, and Chipping Sparrow, are able to recognize a cowbird egg and will toss it out of the nest. Yellow Warblers are well-known discriminators and will build a second nest over a nest containing a cowbird egg. But many species do not recognize the

This fledgling Brown-headed Cowbird is still dependent on a "parent" to feed it. In this case, that parent happened to be a Northern Parula (not in photo), vastly smaller than the cowbird.

A male American Redstart dutifully feeds his "offspring," a nearly fledged Brown-headed Cowbird. The question of why the foster parents are so easily fooled by a bird bearing scant resemblance to them remains of interest. *(Ed Harper)*

foreign egg or the chick that hatches from it and they do foster it to maturity.

More than 140 bird species are known to have fostered cowbirds, and cowbird impact is of concern in situations such as preservation of the endangered Kirtland's Warbler. Cowbird control programs have been used to enhance nesting success of endangered and threatened species. This is because cowbird chicks grow fast and are aggressive, claiming a disproportionate share of the food brought to the young. Bird species that raise a cowbird often lose most if not all of their young. Small birds such as some wood-warbler species and Song Sparrow will readily feed a cowbird fledgling four or five times their size, clearly not able to perceive it as different from their own fledglings. Even in those species that are able to differentiate a cowbird egg from their own clutch, there could be problems. Female cowbirds are known to sometimes return to a nest if their egg has been tossed and to damage the other eggs.

Brown-headed Cowbirds form pair bonds. Several males attempt to court a female until she bonds with the bird that will be her mate. Females respond to the individualized songs of males during a courtship performance that includes exaggerated head bowing with partially open wings and a spread tail. A male alters his harsh, short, nonmusical song until he receives a yes response from a female and the pair bond forms. Males typically mate-guard females. Sometimes males mate with more than one female, and females also on occasion solicit copulations from males to whom they are not mated.

Female Brown-headed Cowbirds usually seek open cup nests and are surreptitious in doing so. Females are somewhat territorial, defending areas from other females that have the same thing in mind, though territories tend to overlap. A female cowbird is capable of laying about 40 eggs during a single nesting season (sometimes more), and well over 90 percent are doomed to fail. Nonetheless, enough succeed to account for the Brown-headed Cowbird being an abundant species. When a female lays her egg, which she does very quickly, usually in the early morning when the female nest owner is foraging, she usually removes one of the eggs in the clutch. The cowbird egg typically hatches before other eggs in the nest, and the nestling is aggressive and fast-growing, able to leave the nest in as short as ten days.

Once a cowbird juvenile is capable of flight it leaves the host species' nest at night and associates with roosting cowbirds. This behavior is nothing short of remarkable, and it is how cowbirds learn self-identity as a species, a critical component to their survival. Otherwise you would predict that a cowbird raised by a Song Sparrow would, upon maturity, assume it is a Song Sparrow and attempt to mate with a Song Sparrow. They never do this. Nor are they influenced by the song of the host species. When a juvenile cowbird hears the chatter call of an adult cowbird, the sound triggers innate recognition in the juvenile. This process has been likened to learning its species password. Thus a juvenile cowbird is taught that it is a cowbird by other cowbirds while still being fledged by a different species.

Cowbirds travel and roost in large flocks, often with other blackbird species such as Red-winged Blackbirds and grackles. The name cowbird refers to the habit of following herds of large animals, such as bovines, that act as beaters, stirring up insects that the birds devour. Cowbirds are mostly vegetarians, augmenting their diet with arthropods. Cattle Egrets also use large animal herds to stir up arthropod food. It is thought by many that the obligate brood parasitism practiced by the Brown-headed Cowbird evolved in response to having followed herds of bison as they nomadically roamed the prairies. As humans cleared more and more land for pasturage and agriculture, cowbirds thrived and the population increased immensely from precolonial times. This is a reasonable supposition, but it does not fully explain how other cowbird species evolved, unless they evolved directly from Brown-headed Cowbirds, inheriting the innate behavior or obligate brood parasitism. In any case, cowbirds represent one of the most remarkable natural histories of any North American bird.

MIGRATORY BEHAVIOR

OBSERVING MIGRATION . . . AND UNDERSTANDING IT

The breadth of bird migration encompasses one of the most impressive spectacles of bird behavior. Hundreds of bird species comprising billions of individual birds, many traveling thousands of miles, effectively relocate twice annually, often returning to the same breeding and wintering areas as occupied the previous year. Such feats of physical endurance, navigational skill, and local adaptation border on mind-boggling, at least with regard to human minds, but birds do it as a matter of routine. Birders crave being afield as often as their time permits during migration, not wanting to miss any of the show.

Many kinds of animals migrate, including various species of fish, sea turtles, butterflies, dragonflies, mammals, and others. *Migration,* defined as the systematic and regular movement of an animal species from one region to another, is a word that, when applied to birds, encompasses multiple forms of behavior. These include the stimuli and timing of migration, whether migration is diurnal or nocturnal, navigation, flocking, and use of stopover sites. Migration is part of a bird's annual cycle (chapter 7). It is influenced by extrinsic signals such as day length, the most essential variable in initiating the urge to migrate, as well as weather systems that affect timing and pace of migration. Migration involves changes in the bird's hormonal system that alter its physiology, in some cases shrinking the organs of digestion while enlarging the reproductive organs during the migratory journey. Migration is lots of behavior wrapped into one word.

There are various ways to observe migration. One way is to keep a year-long calendar of when bird species first arrive and when they leave (though establishing exact departure dates is more difficult). If you do this anywhere in North America and you survey multiple kinds of habitats, you'll soon realize that some form of bird migration is happening through-

This male Blackpoll Warbler, photographed in Michigan in May, has successfully migrated from Amazonia on his way to his boreal forest nesting grounds.

out the year. As the great ornithologist Frank M. Chapman pointed out over a century ago, no matter the month, some bird species are migrating.

I have done most of my birding in the Northeast and so I offer a brief example of an annual migration calendar for northeastern states, in no way detailed, beginning in the depths of winter. The example illustrates that birds of various species are moving every month of the year. We begin at the end of one year and the beginning of another.

Snowy Owls and Rough-legged Hawks arrive around mid-December and January along with species such as Northern Shrike, and Glaucous and Iceland Gulls. Offshore, loons, Northern Gannets, and wintering alcids in varying numbers are moving along the coastline. Flocks of Brant and scoters begin flying toward their northern breeding areas in

OPPOSITE: Tree Swallows assemble in immense flocks for their fall migration.

Black-throated Blue Warblers are typically part of the big influx of May migrants.

early February. Large mixed flocks of Red-winged Blackbirds, grackles, and cowbirds move north in February, continuing into March, as do American Woodcock, Eastern Phoebes, Eastern Bluebirds, and American Robins. March is also the month when many duck and goose species migrate north and westward to their prairie pothole nesting region.

April and May, taken together, are peak months, with major influxes of passerines and other species arriving from southern latitudes. Within that mixture there is a temporal pattern. The first to arrive are Ruby-crowned Kinglets, Blue-headed Vireos, Black-and-white Warblers, Palm Warblers, Yellow-rumped Warblers, Tree Swallows, Northern Flickers, Yellow-bellied Sapsuckers, Hermit Thrushes, Eastern Towhees, and a few others. Orioles, Gray Catbirds, and migrant flocks of Blue Jays appear in mid- to late April. May is the time of densest passerine migration with major influxes of hummingbirds, tyrant flycatchers, wood-warblers, vireos, orioles, tanagers, grosbeaks, buntings, and sparrows. Flocks of Double-crested Cormorants, various shorebirds, and terns are present along coastal regions. Egrets and herons return.

In June and throughout July, as passerines are breeding, many shorebirds are already migrating south, returning from their brief breeding season in the Arctic. July and August feature numerous offshore seabirds such as storm-petrels and shearwa-ters that migrate north from their breeding grounds in the southern Atlantic Ocean to spend northern summers in Atlantic waters. Soon they will be migrating south. Red-necked and Red Phalaropes and various jaeger species returning from northern breeding grounds begin to appear offshore by late July, continuing into October. Waves of shorebirds and passerines migrate southward from July through October. In September, hawk and falcon species are in full migratory mode. Huge concentrations of Tree Swallows assemble in late August and are seen through early October. Hawk migration continues through October and into December, along with numerous sparrow species and often a few vagrant species mostly from the West.

November brings waterfowl, Horned Larks, Lapland Longspurs, and Snow Buntings, and as December arrives, so, occasionally, do irruptive passerine species, Red-breasted Nuthatch, the crossbills, Pine Grosbeak, redpolls, and Bohemian Waxwings. There is no month in the Northeast when it is not possible to find at least some migrating bird species.

Keeping a migration calendar with notes on arrivals and departures is enjoyable and fulfilling anywhere in North America, and it will, of course, differ from place to place, region to region. It is an ideal way to learn the birds of your region, the permanent resident species as well as the migrants. But bear in mind that when you focus only on your region, you are seeing migration through a narrow window. There is much more to bird migration than an annual calendar.

SOME EXAMPLES

To examine more deeply what bird migration actually is, let's take a look at what's happening in mid-May along the Delaware Bayshore, at Reeds Beach, New Jersey. Concentrations of Red Knots (of the *rufa* subspecies) have joined Ruddy Turnstones, Semipalmated Sandpipers, and Sanderlings, all fueling up for their transit to their arctic breeding grounds. It is quite a migration panorama, aggregations of shorebirds gathered sometimes shoulder-to-shoulder along with hundreds of cacophonous Laughing Gulls, all frenetically feeding on the narrow beach. The shorebirds are gathered at a critical staging area, gorging food before beginning the final leg of their northward migration. The timing of the shorebird influx coincides with the shedding of eggs by mating horseshoe crabs (*Limulus polyphemus*). The nutritious horseshoe crab eggs that annually wash up in the high millions provide rich and essential fuel for the northward migration of the shorebirds.

Red Knots soon depart from the shores of Delaware Bay in large flocks that fly northward, high over land, at speeds of 40 to 50 miles per hour and at elevations of up to 15,000 feet. They cross over the Northeast and Great Lakes, dispersing, fanning out, not stopping until they reach their breeding grounds in the far northern Arctic tundra. These same Red Knots began their migration weeks earlier, flying northward from the southern coast of Argentina, overflying parts of Amazonia, and stopping to refuel at various points along the Brazil and northeastern South America coastline. By the time they arrive along the shores of Delaware Bay, some have made stops to feed along the southeastern U.S. coast from Georgia to the Carolinas, but most have flown non-stop across the Atlantic Ocean directly to the Delmarva Peninsula.

A birder may spend a joyous hour or two watching these birds at select spots such as Reeds Beach, New Jersey, just a few miles north of Cape May. But realize that the birds have already devoted weeks of effort to their long flight and still have a precarious flight ahead of them to get to a distant breeding area that they will occupy for only 45 to 50 days at most, while they attempt to accomplish the breeding cycle. After breeding, in August through October, they will make the long migration southward, taking a different route to eventually reach the mudflats of southeastern Argentina. Red Knot migration comprises much of the bird's year, includes flying about 18,000 miles (28,000 km), and requires prodigious energy.

Long-distance migrants must be adapted to recognize and exploit numerous different habitats. If you were to travel with a Gray-cheeked Thrush on its migration north in the spring, you would begin somewhere within the wet lowland rain forests of northern Amazonia, where the wintering thrush shared the forest with hundreds of species of antbirds, antpittas, woodcreepers, tyrant flycatchers, puffbirds, motmots, toucans, tanagers, and others that never migrate, some of which spend long lives (15 years or more) in one small area of rain forest. At around the time of the vernal equinox, migratory restlessness (see "Timing of Migration," page 284) begins and the Gray-cheeked Thrush feels a powerful and irresistible urge to migrate. Among several potential routes, it could fly to the Yucatán Peninsula, an area of dry forest and scrub quite different from where the bird wintered. Some days might be spent there feeding, doing essential refueling before crossing the Gulf of Mexico.

One night, when tailwinds prevail, beginning as darkness falls and continuing through the night and into the next day, the thrush crosses the Gulf of Mexico, flying until it encounters land, perhaps continuing well inland before landing. More refueling is needed, perhaps along the Gulf Coast in Texas or Louisiana, among woods and coastal scrub or further inland in rich bottomland forests. Then it's time to move onward. Days later the thrush might be observed by multiple birders in the woodlands bordering Magee Marsh in Ohio, on the shore of Lake Erie. Always flying after dark and usually with other migrants, making various chip calls, the Gray-cheeked Thrush crosses the Great Lakes or perhaps flies over the Rockies, depending on its specific route north to its final destination. The bird transects the vast Canadian boreal forest, perhaps ending its flight near tree line in northern Alaska, maybe around Nome, its breeding territory.

The entire migration may have required five to six weeks, from leaving winter territory in mid-April until arriving on the breeding site, perhaps around May 25. Risk has accompanied the thrush throughout its migration. It's quite a journey but in no way unique. Many individuals of many species of birds have their own versions of such an ambitious itinerary.

WHERE TO SEE MIGRATION

There are many places where large numbers of bird species congregate during migration. Some are staging areas where birds assemble as they prepare to migrate. Red Knots stage on specific mudflats in Argentina before beginning the long flight north. They restage at the Delaware Bayshore before the final push of their migratory journey. Various shorebird species congregate and feed at key staging areas, such as Cheyenne Bottoms in the Kansas Great Plains and numerous stopover sites along the West, Gulf, and East Coasts. Bolivar Flats, near Galveston, Texas, is one of the best known shorebird stopover sites. National wildlife refuges throughout the United States annually host many migrant species of shorebirds and waterfowl, some of which are present all year, some only during migratory periods or during nonbreeding months. Other spots to observe migration are along well-known migratory routes, where, for example, it is possible to enjoy raptor migration in autumn. Hawk Ridge near Duluth, Minnesota, Hawk Mountain near Kempton, Pennsylvania, and Cape May, the southernmost point in New Jersey, are well known to generations of birders, but there are many other places, particularly along Appalachian ridge lines, where aerial streams of southbound hawks pass during late summer and autumn.

Many places, referred to as migrant traps, offer incredible opportunities to see large numbers of migrant passerines and other species, often up close. Spacious parks in large cities are prime ex-amples. To migrating birds, these parks act as welcome islands of green habitat in a sea of inhospitable concrete and asphalt. Birders have even coined a term, "the Central Park effect," to describe how a diverse array of migrant passerines and other species may, for a few weeks in April and May, abound in city parks, thrilling birders. There are also numerous well-known migrant traps that are well outside of major cities. Places such as Cape May, New Jersey; Point Reyes, California; the Florida Keys; Fort Jefferson Monument off of Key West; and many places along the Gulf Coast from Florida to Texas are major migrant traps. High Island, Bolivar Flats, and Padre Island, all part of coastal Texas, are well-known sites for seeing huge numbers of migrants, particularly in spring. Peveto Woods in coastal Louisiana is one of several critical stopover sites along the Gulf Coast that often teem with spring migrants coming in from their Gulf crossing. In the Midwest and Great Lakes region Magee Marsh in Ohio, Tawas Point in Michigan, and Point Pelee in Ontario are legendary for the numbers of spring migrants that concentrate and refuel there.

Along the Delaware and New Jersey Bayshore, mating horseshoe crabs produce huge numbers of nutrient-rich eggs that serve as essential fuel for migrating shorebirds such as Ruddy Turnstones, Sanderlings, Semipalmated Sandpipers, and Red Knots.

Fallouts are sometimes indicative of very difficult weather conditions in which to migrate. This male Black-and-white Warbler is fluffed, meaning that he is cold, and he is barely awake.

In the western states places such as Crow Valley Campground in eastern Colorado, Westby in eastern Montana, Corn Creek Springs in southern Nevada, and Furnace Creek Ranch in Death Valley, California, are good migrant traps. Local migrant traps or hotspots occur throughout North America and are well known to local birders. Every place has its hotspots.

Migrant traps offer essential stopover sites. Sometimes meteorological conditions are such that nocturnally migrating passerines will be forced down by a weather front, an event that may result in a large fallout of species, many more birds than are usually present even during a good birding day.

What should be understood about fallouts is that while they offer an outstanding opportunity for birders, some fallouts are critically stressful for birds. If the migrant birds face difficult flying conditions—such as persistent headwinds or heavy rain—they drop exhausted, precariously low on fuel, perhaps having metabolized some of their own flight muscle mass in the struggle to reach a safe haven. Some likely have perished en route and others that land exhausted may not survive. This scenario is particularly applicable to the many bird species that cross the Gulf of Mexico in spring if they encounter strong headwinds and rain while over the Gulf. Severe weather events, including tropical storms and hurri-canes, are particularly devastating when migrants become caught up in them. Doppler weather radar observations confirm that some migrants manage to find their way into the eye of a hurricane, where wind is calm, and essentially ride within the hurricane until over land, but hurricanes nonetheless are known to cause considerable mortality to migrating birds (see "Risks and Advantages of Long-Distance Migration," page 291). When severe weather occurs during spring migration along the Gulf Coast, there are on occasion large-scale migrant wrecks, documented by hundreds of dead birds washing ashore on beaches. Under such conditions any large fallout will include many weakened birds desperate to regain strength to continue.

But if the winds are favorable, as they usually are (which is why trans-Gulf migration actually works for birds), birds that have flown about 500 to 600 miles across the Gulf of Mexico and have been aloft for up to 18 hours nonetheless have sufficient fuel to continue over scrubby coastal areas to richer inland habitats, particularly along riverine areas where the spring flush of insects is in full force. Under ideal conditions for birds, there is usually little in the way of a fallout for birders.

The best kind of spring fallout is when the birds had a successful flight but met a moderate rainy cold front, forcing them down into various migrant traps, sometimes for several days as they wait out the weather and feed and regain fat. The birds will feed, refuel, and eventually continue their journeys. Enjoy them.

Staging areas and stopover sites are places of critical importance to assure the continuation of migration. Stopover sites, some habitat specific, provide essential sustenance as well as a relatively safe haven. Whimbrels are uniquely dependent on the rich marshes of coastal Georgia during their spring migration. The Whimbrels refuel in the extensive marshes, gorging on an abundance of fiddler crabs, just as Red Knots are dependent on the horseshoe crab eggs at Delaware Bay. It is essential to understand that migrants require not just breeding and nonbreeding sites but also stopover sites.

The Western Hemisphere Shorebird Reserve Network (WHSRN) was formed in 1986 when ornithologists and conservation biologists recognized the need to safeguard shorebird staging areas and stopover sites. It was the discovery of the unique dependence of Red Knots on horseshoe crab eggs along the Delaware Bayshore that focused attention on saving critical shorebird staging areas and stopover sites. WHSRN has since identified numerous critical stopover sites and effectively promoted their conservation. Thus far there are 106 WHSRN sites in 16 countries that are conserving approximately 15 million hectares of habitat for migrating shorebirds.

It is easy to observe diurnal migratory flights both in spring and autumn. Cranes, various waterfowl (especially swans and geese), cormorants, hawks, falcons, vultures, swallows, swifts, Mourning Doves, Blue Jays, American Robins, American Goldfinches, and various blackbird and sparrow species all migrate by day, sometimes in large flocks. Skeins of geese in spring and fall have long been emblematic of migration.

In contrast, nocturnal migrant species are difficult to observe directly, but birders may hear their multiple chip notes as migrants pass at low altitudes overhead (chapter 13). An old technique for studying nocturnal migration was to watch the face of the full moon with a telescope and count the bird sil-

Whimbrels require food-rich salt marshes along the southeastern coast to refuel during their northward migration.

houettes that passed across the moon's brightly lit face. Mathematical formulas were designed to estimate the total numbers of migrants per hour based on passage of birds over the face of the moon. Nocturnal migrants often fly at low altitude and are seen when illuminated by the lights of towns and cities. However, city lights and tower lights put nocturnal migrants at risk (see page 291). There are many areas along shorelines where, at dawn, nocturnal migrants that find themselves over water can be observed making landfall.

Another way to observe migration is to look online at some of the numerous sites where live Doppler weather radar is used to track the flights of migrants. Most regions throughout North America have websites with radar maps that allow you to follow migration as it is happening and predict where there might be a significant fallout. Radar observations have been used by ornithologists for decades to monitor bird migration and document its density. In recent years Doppler weather radar has been fine-tuned to focus specifically on bird migration in spring and fall. When used with information on frontal systems, radar observations are useful in predicting large-scale migratory movements.

HIGH-TECH TRACKING METHODS: GEOLOCATORS, NANOTAGS, AND SATELLITE TELEMETRY

For many years ornithologists have used radio tracking, sometimes on birds as small as thrushes, in an attempt to follow bird migration. More recently, many studies are being conducted on birds wearing ultra-small communication devices. This allows researchers to track individual birds with high precision. It is somewhat like the bird having its own GPS.

Geolocators are tiny devices usually attached to a bird's back or feet. The slight added weight does not impede the bird. Geolocators measure day length, which varies with both longitude and latitude. Records of daily day length are stored in the geolocator and recovered when the bird is recaptured and the device is removed and read on a computer. The data provide a relatively detailed record of the bird's travels. Recovery of birds wearing geolocators is possible only when geolocators are attached to birds with high natal philopatry.

Nanotags are lightweight, coded radio tags that transmit VHF signals on a common frequency. The signals are captured by various tracking stations that pick up nanotag signals from as far as 20 kilometers. One nanotag system, called Motus, has 325 tracking stations in various places throughout eastern North America and parts of Central and South America.

Satellite tracking has been used for large birds such as Ospreys that carry heavier transmitters. The transmitter is tracked by various satellites that allow researchers to tune in at almost any time and see the bird's exact location.

LONG- AND SHORT-DISTANCE MIGRATION

Species such as the Bar-tailed Godwit, Arctic Tern, Blackpoll Warbler, and Bobolink are among the icons of long-distance migration. Every year of their lives these remarkable species log many thousands of miles from their breeding range to their nonbreeding range and back. Each crosses the equator twice annually, making them birds that basically live their lives in an endless summer. The same may be said about many other species, including various seabirds, shorebirds, hawks, vultures, swifts, and passerines.

The Alaskan subspecies of Bar-tailed Godwit (*Limosa lapponica baueri*) has been shown to fly nonstop for 6,800 miles (11,000 km), requiring about nine days in the air, as it migrates from its nesting grounds in Alaska to its nonbreeding ("wintering") grounds in New Zealand. This is an amazing feat, even by bird standards. Arctic Terns, which are more leisurely migrants than Bar-tailed Godwits, have been measured to fly on average 44,000 miles (70,900 km) in the course of a year from their breeding colonies in Greenland and Iceland to the Weddell Sea in Antarctica and back again. Bobolinks breed in pastures in midwestern and northeastern states and then migrate over the Atlantic Ocean to South America, eventually reaching their nonbreeding area in southern South America, a round-trip journey estimated to be about 12,500 miles (20,100 km).

The distance a species migrates is not related to taxonomic group or to body size. Hummingbirds such as Ruby-throated in the East and Rufous in the West migrate thousands of miles annually, and Ruby-throats cross the Gulf of Mexico in both spring and fall migration. Relative to body length, the Rufous Hummingbird makes the longest migration of any North American bird. Most wood-warblers are long-distance migrants (though most do not cross the equator) and some, such as the Blackpoll and Connecticut Warblers, which breed in the northern boreal forests, fly many hours nonstop over parts of the Atlantic Ocean to reach their wintering grounds in Amazonia. By way of example, let's have a closer look at the Blackpoll.

The Blackpoll Warbler exhibits one of the most remarkable long-distance migratory patterns of any passerine. The species' nesting range includes the vast and wide belt of boreal forest from western Alaska to eastern Canada and northern New England. As far back as 1970, banding records plus field observations indicated that Blackpolls, abundant in New England in late summer and fall, are rare

Bar-tailed Godwits are among the most amazing of the long-distance migrants.

autumnal migrants in the central and southern states, including Florida. Nor are they numerous on various Caribbean islands. On the other hand, as noted above, Blackpolls are regularly observed as common to abundant autumn migrants from New Jersey northward through New England and eastern Canada. These observations collectively suggested that Blackpolls did not fly south through the southern states but instead flew from northeastern coasts out over the Atlantic Ocean, aided by tailwinds from cold fronts driven by westerly winds. The birds, tailwind assisted, eventually encountered trade winds blowing toward the southwest that would eventually take them to South America, where they winter. This route would mean that the majority of fall migrant Blackpolls would fly nonstop from as far north as Nova Scotia to their wintering area in northern and eastern Amazonia. Some Blackpolls are observed in the Lesser Antilles, and these individuals may have been forced down by meteorological conditions, or the birds may be juveniles or older birds that are not sufficiently conditioned to endure the long nonstop flight. But if conditions are satisfactory, tailwinds throughout, the birds do not stop once they leave the North American coast until they reach South America.

The initial claim of such a nonstop flight by a tiny wood-warbler drew sharp criticism from some ornithologists, as it seemed too extreme to be believed. But geolocators attached to migrant Blackpolls have confirmed that the tiny birds do indeed make a nonstop fall migration from east coastal North to South America. The flight requires between 72 and 88 hours aloft, averaging about 27 miles per hour (43 km/h). The birds accomplish a nonstop flight of over 2,000 miles (about 3,200 km). That is the equivalent of about 77 human marathons. During flight, Blackpolls, as well as all other long-distance migrant species, burn fat that was stored during premigratory feeding (which may almost double the body weight). In the case of Blackpolls, it has been calculated that these birds attain a fuel efficiency of about 720,000 miles to the gallon. That is impressive only by human standards. By bird standards, it is just another adaption. All Blackpolls do it, or they fail to survive and reproduce.

But there is more to Blackpoll migration. The birds that nest in the western parts of the range, as in Alaska and the western Canadian provinces, engage in a kind of "premigration" in autumn. Before beginning their transoceanic flight to South America, these westernmost Blackpolls must fly east across the North American continent from as far as western Alaska to Nova Scotia, a distance longer than the transoceanic flight that still lies ahead for them. The birds coming from the west have been shown to have longer wing cord lengths, likely a result of natural selection adapting them to endure a considerably longer migration route from that of eastern birds. Juvenile birds, which depart for the East Coast before adults, fare more poorly than adults and require more time to recover before departing for South America. Many juveniles probably do not survive their first migration.

Blackpolls, as well as many other bird species, follow an entirely different migratory route in the spring, as would be expected from an understanding of the prevailing fronts and wind directions. It is important to realize that wind is a major factor aiding or hindering flight. Because birds have such light body mass, a tailwind greatly enhances flight efficiency, and crosswinds or headwinds act to reduce flight efficiency. In spring, Blackpolls are trans-Gulf migrants that move north through the central and eastern states as they disperse to their various breeding sites in the boreal forest. In the course of the spring migration individuals do not fly nonstop (except over the Gulf of Mexico) as there is no reason for it. They are mostly flying over land and visit numerous stopover sites.

Turkey Vulture, Swainson's Hawk, and Broad-winged Hawk are long-distance diurnal migrants that depend upon warm thermal air currents rising from the ground to keep them aloft, allowing them to soar and save energy during their lengthy southward migration in late summer and autumn. Because there are no thermal currents over the Gulf of Mexico that would allow for soaring, these hawks all converge into immense flocks as they migrate south over eastern Mexico, an event that has been termed an aerial "river of raptors." Normally the birds migrate in broad fronts, but because of geographic realities, immense numbers, often in the millions, concentrate over narrow areas of land, such as Veracruz, Mexico, and the Isthmus of Panama as they move toward South America. The birds fly by day and put down to roost at night.

It appears that Swainson's Hawk, a diurnal hunter, does not feed while migrating, instead storing sufficient food for the journey before departure. Swainson's Hawks feed in open grassland habitat and are not adapted to feed in the dense forests of Amazonia. Swainson's Hawks fly as far south as Argentina, where there is suitable grassland habitat and where they find food. Broad-wings settle to winter in Amazonia, as do Turkey Vultures. Essentially the entire North American populations of Swainson's and Broad-winged Hawks migrate, but in the case of Turkey Vultures, many birds remain in North America. In the spring the vultures and hawks return over a period of weeks, mostly in a broad-front migration, though it is possible to encounter

Basic-plumaged Blackpoll Warblers are often abundant along coastal New England in autumn as they prepare to migrate.

Blue Jays are diurnal and somewhat facultative short-distance migrants.

large concentrations of Turkey Vultures and Broad-winged and Swainson's Hawks in the lower Rio Grande Valley of Texas as they make their way back in spring.

Short-distance migrants are species whose annual migration distance requires far fewer miles than the species discussed above. These species never migrate very far south (if at all) of the southern border of the United States. There are numerous examples, including Killdeer, Northern Saw-whet Owl, Tree Swallow, Blue Jay, Yellow-rumped Warbler, both kinglet species, various wrens, Mourning Dove, Red-winged Blackbird, Common Grackle, Eastern and Say's Phoebes, Pine Warbler, White-crowned and White-throated Sparrows, and American Robin, a list far from inclusive. Short-distance migrants do not log many thousands of miles annually and rarely cross large bodies of water other than lakes. Nonetheless these species are adapted for navigation just as long-distance migrants are. Some migrate diurnally, some nocturnally.

As you might surmise, the distinction between long- and short-distance migration is on a spectrum. There are extreme short-distance migrants and extreme long-distance migrants, but many species or populations within species fall between the extremes. While Western Meadowlarks have a typical short-distance migration, northern populations of Eastern Meadowlarks may migrate from the northern states as far as Mexico. Many populations of Dark-eyed Junco are short-distance migrants, but some, such as those that breed in northern Canada, may migrate a distance of 1,400 miles (about 2,250 km) to winter in Georgia. Many waterfowl species migrate generally from the northwest to the eastern coast or Gulf Coast, covering several thousand miles. Species such as Great and Snowy Egrets from nesting areas as far north as New England and the upper Midwest winter as far south as Mexico. Loon and grebe species likewise are somewhere between long- and short-distance migrants, depending on the population. Snow Buntings, longspurs, and some populations of Horned Larks are moderate to long-distance migrants, many coming south as far as New Jersey and various midwestern and Middle Atlantic states from their breeding areas in the Arctic. A Snow Bunting that has bred in the high Arctic and is wintering on the dunes of Cape Cod, Massachusetts, has migrated close to 2,000 miles (about 3,200 km).

Facultative migrants are species that may or may not migrate, depending on conditions, most particularly food supply. In the northern states and Canadian provinces, Black-capped Chickadees are facultative migrants, flocks moving south when conditions dictate, the northern birds augmenting more southern resident populations. Blue Jays in some areas are facultative migrants. Some Blue Jays migrate annually, but the numbers of jays migrating are variable, typical of facultative migration. Migrant Blue Jays appear to be searching for high-density acorn crops. Crossbills are similar in that their largely nomadic flights are associated with the abundance (or shortage) of the conifer cone crop.

Irruptive species are facultative migrants. These include Rough-legged Hawk, Snowy Owl, Great Gray Owl, Red-breasted Nuthatch, Boreal Chickadee, Northern Shrike, Bohemian Waxwing, Common Redpoll, Pine Siskin, Evening Grosbeak, and Pine Grosbeak. Irruptions are irregular moderate to large-scale movements that appear, at least in the cases of seed consumers, to be associated with lack of food within the normal northern range of the species. Species irrupt only if they are forced to do so, and the magnitude of the irruption is variable. Northern Shrikes, for example, appear every winter south of their breeding range, but numbers vary considerably from one winter to another. Some years there are many, but many years there are few. The same can be said about Bohemian Waxwings, especially their eastern populations. Large-scale influxes of Boreal Chickadees south of their breeding range are rare, but they happen on occasion.

Major Snowy Owl irruptions in recent years have revealed that the birds, mostly juveniles and adult females, migrate south essentially because a high density of lemmings during the previous summer resulted in large clutches of Snowy Owls and thus many potential migrants. The birds are healthy when they arrive in southern Canada and the northern states and most remain healthy, adding weight in winter by feeding on anything from rats to waterfowl. Radio tracking suggests that many if not most of these birds successfully overwinter and subsequently migrate north, back to the Arctic to breed. There is a slight winter survival advantage to being a female Snowy Owl. Females are larger than males and may outcompete them for food when on wintering areas. When starving and emaciated Snowy Owls have been examined, a majority have been males. Females may be able to drive smaller males from optimal winter territories.

Finally, there are bird species that are nonmigratory, though some may exhibit migration in the far northern parts of their range. These permanent residents occupy home ranges that may change seasonally, depending on what phase of the annual cycle they are in. American Crows are permanent residents over much of their range. So are Northern Cardinals, Carolina and Cactus Wrens, White-breasted Nuthatches, Downy and Hairy Woodpeckers, House Sparrows, European Starlings, Song Sparrows, and many other species. Blue Jays and

Black-capped Chickadees are permanent residents over much of their range but, as noted above, local populations may be augmented by migrants in winter. Northern Mockingbirds are mostly permanent residents, but in the northern parts of their range, females and juveniles are migratory while adult males are mostly resident, even in winter. Everywhere in North America some individuals of various bird species present at a given site will be there every day of the year.

MIGRATION AND LOCAL POPULATIONS

Migration is not universal within many species that are, indeed, known to migrate. Rather, it is population specific and in some cases, within a population, some individuals migrate while others do not. Some birds, particularly if they are immatures, may not migrate to breeding areas, remaining on the wintering range until they are older. This is typical of Ring-billed Gulls, for example. Immature Black-bellied Plovers are seen throughout the summer months in the Antilles, where large numbers winter. These birds did not migrate to the northern breeding grounds. Migration is one of the most labile of evolutionary traits (see "Evolution of Migration," page 295). It evolves rapidly and independently among and within species.

The Turkey Vulture is such a species, with some migratory populations and some permanent resident populations. While many thousands of Turkey Vultures migrate to South America in autumn (see "Long- and Short-Distance Migration," page 277), many others, especially from the more southern parts of the range, do not migrate and are permanent residents. Even as far north as southern New England, some but not all Turkey Vultures remain year-round, forming winter roosts. Why some migrate and some do not is a vexing question whose answer is known only to the vultures.

Eastern Towhees from northern regions are short-distance migrants and move from states such as Massachusetts and Pennsylvania to winter in more clement southern states, where they augment nonmigratory populations of Eastern Towhee. Nonetheless, even in northern states, some individual towhees will, for reasons known best to them, not migrate, opting to remain where they presumably bred. Bird feeders may appear to help keep some individuals from migrating by providing winter food, but that is misleading. The decision by individuals not to migrate is made before the weather deteriorates, so feeders come into use only when the overwintering birds become desperate for food.

Mourning Dove migration is also population specific. Northern birds tend to migrate but many do not, remaining where they bred. Southern

Mourning Doves pretty much all stay put and form large winter flocks. The Sandhill Crane population of the Florida panhandle region is also nonmigratory but other, much larger populations of Sandhill Cranes migrate, various populations following different migration routes.

DIFFERENTIAL MIGRATION

Among some short-distance migrant bird species, females and juvenile birds migrate farther south than males. Dark-eyed Juncos, Mourning Doves, and Yellow-bellied Sapsuckers are examples. In the case of the eastern "Slate-colored" Dark-eyed Juncos, because adult males look different from females and juvenile males, it is easy to see that the southernmost birds have a lower proportion of adult males in their flocks. The opposite is true of flocks in the Great Lakes states and New England, where typically up to 70 percent of the wintering juncos are males. Male juncos are somewhat larger than females and may be better able to endure cold weather, a possible example of Bergmann's rule (chapter 10). The males may be most robust and socially dominant. Taken together, these characteristics may enable them to better survive winter in more northern latitudes. They would then have the advantage of a shorter northward migration to reclaim territories in breeding season. Male Mourning Doves from migratory populations also have the shortest migration. Females migrate somewhat farther, and first-winter birds of both sexes generally migrate the farthest.

Northern Flickers, however, show a different pattern. Males, which are slightly larger than females, actually migrate farther south than females, the opposite of what happens in species such as juncos and doves. It is not clear why male Northern Flickers migrate farther south than females, but one unusual characteristic of the species is the uniquely high level of parental care displayed by males. Males do the work of carving out and defending the nesting cavity and then brood the eggs and young more than females do, as well as provide disproportionally more food to the young than do females. This higher reproductive effort is hypothesized to deplete the body resources of males such that they benefit from a more clement winter environment in which

they may regain energy and body mass lost during breeding season. Thus they migrate farther south to reach a more favorable wintering climate. This "fasting endurance" hypothesis (Gow and Weibe) would, if true, also contribute to understanding why most female short-distance migrants winter on average farther south than most males, since females normally devote considerably more energy to nesting and raising young birds than do males.

In observing differential migration such as the above examples, it is important to realize that the differences are not absolute. As you move from north to south, for example, you will encounter proportionally fewer male juncos in the junco flocks you see. Similarly, with flickers, males are most well represented in the more southern parts of the wintering range. There are rarely all-male or all-female flocks anywhere in the wintering range.

Leapfrog migration is also a form of differential migration. In widespread species with various local populations and subspecies, the migratory routes of various populations may be such that those that breed farthest north migrate farthest south. The term *leapfrog* refers to the fact that some of the populations migrate or "leap" over the others during their migratory flights. The White-crowned Sparrow is an example of a leapfrog migrant. The *gambelii* subspecies is the most northern breeder and it winters farthest south. Other subspecies populations of White-crowned Sparrow also show various degrees of leapfrog migration. Western Fox Sparrow populations show a similar pattern. Leapfrog migration is evident in numerous species that have populations spread over wide expanses of latitude.

A different form of differential migration is revealed in looking at wintering habitat diversity of Hooded Warbler and American Redstart as it affects distribution of the sexes. Males and females of these species obviously occupy the same habitat during breeding season, but in general they occupy different habitats on their wintering grounds. Mature male Hooded Warblers typically claim winter territories (which they vigorously defend) in primary tropical forest. Females and immature males are more often found in secondary forest and edge-type habitat. American Redstarts winter in Central America, northern South America, Cuba, Puerto Rico, and parts of the West Indies. Males and females are both territorial on their wintering sites. Adult males typically use mature forest, but females are less represented in such habitat. Instead, females and juvenile males are more abundant (perhaps not by choice) in less-food-rich scrub habitats. This difference in habitat value matters because males and females that overwinter in food-rich forest habitat are more successful breeders, usually gaining weight over winter, arriving

---

OPPOSITE TOP: Eastern Towhee populations may be migratory or stationary, depending upon where they are.

OPPOSITE BOTTOM: Female Northern Flickers (Yellow-shafted shown) migrate a shorter distance than males, perhaps because males contribute more to the nesting effort than females, an unusual situation among bird species.

sooner on the breeding grounds, and having earlier dates of both egg laying and fledging young.

ALTITUDINAL (ELEVATIONAL) MIGRATION

Some species confine migration to changing elevation along a mountain slope, and such a change may strongly contribute to enhanced winter survival. The typical pattern is to breed at high elevation and migrate to lower elevations for the winter months. This pattern has not been as heavily studied for North American bird species as other forms of migration.

Dark-eyed Juncos breed at elevations of up to 4,400 feet (1,341 m) in the southern Appalachian Mountains and migrate to lower elevations to spend the winter. Just over 20 percent of North American bird species show some degree of altitudinal migration. By far the majority of altitudinal migrants are found in the western mountains. These include the rosy-finches, kinglets, Phainopepla, Clark's Nutcracker, Prairie Falcon, Dusky Grouse, Sooty Grouse, Greater Sage-Grouse, Mountain Quail, Chukar, American Dipper, some Dark-eyed Junco populations, Yellow-eyed Junco, Anna's Hummingbird, Townsend's Solitaire, and Orange-crowned Warbler, a list that is not all-inclusive and is indicative of the prevalence of elevational migration in the West, where mountain ranges are extensive. Altitudinal migration is common in many bird species worldwide, including many in Hawaii.

MIGRATION FLYWAYS

Ornithologists have long professed that birds migrate by using various "highways in the sky," termed flyways. In some instances the flyway concept seems useful, but not in all.

Beginning in the West, the Pacific Ocean Flyway is the path taken by seabirds such as storm-petrels, shearwaters, alcids, albatrosses, and others. These oceanic species are dependent on the food riches of the water currents off the California coast. The next flyway is the Pacific Coastal Flyway from Alaska and the western Canadian provinces, through the Pacific Coast states, and into Mexico. This flyway follows the north–south mountain ranges of the Olympic, Cascade, and Sierra Nevada Mountains. Few migrants cross the Great Basin Desert, which is not accommodating for stopover sites.

Next is the Great Plains/Rocky Mountain Flyway that goes from Alberta, Saskatchewan, and Manitoba southward along the various ranges of the Rockies, continuing to Mexico or to the Gulf of Mexico. The Mississippi Flyway, used by numerous waterfowl species, connects part of Manitoba and Ontario with the midwestern and Gulf Coast states, roughly following the direction of the Mississippi River. The Atlantic Flyway begins east and west of

OPPOSITE: The White-crowned Sparrow (*top;* this is the *leucophrys* subspecies) engages in leapfrog migrations. This male Hooded Warbler (*bottom*) foraging among palms in Honduras is typical in that he winters in closed forest.

Hudson Bay, continuing from Quebec and Nova Scotia south through the eastern states. The westernmost part of the flyway follows the Appalachian Mountains (which orient north–south) while the easternmost part follows the eastern coastline. Finally there is the Atlantic Ocean Flyway, which is the route seabirds such as gannets, alcids, phalaropes, and jaegers utilize. Groups of migrants such as various raptors and some waterfowl tend to follow what ornithologists have called leading lines, topographical features such as rivers (think Mississippi) and mountain ranges (think Cascades, Sierra Nevada, Rockies, Appalachians) that mark north–south direction. Migrating hawks follow mountain ranges that provide thermal air currents.

Flyways are not rigid routes. Many bird species cross several flyways during migration, broadly fanning out both in spring and fall migration. It is just as possible to find migrant Blackburnian Warblers and Philadelphia Vireos along the Mississippi Flyway in Michigan as it is to find them along the Atlantic Flyway in West Virginia or somewhere in between. The same holds true for most other migrant passerines. Especially east of the Mississippi, passerines and other species use a broad-front migration during both spring and fall.

TIMING OF MIGRATION

There are three components to the timing of migration. First, what factors trigger the behavior? How does the bird know that it must migrate? Second, does the species migrate diurnally or nocturnally? Third, what is the temporal pacing of migration? In other words, how many miles per day or night does the bird normally travel and how often does it utilize stopover sites? Migrants that nest in more northern areas should time their pace so as not to arrive on their breeding grounds until conditions are satisfactory for survival. Those that arrive early could perish from late-winter storms or be affected by protracted snow cover. Migrants in more southern latitudes time migration to arrive earlier on their breeding grounds because conditions for nesting are suitable much earlier than is the case to the north. Orchard Orioles that nest in Georgia and South Carolina are on territory and nesting when Orchard Orioles that nest in New Jersey and Massachusetts are still migrating, leapfrogging over their southern cousins.

Birds instinctively know to migrate. The migratory urge is initially stimulated by changes in day

length (longer in spring, decreasing in fall), part of the bird's annual cycle (chapter 7). Migrants of many species exhibit migratory restlessness, sometimes called *zugunruhe,* a German word that translates to "travel urge" associated with how birds' hormonal systems physiologically trigger initiation of migration. Day-length change is detected as light passes through the thin bones of the bird's skull, eventually stimulating the pituitary gland and hypothalamus to trigger the urge to migrate. Migratory restlessness is in reality a form of stress, an intense and irresistible urge initiated by hormonal changes satisfied only if the bird migrates. If placed in a cage, migrant passerines experiencing migratory restlessness will vigorously fly against the cage. Not only that, but they fly against the cage in the direction in which they would normally migrate if not caged.

Once a bird enters migratory mode, it accumulates fat (added mostly in the breast area) that will serve as its migratory fuel. Some bird species such as hummingbirds and various wood-warblers are hyperphagic, doubling their weight before beginning their migratory flight. The now-extinct Eskimo Curlew was once called "Dough Bird" because of all of the fat stored around its breast and abdomen during its fall migration from the Arctic.

Weather is an important variable in timing of migration. A bird primed to migrate chooses when to go based on factors such as favorable or unfavorable frontal systems, winds, and temperature. In spring, eastern birders know to look for migrant arrival in conjunction with a northward-moving southern warm front. In autumn, birds generally move with western cold fronts. It is hypothesized that birds also time their movements in relation to barometric pressure changes indicative of frontal changes and oncoming storms. There has been a suggestion that birds may actually be able to hear the ultra-low frequency sounds (infrasounds, too low in frequency for humans to hear) created by approaching storms and take evasive action. This hypothesis, while interesting, requires further verification.

Temperature is an important variable in the timing of migration. As discussed in previous chapters, species such as Eastern Phoebe and American Robin as well as many others tend to align their migrations with isotherms, lines of temperature that move latitudinally as warming spreads northward in spring.

In comparing nocturnal with diurnal migration, meteorological realities have a strong influence. Most migratory passerines as well as many other groups of birds ranging from sapsuckers to waterfowl, herons, egrets, bitterns, and rails are primarily nocturnal migrants. Flying at night offers advantages. The air currents are typically calmest at night,

making it easier for a bird to maintain an accurate heading. Cooler nighttime temperatures aid in avoiding overheating. There is also little vulnerability to predators at night, but that reason is not very compelling since nocturnal migrants must feed by day and may then fall to predators. The need to feed by day is in itself, however, the most compelling reason to fly at night. Passerines and other nocturnal migrants rely on stopover sites for essential refueling. The bird must find suitable habitat and actively search that habitat for food. Foraging plus resting by day makes flying at night adaptive.

Diurnal migration also offers advantages for some groups of birds. Cranes, storks, and raptors rely on thermal currents that rise from land during the heat of the day. These currents allow for effective soaring while migrating, reducing energy cost. Warm rising air currents are the reason why so many hawk species migrate along ridge lines of mountains. The topography of the mountains forces warm air upward, supporting the migrant birds. Other diurnal migrants, such as swallows and swifts, are able to feed as they migrate. They are aerial foragers and find their food by daylight, roosting at night. Still other diurnal migrants, such as various blackbirds, American Robins, and Blue Jays, often migrate relatively short distances, stopping to feed along the way. Mourning Doves, for example, move only about 20 to 35 miles per day as they migrate. These species also use communal roosts at night.

In general, spring migration is more compressed than autumnal migration. There is an obvious urgency to reach the breeding grounds (chapter 14). In autumn there is perhaps less urgency to return to nonbreeding areas, and there are more birds migrating, including inexperienced juveniles. Studies using geolocators (Stuchbury et al.) have shown that individual birds such as Wood Thrush and Purple Martin spend up to several weeks more in autumnal migration than is the case in spring when they migrate north to breed.

But not all spring migration is accomplished in a shorter span compared with fall migration. What matters in many cases is wind direction and food availability. In the case, for instance, of the Alaskan Bar-tailed Godwit, its spring migration includes several stopover sites along the coasts of eastern Australia, Japan, and China as it refuels. Unlike the nonstop September flight over the Pacific Ocean (normally aided by favorable tailwinds), the spring flight along the Asian coastline requires stopovers and thus takes more days to complete.

Millions more migrant birds move southward in late summer and autumn than fly north in spring because birds fledged during the breeding season have now joined the migration of the adult birds that produced them. What is the survival rate over the course of winter? A radar-based study by A. M. Dokter and colleagues concluded that approximately 4 billion birds cross the Canadian border migrating south in late summer and autumn, and 4.7 billion birds cross the southern United States border heading south in autumn. In spring, 3.5 billion birds migrate across the southern United States border and 2.6 billion cross the Canadian border, moving northward. The difference between autumn and spring reflects winter mortality.

Timing of migration varies for age classes among many bird species. It seems logical that after parental care throughout the fledging process, juveniles would join their parents for the upcoming southward migration, but that does not commonly happen. Only in swans, geese, and cranes does migration occur in actual family groups in which juveniles learn the route from their parents. In the case of Whooping Cranes, an endangered species, the reality of juvenile social learning from adults has been used to train juvenile cranes to follow ultralight aircraft in establishing their migratory route.

For many passerine species, adults are the first to migrate and the juveniles are left to find their own way, dependent, of course, on their innate global positioning systems (see below). But in other species juveniles are the first to leave. In either case, juveniles are on their own with no parental guidance during migration.

MOLT MIGRATION

Molt is discussed in chapter 9, but it is important here to note that some migratory patterns include molt strategy. Ducks and geese, for example, usually migrate several hundred miles from northern breeding grounds to marshes where they undergo molt. Drakes migrate before hens. Ducks typically lose enough flight feathers that they are unable to fly (eclipse plumage), and the protection of dense marshes while in molt is essential. Marshes also provide crucial food resources for ducks growing new flight feathers. Many large birds such as hawks and gulls undergo a progressive, sequential wing molt even as they migrate that is obvious when observing the birds in flight.

For passerine birds as well as various other landbirds such as woodpeckers, prebasic molts were largely assumed to occur immediately following breeding and prior to migration. But this is not the

Swainson's Thrush is one of the species known to molt during migration, away from its breeding range.

case for many species. There are numerous species that molt at various migratory staging areas or on their wintering range. Molt is energetically costly, and many species are better suited to molting where they are able to find more concentrated food resources, not always the case on the breeding range. For example, eastern Swainson's Thrushes migrating south are in molt when they are in the Mississippi Valley, well south of their breeding range. Wood Thrushes exhibit a complex pattern in that some appear to molt on the breeding grounds while other individuals molt well into their southbound migration. Tree Swallows typically molt along the Gulf Coast, where many of them winter.

Molt migration underscores the complex energetics of migration interacting with the essential need to molt. What is most remarkable about molt migration is the many patterns displayed not only among but within species. Chipping Sparrows that nest in the Rocky Mountains are molt migrants, spending more time in the molt habitats on the Great Plains east of the mountains than on their breeding areas. But eastern Chipping Sparrows mostly undergo prebasic molt where they breed. It all has to do with the availability of food necessary to support the cost of molt.

AVIAN GLOBAL POSITIONING SYSTEMS

How do birds find their way? Homing pigeons (which are Rock Pigeons) are world-renowned for their uncanny abilities to return in short order to their individual dovecotes. How do they, along with thousands of other bird species, manage such precise navigation?

The answer begins with local habitat memory but soon expands well beyond. Many animals have the ability to learn and remember details of their local environs. Independent classic studies by Nobel laureates Karl von Frisch and Niko Tinbergen demonstrated that insect species from honeybees to digger wasps memorize their local landscapes and directionally orient with high accuracy. So do cats and dogs. After fledging, young Field Sparrows locally disperse prior to migrating, and it is thought that each bird is memorizing a breeding territory for when it returns. It is abundantly clear that birds learn their immediate area in great detail.

Local knowledge helps only a little with migration. Two concepts are involved in determining a precise route from point A to point B: orientation and navigation. A migrant animal must have a compass with which to maintain a direction. We can easily use a compass as a guide to maintain, say, a route south. But navigation also involves a map that provides more specific directions. There are many examples of invertebrates and vertebrates that demonstrate geographic orientation and navigation in

their movements. Spotted Salamanders return to traditional vernal breeding ponds during spring rains. Sea turtles lay eggs on the very beaches where they were hatched years earlier. Salmon spawn in the same streams where they were spawned a generation earlier. Monarch Butterflies congregate in the mountains of western Mexico by the hundreds of thousands in winter. But that said, birds would appear to be the champions of geographic orientation and navigation.

Migrant birds must maintain an accurate heading and be able to compensate for changes in time, latitude, and longitude while migrating. Avian geographic orientation, particularly among long-distance migrants, includes an innate physiological global positioning system genetically programmed for the migration route. Learning strongly augments the accuracy of navigation. Experiments (described in Newton) with European warblers (which are in a different family, Phylloscopidae, from North American wood-warblers, Parulidae) have shown that birds do have a strong innate understanding of various components of their migratory route. Even first-time migrants are known to shift their compass orientation during the migratory journey. They learn their specific breeding and wintering sites once their innately guided migration journey is completed.

There are innumerable examples of migrant birds that have annually returned to their specific breeding territories and also to their specific wintering territories (chapter 14). I well recall the first time I was brought a Wood Thrush that was captured in January in a mist net at our Belize study site. The bird was wearing our band and had been captured in the same net position as one year earlier. It had presumably flown back to North America to breed and returned to its own patch of lowland rain forest in southern Belize. I have watched an Iceland Gull in Plymouth, Massachusetts, that has apparently returned to the same parking lot behind a popular restaurant for over five years. The bird is not banded, but when it first showed up it was in first-cycle (first-winter) plumage. In the succeeding three winters it moved into second-cycle, third-cycle, and adult plumage. Now an adult, it continues to come every winter and was present again in the winter of 2018–2019. I assume it is the same bird. As it did since it was in its first cycle, it joins Ring-billed Gulls aggressively grabbing up food that humans toss to the swarming gulls that daily gather in the parking lot. Multiple examples of birds returning to their wintering patches have been reported by birders and confirmed by banding studies.

Research has confirmed that birds enhance their innate directional accuracy by learning. Classic experiments done in Europe with European Starlings showed that when both adults and juveniles were displaced from their normal migration route by about 400 miles, the adults were able to compensate and correct for the displacement. Very few of the juveniles could do that, showing that the adults, with more experience, had a sense of direction greater than that of the inexperienced birds. More recent work with White-crowned Sparrows (Thorup et al.) transported by air from the state of Washington to New Jersey echoed the results of the earlier starling work. Adults flew from New Jersey to their normal wintering areas; juveniles flew south, the correct direction, but failed to compensate for their continent-wide displacement.

The physiological ability to orient and remember strongly involves the hippocampus, the part of the brain implicated as the center of long-term memory. After fledging, migrant birds learn vital information as they migrate, even though migratory direction is innate. Once they get there, wherever "there" may be, they continue learning and subsequently remember the details of their own patch, their territory. Ornithologists are now beginning to identify actual genes that are associated with the innate migratory sense of direction in some bird species.

The role of instinct in determining migratory route is revealed in comparing migratory routes of two subspecies of Swainson's Thrush, both found in the West. The widespread *swainsoni* subspecies (called "Olive-backed Thrush"), which ranges from pretty much all of Canada to the Rocky Mountains, winters in northern South America, particularly Venezuela and Colombia. The more restricted *ustulatus* subspecies (called "Russet-backed Thrush") breeds along the northern Pacific Coast and migrates from Mexico to winter in northern Central America. The migratory routes of the inland western *swainsoni* and the coastal *ustulatus* are split by the southwestern deserts that present hostile environments to the migrants. But in the Northwest where both subspecies occur, they hybridize. A study (Delmore and Irwin) with geolocators has shown that the hybrids used either of the parental routes or something in between. Some of the hybrids wintered in locations where neither of the parental subspecies normally winter. The distinctly different migration routes of the two parent subspecies of Swainson's Thrush suggests that they may be diverging into separate species (chapter 6). But the diversity of routes taken by the subspecies hybrid birds show how their innate directionality got scrambled because they inherited combinations of two different migratory sets of directions.

Long-distance migrants use multiple environmental cues to maintain orientation during their flights. Ornithologists have shown in various experiments (see Emlen) that birds are able to orient by using nightly star rotation and to orient to the position of the sun by using polarized light, and by using latitudinal and longitudinal changes in Earth's magnetic fields. In addition, birds use geographic cues such as following coastlines or mountain ranges. There is even some suggestion (see Hagstrum) that birds may use low-frequency sound and olfaction in migration. Seabirds use odor to help navigate to their various colonies.

Diurnal migrants, such as swallows, swifts, blackbirds, American Robins, and Blue Jays, all use the position of the sun to maintain a heading in flight. They mentally correct for the movement of the sun from east to west over the course of the day. Their internal circadian clock is essential in this regard. Artificially shifting a bird's photoperiod will cause it to orient to the sun as though the shifted time is the real time. Numerous species of nocturnal migrants use the stars to navigate. The celestial sphere of stars appears to slowly spin over the course of the night. Migrant birds learn the polar stars (which do not rotate to the same degree) and use this information to maintain their compass heading.

Migrating birds are also faced with correcting for wind drift. A bird migrating south in autumn along the Eastern Seaboard must be able to adjust its flight direction if it is encountering a wind blowing from the west. Such a wind would tend to force the bird to drift eastward, perhaps out over the ocean. Birds are sometimes not able to correct for wind drift, and it is not uncommon in coastal areas to see wind-displaced birds at dawn coming in off the ocean during migration.

Magnetic fields vary with latitude and longitude and have long been suspected to be involved in avian navigation. It initially proved difficult to demonstrate magnetic orientation in birds. The first experiments (see Keeton) to clearly do so occurred in the 1970s and showed that Rock Pigeons use magnetic fields to navigate. Since those pioneering efforts, other bird species have demonstrated the ability to sense changes in magnetic fields and apply the information to navigation. Humans have difficulty relating to geomagnetic navigation because we cannot do it internally. We have no conscious sense of alterations in magnetic impulses as we traverse the planet. Apparently, birds do. Birds have particles of magnetite (iron oxide) in the skin of their upper beak and within the nasal cavities. Magnetite acts as a compass and could inform the bird's brain of changes in magnetic fields as the bird flies. There are also magnetoreceptor cells in the retina that could transmit information to the brain, information subsequently stored in the hippocampus, the long-term memory bank. Other than being somewhere within the forebrain, no specific area has as yet been identified where the avian magnetic compass is located.

Geographical orientation is a redundant system in long-distance migrant species. Birds rely on a combination of navigational tools, being able to navigate using diurnal and nocturnal celestial cues, geomagnetic cues, and geographic cues, all supported by a well-developed long-term memory. None of these senses are readily visible to birders. We hear the chip notes of birds migrating at night or we see skeins of waterfowl migrating during the day, but we cannot feel or perceive what the migrant birds are feeling or perceiving. It is a tribute to the efforts of ornithologists that they have learned as much as they have about this remarkable sensory ability demonstrated daily by all manner of bird species, many with brains about the size of a pea.

MIGRATORY CONNECTIVITY

Gray Catbirds breed throughout eastern and midwestern North America, a broad range covering many states. Do Gray Catbirds from New Jersey all migrate to essentially the same geographic region to spend the winter months? Does that region differ from where Gray Catbirds that bred in Iowa go during winter? I banded Gray Catbirds both in New Jersey and in southern Belize and often wondered if the Belize birds might breed in New Jersey. But perhaps the individuals from Belize scatter widely, occupying a broad breeding range. Did one wintering Belizean catbird return to the Midwest and the other to an eastern coastal state, perhaps New Jersey? Did two nesting New Jersey catbirds spend their winter in widely different areas, say one in Belize, one in southern Mexico? What is the connection between where a local bird population breeds and where it goes for the nonbreeding season?

Using geolocators as well as banding data, researchers (see Ryder et al.) learned that a sample group of Gray Catbirds that bred in various midwestern states all wintered in relatively close proximity in Central America. In contrast, a group of catbirds that bred along the East Coast wintered variously in Georgia, Florida, and Cuba. Thus the midwestern and eastern catbird populations appear to have differing wintering areas. The study showed strong regional connections between the breeding and wintering ranges of eastern compared with midwestern populations of Gray Catbirds.

Based on this study, my catbirds that were banded in New Jersey were likely to winter in Florida or perhaps Cuba, and the catbirds I banded in Belize were likely to breed somewhere in the Midwest. A study on Wood Thrushes (see Stanley et al.) showed a similar pattern. Of the birds that bred from North Carolina through the northeastern states, 91 percent wintered from Honduras to Costa Rica. In comparison, 65 percent of Wood Thrushes that bred in the Southeast or central states wintered in the Yucatán Peninsula and Belize through Guatemala and El Salvador.

The studies discussed above show that regional populations from geographically widespread species can exhibit strong connectivity between the breeding and wintering ranges. This is unsurprising given that regional populations are likely to be more similar genetically and thus more apt to follow a similar migratory route.

Connectivity is obviously high in species such as Kirtland's Warbler, in which all of the birds breed in close proximity in and around Michigan and spend the nonbreeding season in close proximity in the Bahamas, mostly on the Turks and Caicos Islands.

All six subspecies of Red Knots have high connectivity between their breeding areas and their wintering areas. The *rufa* subspecies that breeds in Arctic North America winters in southern Argentina, with only a small number overwintering on the Gulf coast. One Red Knot Old World subspecies with strong connectivity between breeding and nonbreeding areas is potentially in danger from climate change. The Red Knot subspecies *canutus* breeds in the high Arctic areas of the Taymyr Peninsula in Russia and winters in Banc d'Arguin in Mauritania in coastal West Africa. Changing climate is reducing resource abundance on the Arctic breeding grounds, resulting in smaller body sizes of Red Knots breeding on Taymyr. Earlier annual snow melt is stimulating earlier insect emergence, and Red Knot chicks are not yet sufficiently grown to obtain enough insect biomass and grow to normal size. Reduced insect biomass means smaller body sizes of fledged birds. When these birds migrate to their wintering area at Banc d'Arguin, they are unable to forage effectively because their smaller body size means they have smaller, shorter bills that fail to reach sufficiently deep in sediment to obtain mollusks. The birds, unable to utilize a rich diet of small clams, have taken to devouring less-nutritious seagrass rhizomes, resulting in reduced fitness and potentially endangering the entire subspecies.

This example has profound conservation implications. When connectivity between breeding area and wintering area is high, deterioration of habitat on either the breeding or wintering grounds could potentially impact the entire population. In the case of the *canutus* Red Knots, a change in the breeding grounds has resulted in reduced foraging success on the wintering grounds.

Colonial nesting birds are by definition highly concentrated. Thus it is unsurprising that many species generally have low connectivity between where they breed and where they winter, spreading out widely during the nonbreeding months. Alcids, such

This Red Knot was banded and tagged in May of 2018 along the New Jersey–Delaware Bayshore and is a member of the *rufa* subspecies. Its leg tag is easily readable and there is an antenna projecting from the bird. The antenna is attached to a nanotag affixed to the bird's back. Nanotags connect to cell phone towers and allow monitoring of the bird's movements.

as Atlantic Puffins, scatter widely over the northern Atlantic Ocean in winter. Many other seabird species do as well.

Common Terns nesting on Bird Island in Buzzards Bay, Massachusetts, have been tracked with geolocators to their wintering areas along the coast of South America. Banding data long ago established that Common Terns migrate south through the West Indies and continue on to South America. Tracking data showed that a group of Common Terns (see Nisbet and Mostello), all of which bred in the Bird Island colony, individually spent the winter months at various widespread areas along South America, ranging from northwest Venezuela to Argentina, an area extending nearly 5,000 miles (8,000 km). Regardless of which site a tern chose to winter in, all returned to Bird Island to breed.

On the other hand, Spectacled Eiders that nest along a broad area in northern Alaska all have a short migration, wintering in dense concentrations on the Bering Sea, an example in which winter density is higher than breeding density. This example is unique in that the eiders cluster tightly together in what are called *polynyas,* openings in the winter pack ice, using their combined body heat to prevent the ice from closing.

Connectivity is a complex issue with much yet to be learned. For example, while many long-distance migrant passerine species appear to remain on a fixed territory in winter, studies with geolocators placed on a small number of Veeries (see Heckscher) captured in Delaware revealed that individuals used a potentially huge wintering area, essentially all of the Amazon Basin. Most Veeries initially settled for some weeks in southern Amazonia in lowland forest. But then individuals moved in various directions north, east, west, widely separated from one another, to spend the remainder of winter before the northward migration. It appears that the Veery requires a broad and ecologically diverse area in Amazonia in which to spend the winter.

RISKS AND ADVANTAGES OF LONG-DISTANCE MIGRATION: AN EVOLUTIONARY TRADE-OFF

Migration is dangerous. No one is sure what the mortality rate is among migrant birds, and it obviously must vary among species. It is clear that the inexperience of first-year migrants dooms many, perhaps most, to failure. But mortality during migration, even for adult birds, is likely to be high. There are numerous hazards along the way (and

Dovekies, small alcids the size of starlings, normally winter offshore in the North Atlantic Ocean but many sometimes are blown along shorelines, beaches, estuaries, and lakes by strong northeastern storms. Such concentrations are called "wrecks."

new ones seem to appear annually) that exact a large annual toll on bird numbers. Hazards fall into two broad categories: weather events and structural hazards. Both account for millions of annual bird deaths.

In October of 2005, Hurricane Wilma, a category 3 hurricane with 120 mph sustained winds (190 km/h), struck parts of Florida and then continued out into the Atlantic Ocean where it re-intensified. Its movement toward the Northeast coincided with the southward migration of many migrant bird species, including the eastern Canadian population of Chimney Swifts. Hundreds of dead swifts were found after the storm. The estimate the following year, based on how many swifts returned to nest, was that the eastern Canadian Chimney Swift population decreased by about 50 percent. This huge loss was attributed to the unfortunate coincidence of the migrant birds having become caught up in the hurricane.

This high swift mortality was relatively unusual because major hurricanes are not routine events, but they do occur. Birders along the East Coast often head for the beaches immediately after the passage of a hurricane in hopes of encountering tropical species such as Sooty and Bridled Terns blown off course when caught up in the storm. Winter storms along the East Coast are sometimes nearly equiva-

lent to hurricanes in their intensity. Nor'easters, as they are called, can result in major wrecks of migrating Dovekies and other seabirds that are migrating offshore in the winter months.

More commonly, weather-related mortality of migrating birds is due to rapidly changing and deteriorating weather occurring at an inopportune time for migrants. Weather is fickle. Many species may incur high mortality in spring if they suddenly encounter prolonged inclement weather. After days of mild weather, migrants may move north only to encounter a sudden severe cold front that remains stable for some days. Swallows and martins that return in spring are dependent on aerial insect food, but should a sudden late cold front appear, even bringing snow on some occasions, aerial foragers will experience high mortality rates because of loss of food. So will other migrant species. Trans-Gulf migrants risk being trapped over water in strong headwinds and rain if they encounter a severe cold front while over the Gulf of Mexico.

Autumnal migration is generally less negatively affected by weather (aside from occasional hurricanes). Birds are migrating southward into increasingly warmer weather and can take advantage of multiple stopover sites to refuel.

Many nocturnal migrant species do not fly at high altitudes. Many fly as low as 500 feet (152 m) and usually no higher than 2,500 feet (762 m), depending on atmospheric conditions. Waterfowl fly as high as 4,000 feet (1,219 m) but may also migrate at altitudes as low as 200 feet (61 m). When migrants fly at low altitudes, there is risk of collision with various human-created structural hazards. According to studies summarized by the National Audubon Society, impact with power lines kills up to 64 million birds annually. Many birds die from impacts with buildings (particularly those with massive windows), power lines, communication towers, and wind turbines (Loss et al.). Many instances of major bird kills have been noted at communications towers. The ground is sometimes littered with dead and dying birds on a morning after a flight. Estimates are that 6.6 million birds die annually from tower impact. Wind turbines appear less lethal to birds during migration but that is an ongoing subject of study. Rotors on wind turbines appear to move more slowly than they actually do. The tips of the largest turbine props can move at about 180 miles per hour, certainly posing a risk to anything attempting to fly through the blade. It is estimated that between 250,000 and 350,000 birds perish annually because of wind turbine impact, but as wind turbines continue to proliferate this number will rise. Collisions with tall city buildings represent the highest impact toll, estimated at about 340 million birds annually. Private residences also contribute strongly to migrant mortality, killing about 253 million birds per year. But why do birds fly into buildings? Let's start with lighthouses.

It has long been observed that lighthouses can be lethal to migrants. Attracted to the bright revolving light, particularly if it is a foggy night, the birds are often blown against the structure and perish. Or they simply fly around the lighthouse, mesmerized and disoriented by the light, until they become exhausted and drop. One of the most lethal buildings to migrants has been the Washington Monument in Washington, DC. The upper part of the monument is brilliantly lit to warn low-flying aircraft of its presence. Particularly on cloudy nights, migrants fly around the lights, seemingly disoriented, and may be killed when windblown against the structure or perish due to exhaustion brought about by confusion as they continue flying around the light, losing energy. It is unclear why nocturnal migrants are attracted to uniquely bright lights. Ornithologists have hypothesized that given the migrants' propensity for celestial navigation, bright lights cause them to lose sight of stars and thus act to quickly disorient the birds, essentially trapping them in the light beams.

A detailed study (Van Doren et al.) was conducted that looked at migrant bird behavior in the two immense beams of lights used in the annual "Tribute in Light" celebration in Lower Manhattan, New York, held to commemorate the tragic attacks of September 11, 2001. This annual light display occurs over nine days in September, a time of high migration. It consists of two concentrated, powerful vertical light beams that reach about four miles and can be seen for a radius of about 60 miles (96.5 km). Radar and acoustic sensors as well as direct observation were used to measure bird densities, and it was soon learned that a high number of migrating birds were drawn into the light beams and, once there, flew around in a manner that was clearly disoriented, trapped by the lights. Over a million birds were estimated to be affected. The two light beams, however, do not result in massive bird mortality. Local volunteer birders help search the lights for birds, and when birds are observed within the lights, the beams are turned off at 20-minute intervals to permit the birds to re-orient and continue with their migration. Radar observations show the birds soon disperse once the beams are turned off. Had this humane policy not been enacted, many birds would fly until too exhausted to continue. Exhaustion, even under normal migratory conditions, is likely to be one of the most common causes of migrant mortality.

Cities, with their skyscrapers draped in light, have posed high risks to low-flying migrant birds, but some cities have made attempts to mitigate the risk. Chicago, Illinois, is one such city. In 1995 Chicago instituted a Lights Out program, encouraging owners and managers of tall buildings to dim decorative lights or turn them off entirely during times of bird migration. The program has worked well and other cities have made similar gestures.

Some risks are not obvious. Concern is mounting about the widespread agricultural use of various insecticides such as neonicotinoids and the effects these esoteric compounds might have on the migratory behavior of birds. Studies (see Larson) have been done using White-crowned Sparrows in which birds were fed doses of imidacloprid (a neonicotinoid) or chlorpyrifos (an organophosphate), both of which are widely used in agriculture. The doses were at a level comparable to what birds would be exposed to in the field. The birds were in migratory mode, experiencing migratory restlessness. Unfortunately, the birds exposed to the insecticides temporarily (for two weeks) lost their abilities to orient in the direction they would follow in migration. Control birds experienced no such difficulties. It is thus possible that seed-consuming birds that feed on seeds laced with insecticides might have greater difficulty and reduced success following their migratory routes.

The increase in numbers of ground predators, and domestic cats in particular, also poses a high risk to migrant (as well as resident) birds. Birds at stopover sites need to feed, and a high predator density does not bode well for them. Feral cats are of unique concern because of their increasing abundance and their effectiveness as bird predators. House cats allowed outdoors to roam pose no less a risk to birds. Cats do what they do and cannot help it—they are not to blame. It is humans who owe it to the safety of their pet felines as well as that of birds to keep cats indoors.

The high risks of mortality when migrating are undeniable, so why would birds, particularly long-distance migrants, evolve such a risky behavior? The answer is because, in spite of all the risks, it works. There are many advantages to being a migrant, and most migrant populations are robust in spite of the annual toll that migration takes. There are significant ecological and adaptive advantages gained from nesting in the temperate zone, even if you must abandon it and fly far away before the onset of winter.

The north temperate zone is bountiful with bird food throughout spring and summer. The northern growing season encompasses a vast geographical range comprising multiple ecosystems and whole biomes such as deciduous forest, boreal forest, prairie grasslands, montane forests, chaparral, and tundra. There is ample environmental diversity and area for breeding birds. Add to that the synchronized insect emergence as spring progresses, a time of peaking insect abundance, supplying essential protein to nestling birds. Beginning in midsummer and extending into autumn, fruits become abundant, as do seeds. Many of the fruits are fat-rich and fuel long-distance migrants, such as thrushes and orioles. The long day lengths throughout summer allow birds to feed for more hours, permitting a larger clutch size as well as a shorter time to fledging. In many cases bird species may be double-brooded, occasionally triple-brooded. Birds work hard during breeding season, but the seasonal characteristics of a northern summer are conducive to birds' nesting success. It should be no surprise that numerous bird species breed in the temperate zone. The benefit is a high offspring yield.

But then there is winter, when dormancy among plants replaces daily growth, insect abundance crashes, and most animals are adapted to overwinter as egg cases or pupae, enter hibernation, or, like migrant birds, leave. Food supplies for birds are at their lowest point of the year. Land-based habitats cannot sustain anywhere near the bird biomass in winter as is sustainable during spring and summer. Add to that the stress of enduring the cold and wind of winter that could easily become lethal, below temperature tolerance limits. Migrating to warmer and more food-rich southern latitudes is obviously adaptive to birds' survival even if it means added mortality risk. If a substantial majority of birds make a successful migration, the gain more than offsets the loss.

There is a trade-off evident when comparing the life cycles of long-distance migrant birds with birds that are permanent residents in equatorial regions between the Tropic of Cancer and the Tropic of Capricorn. The equatorial tropics are rich in species that collectively enjoy a warm and relatively wet climate throughout the year. Many more species of birds reside permanently in the tropics, particularly in the rain and cloud forests, than migrate. The Neotropics (New World tropics) abound with bird species, more permanent resident species than anywhere else on earth. Migrant bird species must insert themselves into these complex avian communities when they arrive in autumn.

Why, as spring approaches, migrate away from such an equatorial paradise? Here's why. Nest predators, ranging from insects to snakes to monkeys to kinkajous to various birds, from raptors to toucans, permeate the tropics. Nest failure rates in the tropics well exceed those in the temperate zone (chapter 15), which, of course, provides a strong advantage to nesting in the temperate zone. There is no predictable insect flush during the tropical year (though there is high insect diversity), and thus the density of protein-rich, easily caught insects, such as caterpillars, is much higher during spring and summer in the temperate zone. Thus, temperate zone nesting birds have an easier time finding protein-rich food for nestlings than would be the case had they remained in the tropics. Day length in the tropics is about 12 hours per day, much shorter than summer days in northern latitudes; thus tropical birds have less foraging time available for feeding nestlings.

Because of these realities, resident tropical bird species have significantly smaller clutch sizes than do species of temperate zone nesters (chapter 15). Most open cup nesting passerines in the temperate zone lay a clutch of four to six eggs, often more. Most in the tropics lay only two eggs.

Given the advantages of nesting in the temperate zone, it would perhaps seem that all tropical birds ought to evolve longer wings and fly north to breed, but there is yet another variable of importance. Tropical resident bird species live longer than their counterparts among long-distance migrants. While data are still somewhat limited on this point, it is likely that the lifetime reproductive success of a sedentary tropical resident bird species (such as a Chestnut-backed Antbird, for example) may equal that of most long-distance migrant species that have larger clutch sizes but many fewer years in which to attempt to reproduce. Most long-distance migrant passerine species have the physiological ability to

Tropical permanent resident species such as this Chestnut-backed Antbird are small in body size but enjoy the potential for a long lifespan, on average longer than migrant and temperate resident species.

live for 8 to 15 or more years, but very few attain such longevity. Not so with many tropical resident species in which a bird's lifespan (once it fledges) often reaches a decade and sometimes considerably more. Banding and other data have demonstrated the impressive longevities of even small passerines such as manakins in the tropics. A tropical bird may enjoy a long life occupying a small home range, which it knows intimately well, experiencing constant high rates of nest failure, but managing over the years to reproduce enough to maintain or grow its population.

Risk factors during migration clearly account for much of why migrant birds have increased mortality, but that may be only part of the story. Intriguing research suggests that, for at least one species, the Dark-eyed Junco, (short-distance) migrant populations may actually age more quickly than resident populations, simply because they migrate. It all has to do with chromosomes and, specifically, the tips of chromosomes. Telomeres are areas of noncoding DNA at the tips of chromosomes that act to protect the coding areas of DNA. Telomeres have been implicated in aging because they deteriorate somewhat with each and every replication of DNA and, as that happens, risk of damage to DNA inexorably increases. It is also true that telomeres shrink with stress, and stress is a factor in migration. In a study

conducted in Virginia by Carolyn M. Bauer and colleagues, individuals of two junco subspecies, one migratory and one nonmigratory, were both captured on a shared wintering area. In birds of the same age, the migrant junco subspecies birds (ssp. *hyemalis*) were found to have significantly shorter telomeres than the resident subspecies (ssp. *carolinensis*), suggesting that the migrants literally were aging more quickly than the residents.

Lifetime reproductive success is what counts. The trade-off between long-distance migrants and tropical resident bird species is one of short-term reproductive success and shorter lives compared with greater longevity, albeit with lower annual reproductive success. Maybe it all balances out. It is worth mentioning that some nonmigrant species, such as permanent resident chickadees in the temperate zone, also live (on average) shorter lives than similarly sized tropical birds, perhaps from the stresses of seasonal change. It is undeniable that many species of long-distance migrants, in spite of all of the risks, are maintaining healthy populations. Migration works.

EVOLUTION OF MIGRATION

Migration is an evolutionary labile trait that often shows various patterns within a single species. For perspective, one could, for example, compare the evolution of migration with that of a component of the avian skeleton, such as the synsacrum, the remarkable pelvic complex of bones fused from the vertebral column and pelvic girdle (chapter 3). All birds have a similarly structured synsacrum. It is an ancient characteristic, highly conserved, universally shared throughout the class Aves. Not so with migration. Migration is also likely an ancient behavior but it is constantly recast, reinvented, re-evolved as species disperse, colonize new ranges, and encounter changing environments and climatic conditions. Migration thus evolves independently in most species and varies among populations within species.

Obligate long-distance migrants such as Bobolinks are relatively invariable. All Bobolinks migrate, and all fundamentally use the same annual migration route. But it gets complicated fast when looking across the full spectrum of examples. The Red Knot, discussed earlier, has six subspecies worldwide and all are long-distance migrants. Each subspecies has a different migratory route from the others. Each subspecies must have evolved its migratory pathway independently of the others. Many bird species, as noted earlier in the chapter, have migrant and non-migrant populations. Migration has, for example, only recently evolved in the northernmost populations of eastern House Finches. As the species dispersed northward, migratory behavior gradually began to appear and now some northern House Finch populations are regular diurnal migrants, moving considerably south in winter. Migration does not take very many generations to evolve.

One key to the relative ease with which migratory behavior seems to evolve may be that bird species typically have a tendency, likely innate, to disperse as populations increase. That may be what stimulated some traditionally southern bird species (chapter 10) to spread to New England during the latter part of the twentieth century. Dispersal is not the same as migration, but the ease of movement accorded birds by their ability to fly allows them to quickly expand their ranges. The Cattle Egret has colonized widely in South and North America since first appearing in South America around 1880 and in North America around 1952. It dispersed initially from Africa and its dispersal behavior has continued. In many areas Cattle Egrets have evolved migration. Dispersal could thus be a significant initial step toward evolution of migration. The Eurasian Collared-Dove, which spread widely in Europe, has also done so in North America after its initial introduction in the Bahamas in the mid-1970s. But not all bird species have equally strong dispersal tendencies. Many species are not commonly found outside of their normal ranges. It is unlikely that Yellow-billed Magpies will ever spread beyond their small range in the Central Valley of California.

In 2018 the Chivi Vireo, widespread in much of Amazonia, was accorded full species status by the American Ornithological Society. Before that it was considered to be a Neotropical subspecies of Red-eyed Vireo, a long-distance migrant that winters in Amazonia. At the time of the split some authorities recognized up to eight subspecies of Red-eyed Vireo occupying much of South America, including Chivi Vireo. What was interesting was that some of these subspecies populations were entirely sedentary, some were migratory within South America, and only one, the Red-eyed Vireo (formerly *Vireo olivaceus olivaceus,* now simply *Vireo olivaceus*), was a North American long-distance migrant. There are two scenarios that might be hypothesized for the evolution of the Red-eyed/Chivi Vireo complex.

One hypothesis might suggest that the population that eventually became the Red-eyed Vireo evolved long-distance migration from South to North America while other populations that comprise the present Chivi Vireo complex remained in South America. But molecular analysis places vireos within a major adaptive radiation that includes the corvids and other groups, all originally radiating out of Australasia across Europe, some to eventually colonize North America. It is thus likely that vireos initially evolved in North America and subsequently colonized South America, not the other way around. This suggestion leads to the second hypothesis, that long-distance migration originated in a north-to-south direction. This would mean that the migratory Red-eyed Vireos or their ancestors likely gave rise to the more sedentary resident populations (now subspecies) found throughout much of South America. The Chivi Vireo is the first to have attained full species status.

Recent studies (Winger et al.) based largely on molecular data and statistical analysis of evolutionary patterns of New World long-distance migrants concluded that all likely evolved migration from a north-to-south direction. One study looked at the Parulidae (wood-warblers), cardinals, buntings, sparrows, blackbirds, orioles, and tanagers (Emberizids).

But some groups of birds may have evolved a south-to-north migration. There are 410 species of tyrant flycatchers, all found only in the New World, and only 33 (8 percent) breed north of the Mexican border. There are 363 extant species of hummingbirds, all found only in the New World, but only 15 (4 percent) breed north of the Mexican border. Those numbers suggest that the major evolution of each group occurred in South America and that the North American species evolved migration from an equatorial origin. But that said, hummingbirds present a bit of a conundrum. Fossils of what appear

Greenland goose species, such as the Greater White-fronted Goose (*top*) and the Pink-footed Goose (*bottom*), are being more frequently encountered during winter in eastern North America, usually among flocks of Canada Geese.

to be hummingbirds have been unearthed in Germany, pretty far away from South America. Did hummingbirds actually evolve in what is now Europe, disperse to North America, then to South America, where they underwent an immense speciation? Also, all hummingbird species lay only two eggs per clutch, their narrow cup nests very cozy even for two nestlings. None of the North American–nesting hummingbirds has as yet evolved a larger clutch size even though, as noted above, clutch size is normally greater among North American nesting birds than it is in their Neotropical counterparts. The best that North American hummingbirds are able to do is to be double-brooded. Female Ruby-throated hummingbirds have been observed brooding a second nest while still feeding young at a first nest.

Perhaps Pleistocene glaciation, as it gradually advanced, provided the stimulus to force dispersal south, soon evolving migration back to the north for many species of North American nesting birds. This process would have required many bird generations, but would nonetheless have had ample time to evolve.

RARITY, VAGRANCY, AND REVERSE MIGRATION

Birders thrive on finding rare birds in their birding patches and, interestingly enough, there are many rarities to be found. A recent book, *Rare Birds of North America*, includes 262 species. Rarity represents a range of probabilities, but it fundamentally may be thought of as referring to a bird that is found outside of its normal range, and thus unexpected when encountered.

Each winter in the eastern states a few Varied Thrushes are discovered, mostly around bird feeders. The increased popularity of bird feeding has likely helped with the detection of many rare and vagrant species, including many sightings of western hummingbird species in the East during late summer and autumn. Even more frequently sighted in the East are species such as Western Kingbird and Ash-throated Flycatcher. Each of these species is a common nesting bird in parts of western North America but a rare bird when discovered in the East, usually in autumn. Similar occurrences happen in the West. In California, eastern species such as Rusty Blackbird, Ovenbird, and Gray Catbird occur as major rarities, well out of their normal range.

Some birds have a strong dispersal tendency. In recent years, for example, more sightings of Roseate Spoonbill have occurred far beyond their normal southeastern range, in places such as Pennsylvania, New Jersey, and Connecticut. Like herons and egrets, juvenile spoonbills are apt to disperse long distances beyond their breeding areas.

In the Northeast, careful scanning of wintering flocks of Canada Geese may result in finding a Greenland rarity such as Greater white-fronted, Barnacle, or Pink-footed Goose. Pink-footed Goose occurs with sufficient infrequency that it is regarded not merely as rare but as a vagrant species. There is a range of probability when characterizing birds as regional rarities versus those considered to be vagrants. No precise definition of a vagrant species exists. It is kind of like saying, "You'll know one when you find one." Any Fork-tailed Flycatcher is so rare that it is best considered a vagrant no matter where it may be found in North America. A Harris's Sparrow, in contrast, is clearly a rare bird when found wintering in eastern Virginia but in Nebraska it is a commonly encountered winter bird.

Species such as Brambling, Little Egret, Northern Lapwing, and Fork-tailed Flycatcher show relatively wide vagrancy, having been seen on occasion in widespread states and Canadian provinces. Brambling, a Eurasian finch species, has been found in 29 states and 8 Canadian provinces, perhaps the most widespread vagrant species in North America. Most Bramblings are discovered in western states, particularly Alaska, as well as British Columbia because the birds that find their way to North America likely originated in Siberia. Other vagrant species show more geographically restricted vagrancy, such as Bahama Swallow, Western Spindalis, and Key West

This Western Kingbird in eastern Massachusetts in November is well out of its wintering range and is cold, as evidenced by fluffed feathers covering its feet. The species is rare in the East but nonetheless some show up every autumn.

The Brambling, a vagrant from Eurasia, is perhaps the most widespread vagrant bird species in North America.

Quail-Dove, each of which have been recorded only in Florida.

Some vagrant species, such as a Wood Stork showing up in California or a male Vermilion Flycatcher perched on an Osprey nest in coastal Maine (and detected by the live-streaming video camera on the nest . . . this actually happened), are from North America. Their vagrant status is recognized by the long distance and unconventional route they have traveled between their normal ranges and where they are as vagrants. But most vagrant species come from other countries. Vagrants from Mexico mostly occur in South Texas and other parts of the Southwest. Vagrants from the Caribbean are generally found in Florida. Many vagrant species from Asia have been found in Alaska and along the Pacific Coast. Vagrants from Europe tend to be most frequently encountered in the Northeast. Vagrant seabird species have been recorded often along both coasts of North America.

The very occurrence of rare and vagrant birds significantly beyond their normal range has implica-

tions for migratory behavior. Is vagrancy a result of weather effects, winds blowing the birds astray? Or might it be biologically based, caused by an error in innate directional programming? These two possibilities are not mutually exclusive.

It is clear that weather systems may displace birds. During migration birds are always subject to wind drift and must correct for it in order to maintain their desired directional heading. Storms produce serious wind drift and sometimes significant geographic displacement. Cave Swallows present an interesting example of this possibility. In recent years whole flocks of Cave Swallows of the subspecies *pallida* from the Southwest have appeared as far north as the Great Lakes states and New England, sometimes in flocks, during the late fall, usually November. The vast majority of the individuals are juveniles. The insect-dependent swallows inevitably encounter cold, sometimes snow, and many, likely most, perish. No firm answer exists as to why Cave Swallows venture to northern states in autumn, but it has been a regular occurrence for nearly two

These Cave Swallows, thoroughly out of their comfort zone, were huddled together in an attempt to stay warm on a cold November day in Chatham, Massachusetts. Most, perhaps all, perished. (Blair Nikula)

decades. The southwestern Cave Swallow population has been increasing its range and population in recent years. One hypothesis is that some birds, particularly juveniles, are caught in their migration by rapidly moving low-pressure systems that take them toward the northeast, exactly the wrong direction. Thus they are flying north against their will, so to speak. If true, the birds are victims of weather systems.

But many instances of rarity and vagrancy appear to be more complex than is accounted for by wind drift and weather patterns. If birds are innately prepared to migrate in a particular direction to a particular location, they could be subject to genetic mutations altering their normal migratory pattern, victims of a genetically faulty GPS. Vagrancy is much more common among juvenile birds than among adults. This could represent occasional mutations in migratory genes.

Vagrancy brings a high mortality risk since it leaves the birds well out of normal range. Do vagrants find their way back to the breeding range? But on the other hand, vagrancy, in rare cases, could prove adaptive in colonizing a new range. It worked for Cattle Egrets!

Migration errors sometimes result in birds going in the opposite direction, or going in the correct direction but overshooting where they ought to stop. When birds go in a direction opposite what would be expected in migration it is termed *reverse migration*. There is much to be learned about the workings of the physiological compass that guides birds on their normal migratory pathway, but reverse migration may represent an example of the internal compass being faulty, the result of genetic mutation. If a bird is incorrectly programmed, so to speak, the bird will fly in the wrong direction without any clue that it is doing so, and that may account for examples of reverse migration. Ornithologists refer to this possibility as *misorientation*. Overshooting is different and may be caused by another kind of error in the bird's mental compass. Both reverse migration and overshooting are much more frequent in juvenile birds making their first migration. But adults are also occasionally subject to both forms of migration error.

When found in New England, thus far usually in winter, Painted Buntings are well out of range, real rarities, but they nonetheless appear here and there most years in very small numbers. Most are juveniles but some are adult males that have presumably made at least two migrations (because males require two years to attain adult plumage). Are they reverse migrants? Possibly. Are they engaging in dispersal because of warming temperature? Not likely, since they are coming north mostly in fall and winter. Were they blown north, caught up in storms or frontal systems? Possibly.

AFTERWORD

As this book was completed and ready to enter production a major study was published in the journal *Science* authored by Ken Rosenberg and colleagues (see references). The study, focused entirely on North American birds, showed a loss of 2.9 billion birds since 1970, representing an overall decline of around 29 percent over about five decades. The databases used included the Christmas Bird Count and Breeding Bird Survey as well as radar observations. Declines were evident over all habitats, but were worse for grassland birds, shorebirds, and boreal forest birds. This paper graphically demonstrates that many bird species are in trouble and those of us who consider birds essential to our lives and to our world need to face the reality that we must be intensely proactive in promoting bird conservation. Birds are likely now to be in more peril than they were back at the end of the nineteenth century, when the Audubon movement began. That appears to be the sad truth.

Having thought about, observed, and studied birds for the vast majority of my life, I know perfectly well what a bird is, anatomically and physiologically, but I still am far from fully understanding what a bird is mentally. I do know this much: these intriguing, indeed beguiling wonders of nature know more than they are telling. What birds may not know is that they are dependent on us humans, those of us who care about them, to ensure their future.

White Ibis come to roost at dusk.

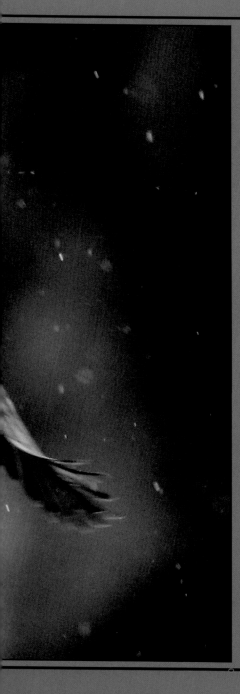

ACKNOWLEDGMENTS

This book represents a culmination of almost seven decades of somewhat fanatical devotion to birds and birding. What began in childhood in the 1950s with robins in a suburban yard and winter juncos flashing their white outer tail feathers has come to fruition in the chapters of this book. Writing a book on bird behavior, when you have spent so much time observing, reading about, teaching about, and thinking about birds—indeed, when birds have essentially consumed a huge chunk of your life—should be easy, or at least you'd think so. You "know" birds.

Well, it isn't easy. Birds are complex, thinking, mental creatures. Their behavioral spectrum ranges widely, from appearing remarkably sentient to seeming stunningly dumb. But I persevered in this writing task (and maintained a high enthusiasm level) in part by thinking back to my lifetime of experiences with birds and of the many folks who helped me to learn how to learn about birds, a long list. It has been satisfying and fun. My early and most influential mentors were well up in years when I met them as a teenager, and all are now gone. I bask in memories of them. Without Florence C. Griscom, Herbert H. Mills, Edward Hicks Parry, Elsie Acker, Esther Heacock, Gertrude Burrell, John Evans, and the entire Wyncote Bird Club (and in particular Jim Lindeman and John Dornan), my curiosity for birds would likely never have become a lifelong passion. Two weeks spent at the Hog Island Audubon Camp in June of 1962, thanks to a generous scholarship from the Wyncote Bird Club, convinced me to follow my interest in birds into a career path in ecology. Not all of my early birding friends were older than me. Some were younger. Leigh Altadonna, Bob Askins, Bruce Carrick, and Joe Wunderle were all my partners in birding way back when the great lifetime adventure was just beginning. They still are.

As I thought about this book, I reached out to friends for thoughts and advice and received no shortage of either. I also asked some friends to read parts of the developing manuscript and offer opinions. No one refused. So I offer a collective and heartfelt thanks to Carol Altadonna, Leigh Altadonna, Bob Askins, the late Pete Bacinski, Bruce Carrick, Ted Davis, Ted Floyd, Sylvia Halkin, Bruce Hallett, Ed Harper, Brian Harrington, Jessie Knowlton, Holly Merker, Kim Peters, Wayne Petersen, Dan Tankersley, Sean Williams, and Joe Wunderle for their comments and suggestions. As should be obvious, in the event that you, the reader, have questions, comments, disagreements, indignation, or cries of moral outrage about any of what I wrote, those are all "on me." I am the author.

There is one person whose opinion I would have sought repeatedly in the course of developing this book. I can only imagine the conversations we would have had. A treasured friend for decades, Professor Edward H. Burtt Jr. passed away in April 2016 after a stellar career in academia and ornithology. Jed Burtt was an inspiration to all who knew him, students and colleagues alike. He certainly was to me. I think there may be "a lot of Jed" in this book.

I am indebted to the following people who generously permitted me to use their photographs: Bob Askins, Frank Caruso, Craig Gibson, John Grant, Bruce Hallett, Ed Harper, Jerry Jackson, Jeff Jones, Kevin Karlson, Mary Keleher, Garry Kessler, Leonard Kessler, Jessie Knowlton, Steve Kress, David Larson, George Martin, Holly Merker, Jon Moulding, Blair Nikula, Van Remsen, Susan Scott, Debi Shearwater, Dan Tankersley, and Jim Zipp. Thanks, my friends, much appreciated. A special thanks goes to Bruce Hallett for his friendship and for showing me how to go about the art of photographing birds. And for all the laughs along the way.

After many years on my own, I finally smartened up and reached out to a literary agent to help me navigate the necessary contractual complexities that accompany authoring a book. I am privileged to have Russell Galen represent me. It has been a pleasure, Russ, thanks so much. I also thank my friend and editor, Lisa White, of Houghton Mifflin Harcourt. I proposed this book to Lisa and she took me up on it. I hope I have fulfilled my end of the bargain; Lisa has certainly fulfilled hers. And Lisa graciously forgave me when I unknowingly and abruptly pushed her out of the way on the crowded boardwalk at Magee Marsh in my pathetic attempt to photograph a roosting Whip-poor-will. My bad. You won't see that photo in the book. I also thank the production team at Houghton Mifflin Harcourt for making a manuscript into a gorgeous book. In particular I wish to thank Beth Burleigh Fuller and Loma Huh. Great working with you both.

Writing this book presented a significant logistical challenge. During the course of its development, I retired from forty-eight years of college teaching, packed up and vacated my office and laboratory, and rode the happy trail into the academic sunset. We then sold and moved out of two properties, downsized dramatically, and moved into a vibrant retirement community where I am now the very picture of blissful elderly contentment. None of this could have transpired without the dogged and stunningly effective efforts of my remarkable and loving wife, Martha Vaughan. She made it happen. She always makes it happen, whatever "it" may be. She is amazing and I owe her so much. She has also frequently clued me in on various cool examples of bird behavior. This book is for Martha.

REFERENCES AND FURTHER READING

PRIMARY REFERENCES

In the preparation of this book, I consulted multiple chapters in the references listed below. I offer some comments with each, should the reader wish to pursue the subject of bird behavior further.

Bent, A. C. 1962–1968. *Life Histories of North American Birds.* New York: Dover. *In the 1960s Dover reprinted and published all of Arthur Cleveland Bent's life histories of North American birds. The works were originally published by Smithsonian. The 26-volume series is now out of print but offers a treasure trove of behavior descriptions. Truly a monumental work and major contribution to North American ornithology.*

Bird, D. M. 2004. *The Bird Almanac: A Guide to Essential Facts and Figures of the World's Birds.* Toronto, ON: Key Porter Books. *Bird presents a plethora of information in the form of various lists and diagrams. Full of fun facts.*

Birds of North America Online. birdsna-org.bna proxy.birds.cornell.edu. *This constantly updated website maintains professionally written monographs on the biology of each species of bird that nests in North America. It is highly detailed and densely referenced and was used extensively in the preparation of this book. It is available by subscription only.*

Birkhead, T., J. Wimpenny, and B. Montgomerie. 2014. *Ten Thousand Birds: Ornithology Since Darwin.* Princeton, NJ: Princeton University Press. *This amazing volume is a broad account of the history of ornithology, beginning from the time of Darwin.*

Dunne, P. 2006. *Pete Dunne's Essential Field Guide Companion: A Comprehensive Resource for Identifying North American Birds.* Boston: Houghton Mifflin. *Dunne does a masterful job of describing the essence of seeing and identifying birds holistically, with much information on behavior.*

Each of the three books by John Eastman listed below offers thorough yet concise accounts of selected bird species that include much on the various species' behaviors. The books focus on species in eastern North America.

Eastman, J. 2000. *Birds of Field and Shore: Grassland and Shoreline Birds of Eastern North America.* Mechanicsburg, PA: Stackpole. *Accounts of 42 bird species.*

———. 1997. *Birds of Forest, Yard, & Thicket.* Mechanicsburg, PA: Stackpole. *Accounts of 68 bird species.*

———. 1999. *Birds of Lake, Pond, and Marsh: Water and Wetland Birds of Eastern North America.* Mechanicsburg, PA: Stackpole. *Accounts of 41 bird species.*

Ehrlich, P. R., D. S. Dobkin, and D. Wheye. 1988. *The Birder's Handbook: A Field Guide to the Natural History of North American Birds.* New York: Simon & Schuster/Fireside. *Only slightly out of date and definitely not a field guide, this cleverly designed book remains an outstanding and concise reference.*

Elphick, C., J. B. Dunning Jr., and D. A. Sibley, eds. 2001. *The Sibley Guide to Bird Life and Behavior.* New York: A. A. Knopf. *An excellent family-by-family overview of bird behavior.*

Elphick, J. 2016. *Birds: A Complete Guide to Their Biology and Behavior.* Buffalo, NY: Firefly Books. *A concise but thorough introduction approaching textbook level.*

Floyd, T. 2019. *How to Know the Birds: The Art and Adventures of Birding.* Washington, DC: National Geographic. *A clever and scholarly overview of birds, bird biology, and birding that features 200 North American bird species, each of which serves as an example of a different "lesson."*

Gill, F. B. 2007. *Ornithology, Third Edition.* New York: W. H. Freeman and Company. *For many years this has been the standard textbook on ornithology used in colleges. It is all encompassing and richly detailed. (A fourth edition of this text was published in 2019, but that new edition was not consulted in the writing of this book.)*

Karlson, K., and D. Rosselet. 2015. *Peterson Reference Guide to Birding by Impression: A Different Approach to Knowing and Identifying Birds.* Boston: Houghton Mifflin Harcourt. *While essentially a book that serves as a broad lesson in how to identify birds, the essence of much of that lesson is to watch their behavior. This book takes the birder from field marks to knowing the whole bird.*

Kaufman, K. 1996. *Lives of North American Birds.* Boston: Houghton Mifflin. *A very useful reference with concise accounts of each species.*

Leahy, C. W. 2004. *The Birdwatcher's Companion to North American Birdlife.* Princeton, NJ: Princeton University Press. *Arranged alphabetically, a sweeping and valuable treatment of everything from anatomy and migration to who Thomas Nuttall was and what the word* scansorial *refers to. A great work.*

Lovette, I. J., and J. W. Fitzpatrick. 2016. *Handbook of Bird Biology, Third Edition.* Ithaca, NY: Cornell Lab of Ornithology and Wiley. *Beautifully designed in full color, this multi-authored, comprehensive, and up-to-date volume would serve well as a college-level text.*

REFERENCES BY CHAPTER

The references listed below are specific for the chapters under which they are listed.

INTRODUCTION

Avellis, G. F. 2011. "Tail Pumping by the Black Phoebe." *Wilson Journal of Ornithology* 123 (4):766–771. *Tail pumping is more frequent when a predator is spotted.*

Hickey, J. J. 1943. *A Guide to Bird Watching.* Oxford: Oxford University Press.

Peterson, R. T. 1934. *A Field Guide to the Birds.* Boston: Houghton Mifflin.

CHAPTER 1
HOW TO USE THIS BOOK

Lack, D. 1970. *The Life of the Robin.* London: Fontana New Naturalist.

Nice, M. M. 1939. *The Watcher at the Nest.* New York: Macmillan.

Tinbergen, N. 1953 (revised 1967). *The Herring Gull's World: A Study of the Social Behaviour of Birds, revised edition.* New York: Doubleday.

CHAPTER 2
LEARNING TO WATCH BIRDS BEHAVE

Peterson, R. T. 1949. *How to Know the Birds: An Introduction to Bird Recognition.* New York: Gramercy.

CHAPTER 3
ON BEING A BIRD: ANATOMY AND PHYSIOLOGY

The books by Gill, Elphick (2016), and Lovette and Fitzpatrick listed among "Primary References" above each contain thorough information on avian anatomy and physiology.

Proctor, N. S., and P. J. Lynch. 1998. *Manual of Ornithology: Avian Structure and Function.* New Haven: Yale University Press. *Outstanding and detailed illustrations of avian anatomy.*

van Grouw, 2013. *The Unfeathered Bird.* Princeton, NJ: Princeton University Press. *Outstanding drawings of multiple bird skeletons and parts of skeletons.*

CHAPTER 4
A PERSPECTIVE ON BIRD BEHAVIOR

Ackerman, J. 2016. *The Genius of Birds.* New York: Penguin. *A sweeping treatment of avian learning potential and cognition with many details of experiments demonstrating avian problem solving, etc.*

Chen, C. C. W., and K. C. Welch Jr. 2013. "Hummingbirds Can Fuel Expensive Hovering Flight Completely with Exogenous Glucose or Fructose." *Functional Ecology* 28:589–600. *The study shows that hummingbirds may ingest and metabolize simple sugars rapidly, sugars that they likely taste.*

de Waal, F. 2017. *Are We Smart Enough to Know How Smart Animals Are?* New York: W. W. Norton. *This book, which includes bird examples, provides a highly readable overview of the reality of animal intelligence in its various manifestations.*

Emory, N. J., and N. S. Clayton. 2004. "The Mentality of Crows: Convergent Evolution of Intelligence in Corvids and Apes." *Science* 306:1903–1907. *A highly thought-provoking comparison between ape and corvid intelligence in which corvids are judged to be equal with apes.*

Ficken, M. S. 1977. "Avian Play." *Auk* 94:573–582.

Gentner, T. Q., et al. 2006. "Recursive Syntactic Pattern Learning by Songbirds." *Science* 440:1204–1207. *This is the study that demonstrates use of recursive grammar by European Starlings.*

Haupt, L. L. 2017. *Mozart's Starling*. NY: Little Brown and Co. *Focused on the fact that Mozart kept a pet starling, the author not only details Mozart's fascination with starlings but researches starling intelligence based on her own pet starling. Includes a lucid discussion of recursive phrasing in starling vocalizations.*

Heinrich, B. 1999. *Mind of the Raven: Investigations and Adventures with Wolf-Birds*. New York: HarperCollins. *Now considered a classic in writings about bird behavior.*

Heinrich, B., and J. Marzluff. 1995. "Why Ravens Share." *American Scientist* 83:342–349. *Discusses how young ravens share food such that dominant adult ravens are unable to take it because of the strength in numbers of the young birds.*

Levey, D. J., R. S. Duncan, and C. F. Levins. 2004. "Use of Dung As a Tool by Burrowing Owls." *Nature* 43:39. *Dung attracts insects that owls devour.*

Marzluff, J., and T. Angell. 2012. *Gifts of the Crow: How Perception, Emotion, and Thought Allow Smart Birds to Behave Like Humans*. New York: Free Press. *A sweeping and lucid account of crow intelligence including clear descriptions of brain anatomy and neural connections linked to intelligence in corvids.*

———. 2005. *In the Company of Crows and Ravens*. New Haven, CT: Yale University Press. *Though the theme of this book is on how crows, ravens, and humans have interacted, there is much insight shared about corvid intelligence and learning abilities.*

Ortega, J. C., and M. Bekoff. 1987. "Avian Play: Comparative Evolutionary and Developmental Trends." *Auk* 104 (2): 338–341.

Pepperberg, Irene. 2002. *The Alex Studies: Cognitive and Communicative Abilities of Grey Parrots*. Cambridge, MA: Harvard University Press. *A summary of the years of research focused on Alex.*

Swift, K. 2018. "Putting the 'Crow' in Necrophilia." *Corvid Research blog*, https://corvidresearch .blog, July 16.

Swift K., and J. M. Marzluff. 2018. "Occurrence and Variability of Tactile Interactions Between Wild American Crows and Dead Conspecifics." *Philosophical Transactions of the Royal Society* B 373 (1754): 20170259. https://doi.org/10.1098/rstb.2017.0259.

———. 2015. "Wild American Crows Gather Around Their Dead to Learn About Danger." *Animal Behavior* 109:187–197.

Wasserman, E. A. 1995. "The Conceptual Abilities of Pigeons." *American Scientist* 83:246–255. *A summary of lab experiments in which Rock Pigeons demonstrated cognitive abilities to sort among objects and categories and to recognize abstract relationships.*

Zickefoose, J., and W. E. Davis Jr. 1998. "Great Blue Heron (*Ardea herodias*) Uses Bread As Bait to Fish." *Colonial Waterbirds* 21:87–88.

CHAPTER 5
A BIRD'S BRAIN AND SENSES

Birkhead, T. 2012. *Bird Sense: What It's Like to Be a Bird*. New York: Walker and Company. *An outstanding account of each of the various senses of birds. Highly recommended.*

Bonadonna, F., and G. A. Nevitt. 2004. "Partner-specific Odor Recognition in an Antarctic Seabird." *Science* 306:825. *The study shows that the Antarctic Prion, a species of seabird, uses odor to locate its partner.*

Brower, L. P. 1969. "Ecological Chemistry." *Scientific American* 220:22–29. *Layman account of the relationship between monarch butterflies and milkweed and how it selects for warning coloration and mimicry.*

Davis, W. E. Jr. 2012. "Northern Cardinal Attacks Its Reflected Image in a Window." *Bird Observer* 40:149–152. *One of dozens of accounts that document bird aggression directed toward their own images as reflected in mirrors or windows.*

Emery, N. 2016. *Bird Brain: An Exploration of Avian Intelligence.* Princeton, NJ: Princeton University Press. *A widely popular and richly illustrated book that reviews the science behind understanding avian intelligence. Highly recommended.*

Gaffney, M. F., and W. Hodos. 2003. "The Visual Acuity and Refractive State of the American Kestrel (*Falco sparverius*)." *Vision Research* 19:2053–2059. *Documents the amazing ability of kestrels to detect small prey items from high above.*

Kress, S. W., and D. Z. Jackson. 2015. *Project Puffin: The Improbable Quest to Bring a Beloved Seabird Back to Egg Rock.* New Haven, CT: Yale University Press. *Describes the behaviors of puffins and techniques used to entice puffins to recolonize islands off the coast of Maine.*

Morell, V. 2018. "Bird Brainiacs." *National Geographic* 233 (2): 108–129. *Engaging popular article that surveys the scope of avian intelligence.*

Nevitt, G. 1999. "Foraging by Seabirds on an Olfactory Landscape." *American Scientist* 87:46–53. *Discusses how seabirds locate prey using chemical cues such as dimethyl sulfate.*

Savoca, M. S., et al. 2016. "Marine Plastic Debris Emits a Keystone Infochemical for Olfactory Foraging Seabirds." *Science Advances* 2 (11): e1600395. https://doi.org/10.1126/sciadv .1600395. *The study shows that decomposing plastics emit the odor of dimethyl sulfate, attracting seabirds that might consume the debris, assuming it to be food.*

Stryker, N. 2014. *The Thing With Feathers: The Surprising Lives of Birds and What They Reveal About Being Human.* New York: Riverhead Books. *Engaging book with many chapters that deal with aspects of avian intelligence and cognition.*

Tewksbury, J., and G. Nabhan. 2001. "Seed Dispersal: Directed Deterrence by Capsaicin in Chilies." *Nature* 412:403–404. *Features discussion of role of Curve-billed Thrasher in seed dispersal.*

Waldvogel, J. A. 1990. "The Bird's Eye View." *American Scientist* 78 (4): 342–353. *Provides a fine account of the range of avian vision.*

CHAPTER 6
UNDERSTANDING BIRD DIVERSITY

Barrowclough, G. F., J. Cracraft, J. Klicka, R. M. Zink. 2016. "How Many Kinds of Birds Are There and Why Does It Matter?" *PLoS ONE* 11 (11): e0166307. https://doi.org/10.1371/journal .pone.0166307. *This provocative paper suggests that as many as 18,000 bird species may inhabit earth.*

Benkman, C. W. 1993. "Adaptation to Single Resources and the Evolution of Crossbill (*Loxia*) Diversity." *Ecological Monographs* 63:305–325. *Benkman has been a leading researcher in the evolution of the red crossbill complex and their various seed trees.*

Benkman, C. W., et al. 2009. "A New Species of Red Crossbill (Fringillidae: *Loxia*) from Idaho." *Condor* 111:169–176.

Block, N. 2015. "An Introduction to Understanding Phylogenies and Their Effects on Taxonomy." *Birding* 47:16–19. *This is a highly lucid explanation of a topic that birders have often found somewhat confusing. Strongly recommended.*

Bronson, C. L., et al. 2005. "Reproductive Success Across the Black-capped Chickadee (*Poecile atricapillus*) and Carolina Chickadee (*P. carolinensis*) Hybrid Zone in Ohio." *Auk* 122:759–772.

Chesser, R. T., et al. 2017. Fifty-eighth Supplement to the American Ornithological Society's *Check-List of North American Birds. The Auk: Ornithological Advances* 134:751–773.

Churchill, M. 2014. "Species & Subspecies: A Brief History." *Birding* 46:20–25.

Curry, R. L. 2005. "Hybridization in Chickadees: Much to Learn from Familiar Birds." *Auk Ornithological Advances* 122:747–758. *Curry is a leading researcher on chickadees, particularly the hybrid zone between Black-capped and Carolina Chickadees.*

DeBenedictus, P. E. 1996. "More Mitochondrial Madness." *Birding* 28:141–145. *Dated but useful primer on how use of mitochondrial DNA analysis affects avian taxonomy.*

———. 1995. "Red Crossbills, One Through Eight." *Birding* 27:494–501.

Donald, P. F. 2007. "Adult Sex Ratios in Wild Bird Populations." *Ibis* 149:671–692.

Gill, F. B. 2014. "Species Taxonomy of Birds: Which Null Hypothesis?" *Auk* 131:150–161. *Provocative and well-reasoned argument that elicited much comment.*

Haiman, A. N. K. 2012. "Evening Grosbeaks: Evolution in Action." *Birding* 44:34–40.

Jamarillo, A. 2006. "Tales from the Cryptic Species: Lumps and Splits." *Birding* 38:30–38.

Jarvis, E. D., et al. 2014. "Whole-genome Analyses Resolve Early Branches in the Tree of Life of Modern Birds." *Science* 346:1320–1331. *This technical paper presents the evidence for why the order of bird families was rearranged in recent field guides.*

Kerr, K. C. R., et al. 2007. "Comprehensive DNA Barcode Coverage of North American Birds." *Molecular Ecology Notes* 7:535–543.

Knue, A. 2013. "What the Heck Is a Tanager?" *Birding* 45:12–23. *A layman-friendly account of the taxonomic conundrum presented by tanagers and things like tanagers.*

Lovette, I. J., et al. 2010. "A Comprehensive Multilocus Phylogeny for the Wood-warblers and Revised Classification of the Parulidae (Aves)." *Molecular Phylogenetics and Evolution* 57:75–770. *This is the paper that changed the entire classification of wood-warblers, including eliminating the genus Dendroica. Birders and ornithologists alike took notice.*

Newton, I., 2003. *The Speciation and Biogeography of Birds.* London: Academic Press. *This book presents an overarching treatment of the topic and should be consulted by anyone with a serious interest in avian biogeography.*

Price, T. 2007. *Speciation in Birds.* Greenwood Village, CO: Roberts & Company. *Outstanding overview of the various concepts and questions about how birds undergo speciation.*

Prum, R. O., et al. 2015. "A Comprehensive Phylogeny of Birds (Aves) Using Targeted Next-generation DNA Sequencing." *Nature* 526:569–573. *A technical paper that presents molecular evidence for taxonomic sequence changes.*

Pyle, P. 2012. "In Memory of the Avian Species." *Birding* 44:34–39.

Shapiro, L. 2005. "More than Meets the Eye: Hybrids of Golden-winged and Blue-winged Warblers." *Birding* 37:278–286.

Vallender, R., R. J. Robertson, V. L. Friesen, and I. J. Lovette. 2007. "Complex Hybridization Dynamics Between Golden-winged and Blue-winged Warblers (*Vermivora chrysoptera* and *Vermivora pinus*) Revealed by AFLP, Microsatellite, Intron and mtDNA Markers." *Molecular Ecology* 16:2017–2029.

Young, M., and T. Spahr. 2017. "Crossbills of North America: Species and Red Crossbill Call Types." *eBird*, https://ebird.org/news/crossbills-of-north-america-species-and-red-crossbill-call-types/.

Young, M., T. Spahr, and A. Spencer. 2018. "Evening Grosbeak Call Types in North America." *eBird*, https://ebird.org/news/evening-grosbeak-call-types-of-north-america.

Zhang, G., et al. 2014. "Comparative Genomics Reveals Insights into Avian Genome Evolution and Adaptation." *Science* 346:1311–1320. *Technical but worth the effort.*

CHAPTER 7
THE ANNUAL CYCLE OF BIRDS

Consult Gill, *Ornithology,* chapter 9, cited above under "Primary References," for a thorough overview of annual cycle.

Blanchard, B. D. 1941. *The White-crowned Sparrows (Zonotrichia leucophrys) of the Pacific Seaboard: Environment and Annual Cycle.* Oakland, CA: University of California Press.

The papers cited below all treat various technical aspects of the biochemistry on the annual cycle.

Dawson, A. 2008. "Control of the Annual Cycle in Birds: Endocrine Constraints and Plasticity in Response to Ecological Variability." *Philosophical Transactions of the Royal Society B: Biological Sciences* 363:1621–1633.

———. 1983. "Plasma Steroid Levels in Wild Starlings (*Sturnus vulgaris*) During the Annual Cycle in Relation to Stages of Breeding." *General and Comparative Endocrinology* 49:286–294.

Dawson, A., and A. R. Goldsmith. 1982. "Prolactin and Gonadotrophin Secretion in Wild Starlings (*Sturnus vulgaris*) During the Annual Cycle and in Relation to Nesting, Incubation, and Rearing Young." *General and Comparative Endocrinology* 48:213–294.

Farner, D. S. 1964. "The Photoperiodic Control of Reproductive Cycles in Birds." *American Scientist* 52:137–156.

Helms, C. W. 1963. "The Annual Cycle and Zugunruhe in Birds." *Proceedings of the International Ornithological Congress* 13:925–939. *Classic paper on migratory restlessness.*

Hendenstrom, A. 2006. "Scaling of Migration and the Annual Cycle of Birds." *Ardea* 94:399–408.

Ketterson, E. D., and J. W. Atwell. 2016. *Snowbird: Integrative Biology and Evolutionary Diversity in the Junco.* Chicago: University of Chicago Press. *A somewhat technical book but one that superbly looks at the genetics and biochemistry of the Junco's annual cycle.*

Mewaldt, L. R., and J. R. King. 1977. "The Annual Cycle of White-crowned Sparrows (*Zonotrichia nuttalli*) in Coastal California." *Condor* 79:445–455.

Wiersma, P., and T. Piersma. 1994. "Effects of Microhabitat, Flocking, Climate and Migratory Goal on Energy Expenditure in the Annual Cycle of Red Knots." *Condor* 96:257–279.

Wolfson, A. 1965. "Circadian Rhythm and the Photoperiodic Regulation of the Annual Reproductive Cycle in Birds." In *Circadian Clocks,* Amsterdam: New Holland, pp. 370–378.

CHAPTER 8
FEATHERS AND FLIGHT

The books by Gill, Elphick (2016), and Lovette and Fitzpatrick listed among "Primary References" each contain thorough information on feathers and flight.

Brown, C. R., and M. B. Brown. "Where Has All the Road Kill Gone?" *Current Biology* 23 (6): https://doi.org/10.1016/j.cub.2013.02.023. *Documents the decrease in wing dimensions of Cliff Swallows that enable the birds to better avoid collisions with vehicles.*

Hill, G. E., and K. J. McGraw. 2006. *Bird Coloration, Volume 1: Mechanisms and Measurement.* Cambridge, MA: Harvard University Press.

Hudon, J., et al. 2017. "Diet Explains Red Flight Feathers in Yellow-shafted Flickers in Eastern North America." *Auk* 134:22–33.

Seneviratne, S. S., and I. L. Jones. 2008. "Mechanosensory Function for Facial Ornamentation in the Whiskered Auklet, a Crevice-dwelling Seabird." *Behavioral Ecology* 19:784–790. *Birds rely on their elaborate filoplumes to navigate dark rocky crevices.*

Terres, J. K. 1968. *How Birds Fly.* New York: Hawthorne Books. *Dated but engaging overview and still very useful.*

CHAPTER 9
MAINTENANCE BEHAVIOR OF BIRDS

Burtt, E. H., Jr., and J. M. Ichida. 2004. "Gloger's Rule, Feather-degrading Bacteria, and Color Variation among Song Sparrows." *Condor* 106: 681–686. *This paper helped establish that feather-degrading bacteria may be a strong selection pressure in explaining Gloger's Rule.*

Clayton, D. H., et al., 2010. "How Birds Combat Ectoparasites." *The Open Ornithology Journal* 3: 41–71.

Floyd, T. 2008. *Smithsonian Field Guide to the Birds of North America.* New York: HarperCollins. *Includes an excellent overview of molt and includes the molt strategy of every species.*

Howell, S. N. G. 2010. *The Peterson Reference Guide to Molt in North American Birds.* Boston: Houghton Mifflin Harcourt. *The best reference currently available. Lucidly explains the complexities of molt and looks at molt in each family of North American birds.*

Humphrey, P. S., and K. C. Parkes. 1959. "An Approach to the Study of Molts and Plumages." *Auk* 76:1–31. *This is the paper that describes the system now in use for molt terminology.*

Kaufman, K. 2011. *Field Guide to Advanced Birding: Understanding What You See and Hear.* Boston: Houghton Mifflin Harcourt. *See Chapter 4.*

Peele, A. M., et al. 2009. "Dark Color of the Coastal Plain Swamp Sparrow (*Melospiza georgiana nigrescens*) May Be an Evolutionary Response to

Occurrence and Abundance of Salt-tolerant Feather-degrading Bacilli in its Plumage." *Auk* 126:531–535.

Potter, E. F. 1970. "Anting in Wild Birds, Its Frequency and Probable Purpose." *Auk* 87:692–713.

Pyle, P. 1997. *Identification Guide to North American Birds, Part I: Columbidae to Ploceidae.* Point Reyes Station, CA: Slate Creek Press.

———. 2008. *Identification Guide to North American Birds, Part II: Anatidae to Alcidae.* Point Reyes Station, CA: Slate Creek Press. *The two books by Pyle offer highly detailed molt information useful especially for bird banders as they attempt to age and sex bird species.*

Scott, S. D., and C. McFarland. 2010. *Bird Feathers: A Guide to North American Species.* Mechanicsburg, PA: Stackpole. *Excellent introduction to bird feather biology and molt as well as species-by-species feather illustrations useful for identifying feathers.*

Simmons, K. E. L. 1957. "The Taxonomic Significance of the Head-scratching Methods in Birds." *Ibis* 99:178–181.

CHAPTER 10
TEMPERATURE AND BIRD BEHAVIOR

Arad, Z., and M. H. Bernstein. 1988. "Temperature Regulation in Turkey Vultures." *Condor* 90:913–919.

Bartholomew, W. R., C. Lasiewski, and E. C. Crawford Jr. 1968. "Patterns of Panting and Gular Fluttering in Cormorants, Pelicans, Owls, and Doves." *Condor* 70:31–34.

Burtt, E. H., Jr. 1999. "Rules to Bird By: Bergmann's Rule." *Birding* 29:267–270. *An excellent overview of Bergmann's rule as it is demonstrated in birds.*

Danner, R. M., et al. 2017. "Habitat-specific Divergence of Air Conditioning Structures in Bird Bills." *Auk* 134:65–75. *This paper discusses the function of respiratory chonchae in the bills of Song Sparrows from two populations, demonstrating the role of greater area in chonchae helping cool the birds and reabsorb moisture.*

Dunn, P. O., and D. W. Winkler. 1999. "Climate Change Has Affected the Breeding Date of Tree Swallows Throughout North America." *Proceed-*

ings of the Royal Society B: Biological Sciences 266 (1437). https://doi.org/10.1098/rspb.1999.0950. *This is one of several papers that show changes in Tree Swallow arrival dates that correlate with warming temperature.*

Greenberg, R. R., et al. 2012. "High Summer Temperature Explains Bill Size Variation in Salt Marsh Sparrows." *Ecography* 35:146–152.

Iliff, M. J. 2017. "A Bad Winter Hangover." *Living Bird* 36:43–44. *Describes the loss of Carolina Wrens in the Northeast as a result of the cold winter of 2015.*

James, F. C. 1970. "Geographic Size Variation in Birds and Its Relationship to Climate." *Ecology* 51:365–390. *An excellent paper demonstrating the effects of environment on body size.*

Marra, P. P., et al. 2014. "Full Annual Cycle: Climate Change Vulnerability Assessment for Migratory Birds of the Upper Midwest and Great Lakes Region." migratoryconnectivityproject.org/climate-change-vulnerability. *This website paper includes an outstanding overview and species-by-species accounts.*

Møller, A. P., W. Fiedler, and P. Berthold. 2010. *Effects of Climate Change on Birds.* Oxford, UK: Oxford University Press. *This book presents a view of the scope of climate change impact on bird populations.*

Root, T. 1988. "Energy Constraints on Avian Distribution and Abundances." *Ecology* 69:330–339. *Describes how the Eastern Phoebe is limited by the −4°C isotherm.*

Schleucher, E. 2004. "Torpor in Birds: Taxonomy, Energetics, and Ecology." *Physiological and Biochemical Zoology* 77:942–949.

Socolar, J. B., et al. 2017. "Phenological Shifts Conserve Thermal Niches in North American Birds and Reshape Expectations for Climate-driven Range Shifts." *Proceedings of the National Academy of Sciences* 114:12976–12981.

Symonds, M. R. E., and G. J. Tattersall. 2010. "Geographical Variation in Bill Size Across Bird Species Provides Evidence of Allen's Rule." *American Naturalist* 176:188–197. *Excellent discussion of factors affecting bill size with a good discussion of Allen's rule.*

Tingley, M. W., et al. 2009. "Birds Track Their Grinnellian Niche Through a Century of Climate Change." *Proceedings of the National Academy of Sciences* 106 (Supplement 2): 19637–19643. www.pnas.org/cgi/doi/10.1073/pnas.0901562106. *This California study shows shifting breeding times in relation to temperature changes over a multiple-year period.*

Van Buskirk, J., R. S. Mulvill, and R. C. Liberman. 2010. "Declining Body Sizes in North American Birds Associated with Climate Change." *Oikos* 119:1047–1055. *Perhaps reflecting Bergmann's rule, this paper documents how some bird species have smaller body dimensions in recent years as temperatures rise.*

CHAPTER 11
SOCIAL BEHAVIOR

Askins, R. A. 2000. *Restoring North America's Birds: Lessons from Landscape Ecology.* New Haven, CT: Yale University Press. *See chapter 9, "Red-cockaded Woodpeckers and the Longleaf Pine Woodland." Other chapters feature other ecosystems and their birds. Outstanding book.*

Bednarz, J. C. 1988. "Cooperative Hunting in Harris' Hawks (*Parabuteo unicinctus*)." *Science* 239:1525–1527.

Caughey, M. 2017. *How to Speak Chicken: Why Your Chickens Do What They Do & Say What They Say.* North Adams, MA: Storey Publications. *Written for lay readers and containing much wit and a clear fondness for the subject matter, the book does a fine job of introducing the behavior of the world's most abundant bird species.*

Conner, R. N., D. C. Rudolph, and J. R. Walters. 2001. *The Red-cockaded Woodpecker. Surviving in a Fire-Maintained Ecosystem.* Austin: University of Texas Press. *Detailed overview of all ecological factors affecting the biology of this endangered species. Technical.*

Ellis, D. H., et al. 1993. "Social Foraging Classes in Raptorial Birds." *Bioscience* 43:14–20.

Firth, J. A., et al. 2017. "Wild Birds Respond to Flockmate Loss by Increasing Their Social Network Associations to Others." *Proceedings of the Royal Society B: Biological Sciences* 284 (1854). https://doi.org/10.1098/rspb.2017.0299. *This study focused on social networking theory as applied to societies of Great and Blue Tits.*

Freeman, B. G., and E. T. Miller. 2018. "Why Do Crows Attack Ravens? The Roles of Predation Threat, Resource Competition, and Social Behavior." *Auk, Ornithological Advances* 135 (4): 857–867. *This study shows that simple acts like aggression by crows toward ravens may have multiple causes.*

Garlinger, R. D., et al. 2017. "Applications of a Social Network Analysis: Investigating Social Behaviors in Resident Carolina Chickadees (*Poecile carolinensis*) and Irruptive Black-capped Chickadees (*P. atricapillus*)." Wilson Ornithological Society Annual Meeting, Florida Gulf Coast University, Ft. Myers, FL, March 9–12.

Koenig, W. D., and R. L. Mumme. 1987. *Population Ecology of the Cooperatively Breeding Acorn Woodpecker.* Monographs in Population Biology, Volume 24. Princeton, NJ: Princeton University Press. *Technical account of the complex social biology of this woodpecker species.*

Miller, J. B. 2017. "Interspecific Aggression by a Common Loon in Winter." *Bird Observer* 45:121–122. *Documents one loon on one pond acting with uncommon aggression directed toward various waterfowl.*

Schjelderup-Ebbe, T. (1975). "Contributions to the Social Psychology of the Domestic Chicken" (M. Schleidt and W. M. Schleidt, translators). In Schein, M. W. (ed), *Social Hierarchy and Dominance: Benchmark Papers in Animal Behavior,* Volume 3. Stroudsburg, PA: Dowden, Hutchinson and Ross, pp. 35–49. Reprinted from *Zeitschrift für Psychologie* 88 (1922): 225–252.

Shunk, S. A. 2016. *Peterson Reference Guide to Woodpeckers of North America.* Boston: Houghton Mifflin Harcourt.

Skutch, A. F. 1985. *Life of the Woodpecker.* Santa Monica, CA: Ibis Publishing Co.

Smith, S. M. 1991. *The Black-capped Chickadee: Behavioral Ecology and Natural History.* Ithaca, NY: Cornell University Press. *Outstanding reference on chickadee behavioral biology, technical but nonetheless easy to read.*

Woolfenden, G. E., and J. W. Fitzpatrick. 1984. *The Florida Scrub Jay: Demography of a Cooperative-breeding Bird.* Monographs in Population Biology, Volume 20. Princeton, NJ: Princeton University Press. *Now classic technical account of the complex social system of the Florida Scrub-Jay.*

CHAPTER 12
PROVISIONING AND PROTECTION

Axelson, G., and V. Greene. 2017. "30 Years of Project FeederWatch Yield New Insights About Backyard Birds." *Living Bird* 36 (Winter): 22–29.

Baicich, P. J., M. A. Barker, and C. L. Henderson. 2015. *Feeding Wild Birds in America: Culture, Commerce, and Conservation*. College Station, TX: Texas A&M University Press. *This book offers a sweeping overview of the history of bird feeding in America and is full of information on behavior at feeders.*

Brittingham, M. C., and S. A. Temple. 1992. "Does Winter Bird Feeding Promote Dependency?" *Journal of Field Ornithology* 63:190–194. *Focused on chickadees, the answer is no.*

Carter, J., et al. 2008. "Subtle Cues of Predation Risk: Starlings Respond to a Predator's Direction of Eye-gaze." *Proceedings of the Royal Society B: Biological Sciences* 275 (1644): 1709–1715. https://doi.org/10.1098/rspb.2008.0095. *Experimental study showing that looking a starling directly in the eye will cause it to avoid you.*

Correia, J., and S. L. Halkin. 2018. "Potential Winter Niche Partitioning Between Tufted Titmice (*Baeolophus bicolor*) and Black-capped Chickadees (*Poecile atricapillus*)." *Wilson Journal of Ornithology* 130:684–695. *Well-crafted study that birders could, themselves, emulate. Shows subtle but important foraging distinctions between the two species during the New England winter.*

Dally, J. M., N. J. Emery, and N. S. Clayton. 2006. "Food-caching Western Scrub-Jays Keep Track of Who Was Watching When." *Science* 312:1662–1665. *An experimental study showing that jays pay close attention to other jays as they cache seeds.*

Darley-Hill, S., and W. C. Johnson. 1981. "Acorn Dispersal by the Blue Jay (*Cyanocitta cristata*)." *Oecologia* 50:231–232. *Focused on pin oak and the amazing dispersal of acorns by blue jays.*

Dunn, E. H., and D. L. Tassaglia-Hymes. 1999. *Birds at Your Feeder: A Guide to Feeding Habits, Behavior, Distribution, and Abundance*. New York: W. W. Norton.

Gallinet, A. S., R. B. Primack, and T. L. Lloyd-Evans. 2019. "Can invasive species replace native species as a resource for birds under climate change? A case study on bird-fruit interactions." *Biological Conservation*, https://doi.org/10.1016/j.biocon.2019.108268.

Goullaud, E. L., D. R. De Zwaan, and K. Martin. 2018. "Predation Risk-induced Adjustments in Provisioning Behavior for Horned Lark (*Eremophila alpestris*) in British Columbia." *Wilson Journal of Ornithology* 130:180–190. *Horned larks alter their behavior when aware of predators, a clear example of how factors such as predation possibility alters optimal foraging.*

Heisman, R. 2017. "Outbreak . . . Then Understanding." *Living Bird* 36:35–41. *Details the spread of eye disease in eastern House Finches.*

Henry, P.-Y., and J.-C. Aznar. 2006. "Tool-use in Charadrii: Active Bait Fishing by a Herring Gull, *Larus argentatus*." *Waterbirds* 29:233–234.

Jablonski, P. G. 1999. "A Rare Predator Exploits Prey Escape Behavior: The Role of Tail-fanning and Plumage Contrast in Foraging of the Painted Redstart (*Myioborus pictus*)." *Behavioral Ecology* 10:7–14.

LaFleur, N. E., M. A. Rubega, and C. S. Elphick. 2007. "Invasive Fruits, Novel Foods, and Choice: An Investigation of European Starling and American Robin Frugivory." *Wilson Journal of Ornithology* 119:429–438.

Ligon, J. D. 1978. "Reproductive Interdependence of Pinon Jays and Pinon Pines." *Ecological Monographs* 48:111–126. *One of the first technical papers to fully document the interdependency of these two species.*

MacArthur, R. H. 1958. "Population Ecology of Some Warblers in Northeastern Coniferous Forests." *Ecology* 39:599–619. *This is one of the most classic studies in avian ecology and it inspired hundreds of other studies.*

McCabe, J. D., and B. J. Olsen. 2015. "Tradeoffs Between Predation Risk and Fruit Resources Shape Habitat Use of Landbirds During Autumn Migration." *Auk* 132 (4): 903–913.

Miller, E. T., et al. 2017. "Fighting over Food Unites the Birds of North America in a Continental Dominance Hierarchy." *Behavioral Ecology* 28 (6): 1454–1463. https://doi.org/10.1093/beheco/arx108.

Moskoff, W. 2001. "Why Some Female Birds Are Larger than Males." *Birding* 33:254–260.

Mumme, R. L. 2014. "White Tail Spots and Tail-flicking Behavior Enhance Foraging Performance in the Hooded Warbler." *Auk* 131:141–149.

Nyffeler, M., et al. 2018. "Insectivorous Birds Consume an Estimated 400–500 Million Tons of Prey Annually." *The Science of Nature* 105 (7–8): 47. https://doi.org/10.1007/s00114-018-1571-z. *This paper also illustrates the strong dependency that birds have for access to insects.*

Robb, G. N., R. A. McDonald, D. E. Chamberlain, and S. Bearhop. 2008. "Food for Thought: Supplementary Feeding As a Driver of Ecological Change in Avian Populations." *Frontiers in Ecology and Environment* 6:476–484.

Sole, Pete. 2017. "Peregrine Falcons Attack a Ross's Gull in Central Coastal California." *Western Birds* 48:150–151.

Stahler, D. B., B. Heinrich, and D. Smith. 2002. "Common Ravens, *Corvus corax,* Preferentially Associate with Grey Wolves, *Canis lupus,* As a Foraging Strategy in Winter." *Animal Behavior* 64:283–290.

Stiles, E. W. 1982. "Expansion of Mockingbirds and Multiflora Roses in the Northeast and Canada." *American Birds* 36:358–364.

Wilcox, J. T. 2018. "Onshore Foraging by an Eared Grebe." *Western Birds* 49:145–148. *Observations of a single bird exhibiting a previously unreported form of foraging behavior.*

Williams, S. M., and E. H. Burtt Jr. 2010. "How Birds' Bills Help Them See." *Birding* 40:32–38. *This study experimentally shows why it is advantageous for some bird species to have dark upper mandibles.*

CHAPTER 13
THE AUDITORY BIRD

Benedict, L. 2008. "Occurrence and Life History Correlates of Vocal Duetting in North American Passerines." *Journal of Avian Biology* 39:57–65. *Comprehensive review.*

Billings, A. C., E. Greene, and S. M. De La Lucia Jensen. 2015. "Are Chickadees Good Listeners? Antipredator Responses to Raptor Vocalizations." *Animal Behaviour* 110:1–8. *Yes, chickadees are very good and highly perceptive listeners.*

Brown, T. J., and P. Hanford. 2002. "Why Birds Sing at Dawn: The Role of Consistent Song Transmission." *Ibis* 145 (1): 120–129. https://doi.org/10.1046/j.1474-919X.2003.00130.x. *One of a great many published works that attempt to explain reasons for the global dawn chorus exhibited by passerines.*

Clark, C. J., and T. J. Feo. 2008. "The Anna's Hummingbird Chirps with Its Tail: A New Mechanism of Sonation in Birds." *Proceedings of the Royal Society B: Biological Sciences* 275. https://doi.org/10.1098/rspb.2007.1619. *Another example of a bird species that sonates with its feathers, in this case specific tail feathers.*

de Kiriline, L. 1954. "The Voluble Singer of the Treetops." *Audubon* 56:109–111. *The account of the incessant singing of a single Red-eyed Vireo. THE classic account of how many times in one day a single Red-eyed Vireo sang.*

Evans, W., and M. O'Brien. 2002. "Flight Calls of Migratory Birds." www.oldbird.org/pubs/EvansOBrien2002.html.

Ficken, M. S., R. W. Ficken, and S. R. Witkin. 1978. "Vocal Repertoire of the Black-capped Chickadee." *Auk* 95:34–48. *Historic account that began documentation of "chickadee language."*

Foote, J. R., et al. 2008. "Male Chickadees Match Neighbors Interactively at Dawn: Support for the Social Dynamics Hypothesis." *Behavioral Ecology* 19 (6): 1192–1199. https://doi.org/10.1093/beheco/arn087. *Chickadees hear each other as individuals and match songs, even during the cacophony of the dawn chorus.*

Gentner, T. C., et al. 2006. "Recursive Syntactic Pattern Learning by Songbirds." *Science* 440:1204–1207. *This major work documents use of recursive grammar in European Starlings.*

Gentry, K. E., et al. 2017. "Evidence of Suboscine Song Plasticity in Response to Traffic Noise Fluctuations and Temporary Road Closures." *Bioacoustics* 27 (2): 165–181. https://doi.org/10.1080/09524622.2017.1303645. *Eastern Wood Pewees alter song in various ways in response to traffic noise.*

Hall, M. L. 2009. "A Review of Vocal Duetting in Birds." In Naguib, M., et al., eds., *Advances in the Study of Behavior,* Volume 40, 67–121 (chapter 3). New York: Elsevier.

Holthuijzen, W. A. 2018. "Stranger Danger: Acoustic Response of the Veery (*Catharus fuscescens*) via Heterospecific Eavesdropping on the Tufted Titmouse (*Baeolophus bicolor*)." *Wilson Journal of Ornithology* 130:168–179. *Veeries learn to "speak titmouse" for their own safety.*

Howard, H. E. 1920. *Territory in Bird Life.* London: Atheneum. *Outstanding classic still worth reading and enjoying.*

Johnsgard, P. A. 1998. *Crane Music: A Natural History of American Cranes.* Lincoln, NE: University of Nebraska Press. *Excellent account, nontechnical, written for the general reader.*

The next three volumes, by Donald Kroodsma, all written for lay readers, each provide a wonderful education on the function of bird songs and calls and how much you can enjoy and learn by listening.

Kroodsma, D. E. 2016. *Listening to a Continent Sing: Birdsong by Bicycle from the Atlantic to the Pacific.* Princeton, NJ: Princeton University Press.

———. 2009. *Birdsong by the Seasons: A Year of Listening to Birds.* Boston: Houghton Mifflin Harcourt.

———. 2005. *The Singing Life of Birds: The Art and Science of Listening to Birdsong.* Boston: Houghton Mifflin.

Kroodsma, D. E., and E. H. Miller, eds. 1996. *Ecology and Evolution of Acoustic Communication in Birds.* Ithaca, NY: Cornell University Press. *A technical multi-authored volume that covers much of the detail of bird vocalization.*

Kroodsma, D. E., and L. D. Parker. 1977. "Vocal Virtuosity in the Brown Thrasher." *Auk* 94:783–785. *Documents the richness of the thrasher repertoire. Amazing.*

Luther, D., and L. Baptista. 2010. "Urban Noise and the Cultural Evolution of Bird Songs." *Proceedings of the Royal Society B: Biological Sciences* 277:473. *Technical overview of how birdsong is changing due to noise stimuli in urban areas.*

Marler, P., and H. Slabbekoorn, eds. 2004. *Nature's Music: The Science of Birdsong.* New York: Elsevier Academic Press. *A "textbook" on birdsong, complete and dense with information.*

Mathews, F. S. 1921. *Field Book of Wild Birds and Their Music.* New York: Putnam's Sons (reprinted by Dover Publications, NY, in 1967). *A quaint but engaging attempt to express bird songs with musical signatures.*

Odom, K. J., and L. Benedict. 2018. "A Call to Document Female Bird Songs: Applications for Diverse Fields." *Auk Ornithological Advances* 135:314–325. *Female birdsong has been known for many years in some species but this major contribution focuses sharply on how understudied the subject has been.*

The two volumes listed next are designed as field guides to help you learn birdsong through the use of sonograms.

Pieplow, N. 2019. *Peterson Field Guide to Bird Sounds of Eastern North America.* Boston: Houghton Mifflin Harcourt.

———. 2017. *Peterson Field Guide to Bird Sounds of Western North America.* Boston: Houghton Mifflin Harcourt.

Price, J. J., S. M. Lanyon, and K. E. Omland. 2009. "Losses of Female Song with Changes from Tropical to Temperate Breeding in the New World Blackbirds." *Proceedings of the Royal Society B: Biological Sciences* 276 (1664): 1971–1980. https://doi.org/10.1098/rspb.2008.1626. *Technical account suggesting that long-distance migrant females "lost" their innate urge to sing with the evolution of migration.*

Searcy, W. A., and S. Nowicki. 2008. "Bird Song and the Problem of Honest Communication." *American Scientist* 96:114–121. *Excellent discussion of song repertoire and soft song.*

Templeton, C. N., and E. Greene. 2007. "Nuthatches Eavesdrop on Variations in Heterospecific Chickadee Mobbing Alarm Calls." *Proceedings of the National Academy of Sciences* 104 (13): 5479–5482. *Nuthatches also understand at least some chickadee language.*

Templeton, C. N., E. Greene, and K. Davis. 2005. "Allometry of Alarm Calls: Black-capped Chickadees Encode Information About Predator Size." *Science* 308:1934–1937. *Another fascinating study of chickadee language.*

CHAPTER 14
REAL ESTATE, MATE ATTRACTION, AND PAIR BONDING

Ardrey, R. 1961. *African Genesis: A Personal Investigation into the Animal Origins of Property and Nations.* New York: Dell.

———. 1966. *The Territorial Imperative: A Personal Inquiry into the Animal Origins of Property and Nations.* New York: Dell.

Birkhead, T. R. 1996. "Mechanisms of Sperm Competition in Birds." *American Scientist* 84:254–262. *Birkhead is the ranking expert on the subject of sperm competition in birds.*

Birkhead, T. R., and A. P. Møller. 1992. *Sperm Competition in Birds: Evolutionary Causes and Consequences.* London: Academic Press.

Brooke, M. 2018. *Far from Land: The Mysterious Lives of Seabirds.* Princeton, NJ: Princeton University Press. *See page 54, about separate wintering ranges of a pair of Sabine's Gulls. Very engaging book with wonderful detailed accounts of seabird natural history.*

Brush, A. H., and D. M. Power. 1976. "House Finch Pigmentation: Carotenoid Metabolism and the Effect of Diet." *Auk* 93:725–739. *Pigment intake matters, at least to male house finches.*

Campagna, L. 2018. "Among Ruffs, Some 'Fight-loving Fighters' Don't Like to Fight." *Living Bird* (Summer).

Darwin, C. 1871. *The Descent of Man, and Selection in Relation to Sex.* London: John Murray. *Arguably Darwin's second most important book, with lots on sexual selection of birds.*

Davis, S. E., M. Maftei, and M. L. Mallory. 2016. "Migratory Connectivity at High Latitudes: Sabine's Gulls (Xema sabini) from a Colony in the Canadian High Arctic Migrate to Different Oceans." *PLOS One* 11 (12): e0166043. https://doi.org/10.1371/journal.pone.0166043.

Garrett, M. G., J. W. Watson, and R. G. Anthony. 1993. "Bald Eagle Home Range and Habitat Use in the Columbia River Estuary." *Journal of Wildlife Management* 57:19–27.

Hess, P. 2009. "Ruffs: A Unique Male Trio." *Birding* 39:28.

Hill G. E. 1990. "Female House Finches Prefer Colourful Males: Sexual Selection for a Condition-dependent Trait." *Animal Behavior* 40:563–572.

Hill, G. E., and K. J. McGraw. 2006. *Bird Coloration, Volume 2: Function and Evolution.* Cambridge, MA: Harvard University Press. *Excellent and detailed account of research into sexual selection and how bird coloration influences it.*

Howard, H. E. 1920. *Territory in Bird Life.* London: Atheneum.

Johnsgard, P. A. 1994. *Arena Birds: Sexual Selection and Behavior.* Washington, DC: Smithsonian Institution Press. *Fine overview of sexually selected lekking species, richly illustrated.*

Jukema, J., and T. Piersma. 2006. "Permanent Female Mimics in a Lekking Shorebird." *Biological Letters* 2 (2): 161–164. https://doi.org/10.1098/rsbl.2005.0416. *This paper describes the male ruff morph that mimics females.*

Ketterson, E. D., and J. W. Atwell, eds. 2016. *Snowbird: Integrative Biology and Evolutionary Diversity in the Junco.* Chicago: University of Chicago Press. *Chapters by multiple authors, all technical, very information rich.*

Keyser, A. J., and G. E. Hill. 2000. "Structurally Based Plumage Coloration Is an Honest Signal of Quality in Male Blue Grosbeaks." *Behavioral Ecology* 11:202–209.

Kress, S. W., and D. Z. Jackson. 2015. *Project Puffin: The Improbable Quest to Bring a Beloved Seabird Back to Egg Rock.* New Haven, CT: Yale University Press. *Charming and informative book on one of the most important long-term projects in seabird biology.*

Møller, A. P. 1987. "Variation in Badge Size in Male House Sparrows *Passer domesticus:* Evidence for Status Signaling." *Animal Behavior* 35:1637–1644. *The bigger the badge, the more dominant the male House Sparrow.*

———. 1994. *Sexual Selection and the Barn Swallow.* Oxford: Oxford University Press. *Males with long deeply forked tails are most successful at attaining mates. Based on work in Europe.*

Nakagawa, S., et al. 2007. "Assessing the Function of House Sparrows' Bib Size Using a Flexible

Meta-analysis Method." *Behavioral Ecology* 18 (5): 831–840. https://doi.org/10.1093/beheco/arm050.

Nero, R. W. 1984. *Redwings.* Washington, DC: Smithsonian Institution Press. *A fine natural history written for lay readers.*

Piper, W., J. Mager, and C. Walcott. 2011. "Marking Loons, Making Progress." *American Scientist* 99:220–227.

Prum, R. O. 2017. *The Evolution of Beauty: How Darwin's Forgotten Theory of Mate Choice Shapes the Animal World—and Us.* New York: Doubleday. *Controversial book, as Prum revisits Darwin's original hypothesis for sexual selection, namely female whimsical preference.*

Safran, R. J., and K. J. McGraw. 2004. "Plumage Coloration, Not Length or Symmetry of Tail-streamers, Is a Sexually Selected Trait in North American Barn Swallows." *Behavioral Ecology* 15:455–461. *Intensity of breast color matters more than tail length.*

Searcy, W. A., and K. Yusukawa. 1983. "Sexual Selection and Red-winged Blackbirds." *American Scientist* 71:166–174.

Stutchbury, B. 2010. *The Private Lives of Birds: A Scientist Reveals the Intricacies of Avian Social Life.* New York: Walker & Company. *The author, an accomplished field ornithologist, deftly describes many aspects of bird behavior from not only her own research but that of others.*

Tuttle, E. M., et al. 2016. "Divergence and Functional Degradation of a Sex Chromosome-like Supergene." *Current Biology* 26:344–350. *This paper deals with the White-throated Sparrow morphs.*

van Leeuwen, C. H. A., and S. E. Jamieson. 2018. "Strong Pair Bonds and High Site Fidelity in a Subarctic-breeding Migratory Shorebird." *Wilson Journal of Ornithology* 130:140–151. *Documents high philopatry among both male and female Dunlins.*

Veiga, J. P. 1993. "Badge Size, Phenotypic Quality, and Reproductive Success in the House Sparrow: A Study on Honest Advertisement." *Evolution* 47:1161–1170. *Another paper documenting that badge size indeed matters . . . in male House Sparrows.*

Wolfenbarger, L. L. 1999. "Red Coloration of Male Northern Cardinals Correlates with Mate Quality and Territory Quality." *Behavioral Ecology* 10:80–90.

Woods, P. E. 1999. "A Familiar Mystery: Polymorphism and the White-throated Sparrow." *Birding* 31:263–266. *Fine overview of the complex white and brown morph relationship in White-throated Sparrows.*

CHAPTER 15
NESTING BEHAVIOR

Baicich, P. J., and C. J. O. Harrison. 1997. *A Guide to the Nests, Eggs, and Nestlings of North American Birds, Second Edition.* New York: Academic Press.

Beals, S. 2011. *Nests: Fifty Nests and the Birds That Built Them.* San Francisco: Chronicle Books.

Birkhead, T. R. 2017. *The Most Perfect Thing: Inside (and Outside) a Bird's Egg.* London: Bloomsbury Publishing.

Birkhead, T. R., and A. P. Møller. 1992. *Sperm Competition in Birds: Evolutionary Causes and Consequences.* London: Academic Press.

Brawn, J., and S. K. Robinson. 1996. "Source–Sink Population Dynamics May Complicate the Interpretation of Long-term Census Data." *Ecology* 77:3–12. *Technical and important study showing that singing males do not necessarily indicate a healthy reproducing population.*

Coon, J. J., et al. 2018. "An Observation of Parental Infanticide in Dickcissels (*Spiza americana*): Video Evidence and Potential Mechanisms." *Wilson Journal of Ornithology* 130:341–345.

Cutler, T. L., and D. E. Swann. 1999. "Using Remote Photography in Wildlife Ecology: A Review." *Wildlife Society Bulletin* 27:571–581. *Describes use of artificial nests in evaluating nest failures due to predators.*

Danner, J. E., et al. 2018. "Temporal Patterns of Extra-pair Paternity in a Population of Grasshopper Sparrows (*Ammodramus savannarum*) in Maryland." *Wilson Journal of Ornithology* 130:40–51.

Goodfellow, P. 2011. *Avian Architecture: How Birds Design, Engineer, and Build.* Princeton, NJ: Princeton University Press.

Harrison, H. H. 1975. *A Field Guide to Birds' Nests.* Boston, MA: Houghton Mifflin.

Hauber, M. E. 2014. *The Book of Eggs: A Life-Size Guide to the Eggs of Six Hundred of the World's Bird Species.* Chicago: University of Chicago.

Hauber, M. E., S. A. Russo, and P. W. Sherman. 2001. "A Password for Species Recognition in a Brood-parasitic Bird." *Proceedings of the Royal Society B: Biological Sciences* 268:1041–1048. *How nestling and juvenile Brown-headed Cowbirds learn the calls of their species.*

Hawkins, G. L., G. E. Hill, and A. Mercadante. 2011. "Delayed Plumage Maturation and Delayed Reproductive Investments in Birds. *Biological Reviews* 87:257–274. *Very thorough review of a subject that has challenged ornithologists to explain.*

Hunt, J. D., and R. M. Evans. 1997. "Brood Reduction and the Insurance Egg Hypothesis in Double-crested Cormorants." *Colonial Waterbirds* 20:485–491.

Kennedy, E. D. 1991. "Determinate and indeterminate egg-laying patterns: a review." *The Condor* 93:106–124.

Johnsgard, P. A. 1994. *Arena Birds: Sexual Selection and Behavior.* Washington, DC: Smithsonian Institution Press. *This well-illustrated book provides a survey of birds lekking ("arena birds"), ranging from grouse to tropical cotingas and manakins.*

Lanyon, W. E. 1963. *Biology of Birds.* Garden City, NY: The Natural History Press. *This classic book is still valuable for the succinct overview it provides about bird biology and it details the experiment in which a Northern Flicker was demonstrated to be an indeterminate egg layer.*

Lifjeld, J. T., and R. J. Robertson. 1992. "Female Control of Extra-pair Fertilization in Tree Swallows." *Behavioral Ecology and Sociobiology* 31:89–96.

Louder, I. M., et al. 2015. "Out on Their Own: A Test of Adult-assisted Dispersal in Fledgling Brood Parasites Reveals Solitary Departures from Hosts." *Animal Behavior* 110:29–37. *Describes how juvenile cowbirds leave their host nest at night to associate with other cowbirds.*

Mauck, R. A., T. A. Waite, and P. G. Parker. 1995. "Monogamy in Leach's Storm-Petrel: DNA Fingerprinting Evidence." *Auk* 112:473–482.

Mock, D. W., H. Drummond, and C. H. Stinson. "Avian Siblicide." *American Scientist* 78:438–449. *Looks at adaptive reasons why siblicide occurs in some species.*

Møller, A. P., and T. R. Birkhead. 1994. "The Evolution of Plumage Brightness in Birds in Relation to Extra-pair Paternity." *Evolution* 48:1089–1100.

Moskoff, W. 1998. "DNA Fingerprinting: Revealing Avian Reproductive and Social Behaviors." *Birding* 30:124–131. *One of the papers that changed the concept of how ornithologists regard monogamy in birds.*

Parsons, J. 1971. "Cannibalism in Herring Gulls." *British Birds* 64:528–537. *Predation of juveniles by unrelated adults at colonies.*

Robinson, S. K., et al. 1995. "Regional Forest Fragmentation and the Nesting Success of Migratory Birds." *Science* 31:1987–1990. *Fragmentation reduces nesting success in many bird species.*

Rohwer, S., and G. S. Butcher. 1988. "Winter Versus Summer Explanations of Delayed Plumage Maturation in Temperate Passerine Birds." *American Naturalist* 131:556–572.

Rohwer, S., S. D. Fretwell, and D. M. Miles. 1980. "Delayed Plumage Maturation in Passerine Plumages and the Deceptive Acquisition of Resources." *American Naturalist* 115:400–437.

Rohwer, S., W. P. Klein Jr., and S. Heard. 1983. "Delayed Plumage Maturation and the Presumed Prealternate Molt in American Redstarts." *Wilson Bulletin* 95:199–208.

Rothstein, S. I. 2004. "Brown-headed Cowbird: Villain or Scapegoat?" *Birding* 36:374–384. *Well-argued article of how birders "judge" the worth of birds.*

Sheldon, B. C. 1994. "Male Phenotype, Fertility, and the Pursuit of Extra-pair Copulations by Female Birds." *Proceedings of the Royal Society B: Biological Sciences* 257:25–30.

Stoddard, M. C., et al. 2017. "Avian Egg Shape: Form, Function, and Evolution." *Science* 356:1249–1254. *This technical paper makes a broad argument for egg form being mostly related* to the challenges of flight, as they vary among bird species.

Suarez-Rodriguez, M., I. Lopez-Rull, and C. M. Garcia. 2012. "Incorporation of Cigarette Butts into Nests Reduces Nest Ectoparasite Load in Urban Birds: New Ingredients for an Old Recipe?" *Biology Letters* 9 (1). https://doi.org/10.1098/rsbl.2012.0931.

Tittler, R., L. Fahrig, and M.-A. Villard. 2006. "Evidence of Large-scale Source–Sink Dynamics and Long-distance Dispersal Among Wood Thrush Populations." *Ecology* 87:3029–3036.

Westneat, D. F., P. W. Sherman, and M. L. Morton. 1990. "The Ecology and Evolution of Extra-pair Copulations in Birds." *Current Ornithology* 7:331–369. *A technical review that established the widespread reality of EPCs in many bird species.*

Woods, P. 1999. "Delayed Plumage Maturation in Orioles." *Birding* 31:416–420.

Zickefoose, J. 2016. *Baby Birds: An Artist Looks into the Nest.* Boston: Houghton Mifflin Harcourt. *This is a magnificently illustrated book combining much information about nesting behavior with outstanding artwork.*

———. 2012. *The Bluebird Effect: Uncommon Bonds with Common Birds.* Boston: Houghton Mifflin Harcourt. *Beautifully illustrated with detailed nesting accounts of many common species.*

CHAPTER 16
MIGRATORY BEHAVIOR

Able, K. P., ed. 1999. *Gatherings of Angels: Migrating Birds and Their Ecology.* Ithaca, NY: Comstock Books. *Able has chosen ten of the best examples of bird migration for his various chapters, each authored by an expert on the subject.*

Able, K. P., and J. R. Belthoff. 1998. "Rapid 'Evolution' of Migratory Behaviour in the Introduced House Finch of Eastern North America." *Proceedings of the Royal Society B: Biological Sciences* 265:2063–2071. *Northern populations quickly evolved winter migration.*

Baird, J. 1999. "The Epic Autumn Flight of the Blackpoll Warbler." *Birding* 31:65–73. *This is the classic study that suggested Blackpolls actually are transoceanic migrants flying nonstop from northeastern North America to Amazonia.*

Bartell, P., and A. Moore. 2013. "Avian Migration: The Ultimate Red-eye Flight." *American Scientist* 101:46–55. *Good overview for the general reader.*

Barton, G. G., and B. K. Sandercock. 2018. "Long-term Changes in the Seasonal Timing of Landbird Migration on the Pacific Flyway." *Condor: Ornithological Applications* 120:30–46.

Bauer, C. M., et al. 2016. "A Migratory Lifestyle Is Associated with Shorter Telomeres in a Songbird (*Junco hyemalis*). *Auk, Ornithological Advances* 133:649–653. *Preliminary but potentially very important study that suggests migratory birds may actually age more quickly than sedentary birds.*

Beehler, B. M. 2018. *North with the Spring: Travels with the Songbird Migration of Spring.* Washington, DC: Smithsonian Books. *A general and nontechnical overview of migration featuring some of the best areas to witness bird migration in spring.*

Berthold, P. 1991. "Genetic Control of Migratory Behavior in Birds." *Trends in Ecology and Evolution* 6:254–257. *Technical.*

Berthold, P., and A. J. Helbig. 1992. "The Genetics of Bird Migration: Stimulus, Timing, and Direction." *Ibis* 134:S35–S40. *Technical.*

Berthold, P., and U. Querner. 1981. "Genetic Basis of Migratory Behavior in European Warblers." *Science* 212:77–79. *Demonstrates that birds inherit knowledge of migratory pathways, at least in a general sense.*

Boyle, W. A. 2017. "Altitudinal Bird Migration in North America." *Auk* 134:443–465.

Chang, A. M., and K. L. Wiebe. 2016. "Body Condition in Snowy Owls Wintering on the Prairies Is Greater in Females and Older Individuals and May Contribute to Sex-based Mortality." *Auk, Ornithological Advances* 133:738–746.

Chapman, F. M. 1932. *Handbook of Birds of Eastern North America.* New York: D. Appleton and Company. *This is a classic of American ornithology, still useful in many ways, including Chapman's outstanding overview of avian natural history and account of migration patterns. It was first published in 1895.*

Chu, M. 2006. *Songbird Journeys: Four Seasons in the Lives of Migratory Birds.* New York: Walker & Company. *Well-crafted book for lay readers, covers mostly passerine migration.*

Cramer, D. 2015. *The Narrow Edge: A Tiny Bird, an Ancient Crab & an Epic Journey.* New Haven, CT: Yale University Press. *A widely acclaimed book chronicling the annual migratory journey of the* rufa *subspecies of Red Knot.*

DeGraaf, R. M., and J. H. Rappole. 1995. *Neotropical Migratory Birds: Natural History, Distribution, and Population Change.* Ithaca, NY: Comstock Publishing Associates.

Delmore, K. E., and D. E. Irwin. 2014. "Hybrid Songbirds Employ Intermediate Routes in a Migratory Divide." *Ecology Letters* 17:1211–1218. *Describes the migration of hybrids between two subspecies of western Swainson's Thrushes.*

Dionne, M., et al. 2008. "Impact of Hurricane Wilma on Migrating Birds: The Case of the Chimney Swift." *Wilson Journal of Ornithology* 120:784–792. *Suggests that eastern Canadian swift populations were devastated by being caught in a hurricane while migrating.*

Dokter, A. M., et al. 2018. "Seasonal Abundance and Survival of North America's Migratory Avifauna Determined by Weather Radar." *Nature Ecology & Evolution* 2:1603–1609.

Dunne, P., and K. T. Karlson. 2016. *Birds of Prey: Hawks, Eagles, Falcons, and Vultures.* Boston: Houghton Mifflin Harcourt. *Includes descriptions of migratory behaviors of various species such as Swainson's Hawk and Turkey Vulture.*

Elphick, J., ed. 2011. *Atlas of Bird Migration: Tracing the Great Journeys of the World's Birds.* Buffalo, NY: Firefly. *Provides a global account of bird migration.*

Emlen, S. T. 1970. "Celestial Rotation: Its Importance in Development of Migratory Orientation." *Science* 170:1198–1201. *Emlen is one of the ornithologists who discovered that birds orient to celestial patterns.*

Gauthreaux, S. A., and C. G. Belser. 2003. "Bird Movements on Doppler Weather Surveillance Radar." *Birding* 35:616–628. *Gauthreaux is the ranking expert of use of radar to study avian migration. This nontechnical account is highly recommended.*

Gow, E. A., and K. L. Weibe. 2014. "Males Migrate Farther than Females in a Differential Migrant: An Examination of the Fasting Endurance Hypothesis." *Royal Society Open Science* 1 (4).

https://doi.org/10.1098/rsos.140346. *Male Northern Flickers migrate longer distances than females.*

Greenbery, R., and P. Marra. 2005. *Birds of Two Worlds: The Ecology and Evolution of Migration.* Baltimore, MD: Johns Hopkins University Press. *This book, with a global perspective, provides an outstanding overview of all aspects of bird migration.*

Hagan, J. M., III, and D. W. Johnston, eds. 1992. *Ecology and Conservation of Neotropical Migrant Landbirds.* Washington, DC: Smithsonian Institution Press.

Hagstrum, J. T. 2000. "Infrasound and avian navigational map." *Journal of Experimental Biology* 203: 1103–1111.

Harrington, B. 1996. *The Flight of the Red Knot.* New York: W. W. Norton & Company. *Harrington was one of the discoverers of the use of Reed Beach on the Delaware Bayshore as a critical provisioning area for Red Knots and other species. This book describes Red Knot migration and Harrington's work.*

Heckscher, C. M., et al. 2011. "Veery (*Catharus fuscescens*) Wintering Locations, Migratory Connectivity, and a Revision of Its Wintering Range Using Geolocator Technology." *Auk* 128:531–542. *Kind of a game changer in understanding patterns of overwintering among lost distance migrants to Amazonia.*

Hess, P. 2008. "First-year Migrants May Need Experience." *Birding* 40:34.

Howell, S. N. G., I. Lewington, and W. Russell. 2014. *Rare Birds of North America.* Princeton, NJ: Princeton University Press. *Outstanding overview and discussion of vagrancy and its hypothesized causation with detailed and well-illustrated species accounts.*

Jønsson, K. A., et al. 2011. "Major Global Radiation of Corvoid Birds Originated in the Proto-Papuan Archipelago." *Proceedings of the National Academy of Sciences* 108:2328–2333. *Describes how birds such as vireos originated from a major adaptive radiation of crowlike birds.*

Kaufman, K. 2019. *A Season on the Wind: Inside the World of Spring Migration.* Boston: Houghton Mifflin Harcourt. *Engaging account of birding over the course of a year in central Ohio, near*

Magee Marsh, with an emphasis on migration, conservation, and the joy of birding.

Keeton, W. T. 1971. "Magnets interfere with pigeon homing." *PNAS 68*: 102–106. *This is the classic study that showed clearly how birds are able to navigate using earth's magnetic fields.*

Kerlinger, P., J. Gehring, and R. Curry. 2011. "Understanding Bird Collisions at Communication Towers and Wind Turbines: Status of Impacts and Research." *Birding 43:44–51.*

Kricher, J. 2017. "Masterful Migrant: How Tiny Blackpoll Warbler Manages to Cross a Continent and an Ocean Each Fall." *BirdWatching 31:22–26.*

Larson, D. M. 2018. "Getting Lost." *Bird Observer 46:36–38. Written for a lay audience, an overview of the work showing how some insecticides may alter the ability of migrants to orient.*

Lehman, P. 2003. "A Weather Primer for Birders." *Birding 35:596–605. Very useful to birders attempting to understand how migration relates to weather patterns.*

Lincoln, F. C., revised by J. L. Zimmerman. 1998. *Migration of Birds, Circular 16.* Washington, DC: U.S. Fish and Wildlife Service. *This is an updated edition of a classic small book on bird migration and remains a valuable and concise reference.*

The next three papers by Loss and others document mortality factors that migrants encounter all related to humans and their buildings.

Loss, S. R., T. Will, and P. P. Marra. 2012. "Direct Human-caused Mortality of Birds." *Annual Review of Ecology, Evolution, and Systematics 46:99–120.*

———. 2012. "Direct Human-caused Mortality of Birds: Improving Quantification of Magnitude and Assessment of Population Impact." *Frontiers in Ecology and Environment 10:357–364.*

Loss, S. R., T. Will, S. Loss, and P. Marra. 2014. "Bird–Building Collisions in the United States: Estimates of Annual Mortality and Species Vulnerability." *Condor 116:8–23.*

Marra, P., and C. Santella. 2016. *Cat Wars: The Devastating Consequences of a Cuddly Killer.* Princeton, NJ: Princeton University Press. *This sobering book documents in detail the extensive losses to birds caused by cats.*

McCaffery, B. J. 2008. "On Scimitar Wings: Long-distance Migration by the Bar-tailed Godwit and Bristle-thighed Curlew." *Birding 40:50–59.*

Morris, A. R., et al. 2016. "Fall Migratory Patterns of the Blackpoll Warbler at a Continental Scale." *Auk, Ornithological Advances 133:41–51. Focuses on the cross-continent migration that many Blackpolls must do before migrating across the Atlantic Ocean.*

Mueller, T., et al. 2013. "Social Learning of Migratory Performance." *Science 341:999–1002. Describes the use of ultralight aircraft to teach migrant cranes their migration route.*

Newton, I. 2008. *The Migration Ecology of Birds.* Amsterdam: Elsevier. *This is the most comprehensive work on bird migration currently available.*

Nisbet, I. T. C. 1970. "Autumn Migration of the Blackpoll Warbler: Evidence for Long Flight Provided by Regional Survey." *Journal of Field Ornithology 41:207–240.*

Nisbet, I. T. C., and C. S. Mostello. 2015. "Winter Quarters and Migration Routes of Common and Roseate Terns Revealed by Tracking with Geolocators." *Bird Observer 43:222–231.*

Perlet, N. G. 2018. "Prevalent Transoceanic Fall Migration by a 30-gram Songbird, the Bobolink." *Auk, Ornithological Advances 135 (4): 992–997.*

Price, J. 2003. "A Birder's Guide to Climate Change." *Birding 35:630–639. Valuable but just scratches the surface of this complex topic.*

Pyle, P., J. F. Saracco, and D. F. DeSante. 2018. "Evidence of Widespread Movements from the Breeding to Molting Grounds by North American Landbirds." *Auk 135:506–520. This comprehensive study, based on thousands of records over many years from banding stations, demonstrates that molt is common during migration and at various sites away from the breeding grounds.*

Rappole, J. H. 1995. *The Ecology of Migrant Birds: A Neotropical Perspective.* Washington, DC: Smithsonian Institution Press. *Concise species-by-species accounts of where the birds breed and where they spend the winter months.*

Robertson, B. A. 2004. "Stable Isotope Analysis: Forging New Links in Bird Migration." *Birding* 36:142–145. *This methodology contributed major breakthroughs in connectivity research. The methodology is well explained in this popular article.*

Ryder, T. B., J. W. Fox, and P. M. Marra. 2011. "Estimating Migratory Connectivity of Gray Catbirds (*Dumetella carolinensis*) Using Geolocator and Mark-recapture Data." *Auk* 128:448–453.

Sherony, D. F. 2008. "Greenland Geese in North America." *Birding* 40:46–56. *Increasing sightings of Greater White-fronted, Barnacle, and Pink-footed Geese.*

Skagen, S. K. 2008. "Shorebird Journeys Across the North American Prairie." *Birding* 40:48–54. *Focuses on the importance of shorebird staging areas such as Cheyenne Bottoms.*

Stanley, C. Q., et al. 2015. "Connectivity of Wood Thrush Breeding, Wintering, and Migration Sites Based on Range-wide Tracking." *Conservation Biology* 29:164–174.

Stuchbury, B. 2007. *Silence of the Songbirds: How We Are Losing the World's Songbirds and What We Can Do to Save Them.* New York: Walker & Company. *Outstanding overview of migration of songbirds, ecology on the wintering grounds, and conservation issues.*

Stuchbury, B., et al. 2009. "Tracking Long-distance Songbird Migration by Using Geolocators." *Science* 323:896. *One of the first major papers on use of geolocators.*

Thorup, K., et al. 2007. "Evidence for navigational map stretching across the continental U.S. in a migratory songbird." *PNAS* 104 (46): 18115–18119. *This paper describes research on displaced White-crowned Sparrows as it affects their migratory pathway.*

Usui, T., S. H. M. Butchart, and A. B. Phillimore. 2016. "Temporal Shifts and Temperature Sensitivity of Avian Migratory Phenology: A Phylogenetic Meta-analysis." *Journal of Animal Ecology* 86:250–261. *Documents shifting patterns in migration timing that correlate with increasing temperatures.*

Van Doren, B. M., et al. 2017. "High-intensity Urban Light Installation Dramatically Alters Bird Migration." *Proceedings of the National Academy of Sciences* 114 (42): 11175–11180. https://doi.org/10.1073/pnas.1708574114. *This is the study of the effects of "Tribute in Light" on migratory birds flying over New York City.*

van Gils, J. A., et al. 2016. "Body Shrinkage Due to Arctic Warming Reduces Red Knot Fitness in Tropical Wintering Range." *Science* 352:819–821.

Weidensaul, S. 1999. *Living on the Wind: Across the Hemisphere with Migratory Birds.* New York: North Point Press. *Essential, indeed eloquent reading for anyone with any interest in bird migration. One of the best books ever written on the subject of bird migration.*

Winger, B. M., F. K. Barker, and R. H. Ree. 2014. "Temperate Origins of Long-distance Seasonal Migration in New World Songbirds." *Proceedings of the National Academy of Sciences* 111:12115–12120. *This is a technical paper that requires understanding of the techniques used in the study but the overall discussion is recommended.*

Winger, B. M., I. J. Lovette, and D. W. Winkler. 2012. "Ancestry and Evolution of Seasonal Migration in the Parulidae." *Proceeding of the Royal Society B: Biological Sciences* 279:610–618. *See comments for previous paper.*

AFTERWORD

Axelson, G. 2019. "Vanishing." *Living Bird* 38:44–53. *This article provides a succinct overview of the details described in the paper (below) by Rosenberg et al.*

Rosenberg, K. V., et al. 2019. "Decline of the North American Avifauna." *Science* 10.1126/science.aaw1313. *This is a major paper that documents (using several databases) the catastrophic decline in numbers of birds since 1970. Very sobering.*

LIST OF SCIENTIFIC NAMES
OF BIRDS MENTIONED IN THE TEXT

Acadian Flycatcher, *Empidonax virescens*
Acorn Woodpecker, *Melanerpes formicivorus*
African Gray Parrot, *Psittacus erithacus*
African Skimmer, *Rynchops flavirostris*
Alder Flycatcher, *Empidonax alnorum*
American Avocet, *Recurvirostra americana*
American Bittern, *Botaurus lentiginosus*
American Black Duck, *Anas rubripes*
American Coot, *Fulica americana*
American Crow, *Corvus brachyrhynchos*
American Dipper, *Cinclus mexicanus*
American Flamingo, *Phoenicopterus ruber*
American Goldfinch, *Spinus tristis*
American Kestrel, *Falco sparverius*
American Oystercatcher, *Haematopus palliatus*
American Redstart, *Setophaga ruticilla*
American Robin, *Turdus migratorius*
American Tree Sparrow, *Spizelloides arborea*
American White Pelican, *Pelecanus
    erythrorhynchos*
American Wigeon, *Mareca americana*
American Woodcock, *Scolopax minor*
Anhinga, *Anhinga anhinga*
Anna's Hummingbird, *Calypte anna*
Aplomado Falcon, *Falco femoralis*
Arctic Tern, *Sterna paradisaea*
Ash-throated Flycatcher, *Myiarchus cinerascens*
Atlantic Puffin, *Fratercula arctica*
Bahama Swallow, *Tachycineta cyaneoviridis*
Bald Eagle, *Haliaeetus leucocephalus*
Baltimore Oriole, *Icterus galbula*
Bank Swallow, *Riparia riparia*
Barn Owl, *Tyto alba*
Barn Swallow, *Hirundo rustica*
Barnacle Goose, *Branta leucopsis*
Barred Owl, *Strix varia*
Bar-tailed Godwit, *Limosa lapponica*
Bay-breasted Warbler, *Setophaga castanea*
Bee Hummingbird, *Mellisuga helenae*
Belted Kingfisher, *Megaceryle alcyon*
Bewick's Wren, *Thryomanes bewickii*
Bicknell's Thrush, *Catharus bicknelli*
Blackburnian Warbler, *Setophaga fusca*
Black-and-white Warbler, *Mniotilta varia*
Black-bellied Plover, *Pluvialis squatarola*

Black-billed Cuckoo, *Coccyzus erythropthalmus*
Black-capped Chickadee, *Poecile atricapillus*
Black-crowned Night-Heron, *Nycticorax nycticorax*
Black-footed Albatross, *Phoebastria nigripes*
Black Grouse, *Tetrao tetrix*
Black-headed Grosbeak, *Pheucticus
    melanocephalus*
Black-legged Kittiwake, *Rissa tridactyla*
Black Phoebe, *Sayornis nigricans*
Blackpoll Warbler, *Setophaga striata*
Black Skimmer, *Rynchops niger*
Black Tern, *Chlidonias niger*
Black-throated Blue Warbler, *Setophaga
    caerulescens*
Black-throated Green Warbler, *Setophaga virens*
Black-throated Mango, *Anthracothorax nigricollis*
Black Vulture, *Coragyps atratus*
Blue-Gray Gnatcatcher, *Polioptila caerulea*
Blue Grosbeak, *Passerina caerulea*
Blue Jay, *Cyanocitta cristata*
Blue-winged Warbler, *Vermivora cyanoptera*
Boat-tailed Grackle, *Quiscalus major*
Bobolink, *Dolichonyx oryzivorus*
Bohemian Waxwing, *Bombycilla garrulus*
Boreal Chickadee, *Poecile hudsonicus*
Boreal Owl, *Aegolius funereus*
Brambling, *Fringilla montifringilla*
Brant, *Branta bernicla*
Brewer's Blackbird, *Euphagus cyanocephalus*
Bristle-thighed Curlew, *Numenius tahitiensis*
Broad-tailed Hummingbird, *Selasphorus
    platycercus*
Broad-winged Hawk, *Buteo platypterus*
Bronzed Cowbird, *Molothrus aeneus*
Brown Creeper, *Certhia americana*
Brown-headed Cowbird, *Molothrus ater*
Brown-headed Nuthatch, *Sitta pusilla*
Brown Pelican, *Pelecanus occidentalis*
Brown Thrasher, *Toxostoma rufum*
Buff-breasted Sandpiper, *Calidris subruficollis*
Bufflehead, *Bucephala albeola*
Bullock's Oriole, *Icterus bullockii*
Burrowing Owl, *Athene cunicularia*
Bushtit, *Psaltriparus minimus*
Cactus Wren, *Campylorhynchus brunneicapillus*

California Condor, *Gymnogyps californianus*
California Scrub-Jay, *Aphelocoma californica*
California Towhee, *Melozone crissalis*
Calliope Hummingbird, *Selasphorus calliope*
Canada Goose, *Branta canadensis*
Canada Jay, *Perisoreus canadensis*
Canada Warbler, *Cardellina canadensis*
Canyon Wren, *Thryothorus ludovicianus*
Cape May Warbler, *Setophaga tigrina*
Carolina Chickadee, *Poecile carolinensis*
Carolina Wren, *Thryothorus ludovicianus*
Caspian Tern, *Hydroprogne caspia*
Cassia Crossbill, *Loxia sinesciuris*
Cattle Egret, *Bubulcus ibis*
Cave Swallow, *Petrochelidon fulva*
Cedar Waxwing, *Bombycilla cedrorum*
Cerulean Warbler, *Setophaga cerulea*
Chestnut-backed Antbird, *Poliocrania exsul*
Chicken (domestic), *Gallus gallus domesticus*
Chimney Swift, *Chaetura pelagica*
Chipping Sparrow, *Spizella passerina*
Chivi Vireo, *Vireo chivi*
Chuck-will's-widow, *Antrostomus carolinensis*
Chukar, *Alectoris chukar*
Clapper Rail, *Rallus crepitans*
Clark's Grebe, *Aechmophorus clarkii*
Clark's Nutcracker, *Nucifraga columbiana*
Cliff Swallow, *Petrochelidon pyrrhonota*
Common Eider, *Somateria mollissima*
Common Gallinule, *Gallinula galeata*
Common Goldeneye, *Bucephala clangula*
Common Grackle, *Quiscalus quiscula*
Common Loon, *Gavia immer*
Common Merganser, *Mergus merganser*
Common Murre, *Uria aalge*
Common Nighthawk, *Chordeiles minor*
Common Ostrich, *Struthio camelus*
Common Pauraque, *Nyctidromus albicollis*
Common Poorwill, *Phalaenoptilus nuttallii*
Common Raven, *Corvus corax*
Common Redpoll, *Acanthis flammea*
Common Swift, *Apus apus*
Common Tern, *Sterna hirundo*
Common Yellowthroat, *Geothlypis trichas*
Connecticut Warbler, *Oporornis agilis*
Cooper's Hawk, *Accipiter cooperii*
Cory's Shearwater, *Calonectris diomedea*
Curve-billed Thrasher, *Toxostoma curvirostre*
Dark-eyed Junco, *Junco hyemalis*
Dickcissel, *Spiza americana*
Double-crested Cormorant, *Phalacrocorax auritus*
Dovekie, *Alle alle*
Downy Woodpecker, *Dryobates pubescens*
Dunlin, *Calidris alpina*
Dusky Grouse, *Dendragapus obscurus*
Eared Grebe, *Podiceps nigricollis*
Eastern Bluebird, *Sialia sialis*
Eastern Kingbird, *Tyrannus tyrannus*

Eastern Meadowlark, *Sturnella magna*
Eastern Phoebe, *Sayornis phoebe*
Eastern Screech-Owl, *Megascops asio*
Eastern Towhee, *Pipilo erythrophthalmus*
Eastern Whip-poor-will, *Antrostomus vociferus*
Eastern Wood-Pewee, *Contopus virens*
Eskimo Curlew, *Numenius borealis*
Eurasian Blackbird, *Turdus merula*
Eurasian Blue Tit, *Cyanistes caeruleus*
Eurasian Collared-Dove, *Streptopelia decaocto*
Eurasian Eagle-Owl, *Bubo bubo*
Eurasian Wren, *Troglodytes troglodytes*
European Robin, *Erithacus rubecula*
European Starling, *Sturnus vulgaris*
Evening Grosbeak, *Coccothraustes vespertinus*
Field Sparrow, *Spizella pusilla*
Fish Crow, *Corvus ossifragus*
Florida Scrub-Jay, *Aphelocoma coerulescens*
Fork-tailed Flycatcher, *Tyrannus savana*
Fox Sparrow, *Passerella iliaca*
Franklin's Gull, *Leucophaeus pipixcan*
Gadwall, *Mareca strepera*
Glaucous Gull, *Larus hyperboreus*
Glaucous-winged Gull, *Larus glaucescens*
Golden-crowned Kinglet, *Regulus satrapa*
Golden-crowned Sparrow, *Zonotrichia atricapilla*
Golden Eagle, *Aquila chrysaetos*
Golden-winged Warbler, *Vermivora chrysoptera*
Grasshopper Sparrow, *Ammodramus savannarum*
Gray Catbird, *Dumetella carolinensis*
Gray-cheeked Thrush, *Catharus minimus*
Gray Kingbird, *Tyrannus dominicensis*
Great Black-backed Gull, *Larus marinus*
Great Blue Heron, *Ardea herodias*
Great Crested Flycatcher, *Myiarchus crinitus*
Great Egret, *Ardea alba*
Great Gray Owl, *Strix nebulosa*
Great Horned Owl, *Bubo virginianus*
Great Shearwater, *Ardenna gravis*
Great Tit, *Parus major*
Greater Roadrunner, *Geococcyx californianus*
Greater Sage-Grouse, *Centrocercus urophasianus*
Greater Scaup, *Aythya marila*
Greater White-fronted Goose, *Anser albifrons*
Greater Yellow-headed Vulture, *Cathartes melambrotus*
Greater Yellowlegs, *Tringa melanoleuca*
Green Heron, *Butorides virescens*
Groove-billed Ani, *Crotophaga sulcirostris*
Hairy Woodpecker, *Dryobates villosus*
Harris's Hawk, *Parabuteo unicinctus*
Harris's Sparrow, *Zonotrichia querula*
Hermit Thrush, *Catharus guttatus*
Hermit Warbler, *Setophaga occidentalis*
Herring Gull, *Larus argentatus*
Hill Myna, *Gracula religiosa*
Hoary Redpoll, *Acanthis hornemanni*
Hooded Merganser, *Lophodytes cucullatus*

Hooded Warbler, *Setophaga citrina*
Horned Grebe, *Podiceps auritus*
Horned Lark, *Eremophila alpestris*
House Finch, *Haemorhous mexicanus*
House Sparrow, *Passer domesticus*
House Wren, *Troglodytes aedon*
Iceland Gull, *Larus glaucoides*
Indian Peafowl, *Pavo cristatus*
Indian Skimmer, *Rynchops albicollis*
Indigo Bunting, *Passerina cyanea*
Island Scrub-Jay, *Aphelocoma insularis*
Ivory Gull, *Pagophila eburnea*
Kea, *Nestor notabilis*
Key West Quail-Dove, *Geotrygon chrysia*
Killdeer, *Charadrius vociferus*
King Penguin, *Aptenodytes patagonicus*
King Vulture, *Sarcoramphus papa*
Kirtland's Warbler, *Setophaga kirtlandii*
Kiwi (several species), *Apteryx spp.*
Lake Duck, *Oxyura vittata*
Lapland Longspur, *Calcarius lapponicus*
Laughing Gull, *Leucophaeus atricilla*
Laysan Albatross, *Phoebastria immutabilis*
Lazuli Bunting, *Passerina amoena*
Leach's Storm-petrel, *Oceanodroma leucorhoa*
Least Bittern, *Ixobrychus exilis*
Least Flycatcher, *Empidonax minimus*
Least Sandpiper, *Calidris minutilla*
Least Tern, *Sternula antillarum*
Lesser Scaup, *Aythya affinis*
Lewis's Woodpecker, *Melanerpes lewis*
Limpkin, *Aramus guarauna*
Little Blue Heron, *Egretta caerulea*
Little Egret, *Egretta garzetta*
Loggerhead Shrike, *Lanius ludovicianus*
Long-billed Curlew, *Numenius americanus*
Long-billed Dowitcher, *Limnodromus scolopaceus*
Long-eared Owl, *Asio otus*
Lucy's Warbler, *Leiothlypis luciae*
Mallard, *Anas platyrhynchos*
Marbled Godwit, *Limosa fedoa*
Marsh Warbler, *Acrocephalus palustris*
Marsh Wren, *Cistothorus palustris*
Merlin, *Falco columbarius*
Mississippi Kite, *Ictinia mississippiensis*
Mountain Bluebird, *Sialia currucoides*
Mountain Quail, *Oreortyx pictus*
Mourning Dove, *Zenaida macroura*
Mute Swan, *Cygnus olor*
Nelson's Sparrow, *Ammospiza nelsoni*
New Caledonian Crow, *Corvus moneduloides*
Northern Bobwhite, *Colinus virginianus*
Northern Cardinal, *Cardinalis cardinalis*
Northern Flicker, *Colaptes auratus*
Northern Fulmar, *Fulmarus glacialis*
Northern Gannet, *Morus bassanus*
Northern Goshawk, *Accipiter gentilis*
Northern Harrier, *Circus hudsonius*

Northern Lapwing, *Vanellus vanellus*
Northern Mockingbird, *Mimus polyglottos*
Northern Parula, *Setophaga americana*
Northern Pygmy-Owl, *Glaucidium gnoma*
Northern Rough-winged Swallow, *Stelgidopteryx serripennis*
Northern Saw-whet Owl, *Aegolius acadicus*
Northern Shoveler, *Spatula clypeata*
Northern Shrike, *Lanius borealis*
Northern Waterthrush, *Parkesia noveboracensis*
Northern Wheatear, *Oenanthe oenanthe*
Orange-crowned Warbler, *Leiothlypis celata*
Orchard Oriole, *Icterus spurius*
Osprey, *Pandion haliaetus*
Ovenbird, *Seiurus aurocapilla*
Pacific Wren, *Troglodytes pacificus*
Painted Bunting, *Passerina ciris*
Painted Redstart, *Myioborus pictus*
Palm Warbler, *Setophaga palmarum*
Parasitic Jaeger, *Stercorarius parasiticus*
Pectoral Sandpiper, *Calidris melanotos*
Peregrine Falcon, *Falco peregrinus*
Phainopepla, *Phainopepla nitens*
Philadelphia Vireo, *Vireo philadelphicus*
Pied-billed Grebe, *Podilymbus podiceps*
Pileated Woodpecker, *Dryocopus pileatus*
Pine Grosbeak, *Pinicola enucleator*
Pine Siskin, *Spinus pinus*
Pine Warbler, *Setophaga pinus*
Pink-footed Goose, *Anser brachyrhynchus*
Pinyon Jay, *Gymnorhinus cyanocephalus*
Piping Plover, *Charadrius melodus*
Prairie Falcon, *Falco mexicanus*
Prothonotary Warbler, *Protonotaria citrea*
Purple Gallinule, *Porphyrio martinica*
Purple Martin, *Progne subis*
Pygmy Nuthatch, *Sitta pygmaea*
Razorbill, *Alca torda*
Red-bellied Woodpecker, *Melanerpes carolinus*
Red-breasted Merganser, *Mergus serrator*
Red-breasted Nuthatch, *Sitta canadensis*
Red-cockaded Woodpecker, *Dryobates borealis*
Red Crossbill, *Loxia curvirostra*
Reddish Egret, *Egretta rufescens*
Red-eyed Vireo, *Vireo olivaceus*
Redhead, *Aythya americana*
Red-headed Woodpecker, *Melanerpes erythrocephalus*
Red Knot, *Calidris canutus*
Red-necked Grebe, *Podiceps grisegena*
Red-necked Phalarope, *Phalaropus lobatus*
Red Phalarope, *Phalaropus fulicarius*
Red-shouldered Hawk, *Buteo lineatus*
Red-tailed Hawk, *Buteo jamaicensis*
Red-winged Blackbird, *Agelaius phoeniceus*
Ring-billed Gull, *Larus delawarensis*
Ring-necked Pheasant, *Phasianus colchicus*
Rock Pigeon, *Columba livia*

Rock Wren, *Salpinctes obsoletus*
Roseate Spoonbill, *Platalea ajaja*
Rose-breasted Grosbeak, *Pheucticus ludovicianus*
Rose-throated Becard, *Pachyramphus aglaiae*
Ross's Gull, *Rhodostethia rosea*
Rough-legged Hawk, *Buteo lagopus*
Royal Tern, *Thalasseus maximus*
Ruby-crowned Kinglet, *Regulus calendula*
Ruby-throated Hummingbird, *Archilochus colubris*
Ruddy Duck, *Oxyura jamaicensis*
Ruddy Turnstone, *Arenaria interpres*
Ruff, *Calidris pugnax*
Ruffed Grouse, *Bonasa umbellus*
Rufous Hummingbird, *Selasphorus rufus*
Rusty Blackbird, *Euphagus carolinus*
Sabine's Gull, *Xema sabini*
Saltmarsh Sparrow, *Ammospiza caudacuta*
Sanderling, *Calidris alba*
Sandhill Crane, *Antigone canadensis*
Savannah Sparrow, *Passerculus sandwichensis*
Say's Phoebe, *Sayornis saya*
Scarlet Tanager, *Piranga olivacea*
Seaside Sparrow, *Ammospiza maritima*
Sedge Wren, *Cistothorus platensis*
Semipalmated Plover, *Charadrius semipalmatus*
Semipalmated Sandpiper, *Calidris pusilla*
Sharp-shinned Hawk, *Accipiter striatus*
Sharp-tailed Grouse, *Tympanuchus phasianellus*
Shiny Cowbird, *Molothrus bonariensis*
Short-billed Dowitcher, *Limnodromus griseus*
Short-eared Owl, *Asio flammeus*
Snail Kite, *Rostrhamus sociabilis*
Snow Bunting, *Plectrophenax nivalis*
Snow Goose, *Anser caerulescens*
Snowy Egret, *Egretta thula*
Snowy Owl, *Bubo scandiacus*
Solitary Sandpiper, *Tringa solitaria*
Song Sparrow, *Melospiza melodia*
Sooty Grouse, *Dendragapus fuliginosus*
Sooty Shearwater, *Ardenna grisea*
Sora, *Porzana carolina*
Spectacled Eider, *Somateria fischeri*
Spotted Sandpiper, *Actitis macularius*
Spotted Towhee, *Pipilo maculatus*
Spruce Grouse, *Falcipennis canadensis*
Steller's Jay, *Cyanocitta stelleri*
Summer Tanager, *Piranga rubra*
Superb Lyrebird, *Menura novaehollandiae*
Surf Scoter, *Melanitta perspicillata*
Swainson's Hawk, *Buteo swainsoni*
Swainson's Thrush, *Catharus ustulatus*
Swallow-tailed Kite, *Elanoides forficatus*
Swamp Sparrow, *Melospiza georgiana*
Tennessee Warbler, *Leiothlypis peregrina*
Thayer's Gull, *Larus thayeri* (no longer considered a separate species)

Townsend's Solitaire, *Myadestes townsendi*
Townsend's Warbler, *Setophaga townsendi*
Tree Swallow, *Tachycineta bicolor*
Tricolored Heron, *Egretta tricolor*
Tufted Titmouse, *Baeolophus bicolor*
Tundra Swan, *Cygnus columbianus*
Turkey Vulture, *Cathartes aura*
Varied Bunting, *Passerina versicolor*
Varied Thrush, *Ixoreus naevius*
Veery, *Catharus fuscescens*
Verdin, *Auriparus flaviceps*
Vermilion Flycatcher, *Pyrocephalus rubinus*
Warbling Vireo, *Vireo gilvus*
Western Bluebird, *Sialia mexicana*
Western Capercaillie, *Tetrao urogallus*
Western Gull, *Larus occidentalis*
Western Kingbird, *Tyrannus verticalis*
Western Meadowlark, *Sturnella neglecta*
Western Screech-Owl, *Megascops kennicottii*
Western Spindalis, *Spindalis zena*
Western Tanager, *Piranga ludoviciana*
Whimbrel, *Numenius phaeopus*
Whiskered Auklet, *Aethia pygmaea*
White-breasted Nuthatch, *Sitta carolinensis*
White-crowned Sparrow, *Zonotrichia leucophrys*
White-eyed Vireo, *Vireo griseus*
White-faced Ibis, *Plegadis chihi*
White Ibis, *Eudocimus albus*
White-tailed Ptarmigan, *Lagopus leucura*
White-throated Sparrow, *Zonotrichia albicollis*
White-winged Crossbill, *Loxia leucoptera*
Wild Turkey, *Meleagris gallopavo*
Willet, *Tringa semipalmata*
Williamson's Sapsucker, *Sphyrapicus thyroideus*
Willow Flycatcher, *Empidonax traillii*
Wilson's Snipe, *Gallinago delicata*
Winter Wren, *Troglodytes hiemalis*
Woodhouse's Scrub-Jay, *Aphelocoma woodhouseii*
Wood Duck, *Aix sponsa*
Woodpecker Finch, *Camarhynchus pallidus*
Wood Stork, *Mycteria americana*
Wood Thrush, *Hylocichla mustelina*
Worm-eating Warbler, *Helmitheros vermivorum*
Wrentit, *Chamaea fasciata*
Yellow-bellied Sapsucker, *Sphyrapicus varius*
Yellow-billed Cuckoo, *Coccyzus americanus*
Yellow-billed Magpie, *Pica nuttalli*
Yellow-crowned Night-Heron, *Nyctanassa violacea*
Yellow-eyed Junco, *Junco phaeonotus*
Yellow-headed Blackbird, *Xanthocephalus xanthocephalus*
Yellow-rumped Warbler, *Setophaga coronata*
Yellow-throated Longclaw, *Macronyx croceus*
Yellow-throated Warbler, *Setophaga dominica*

INDEX OF NAMES

Page references in **bold** refer to text photos.

Purple Gallinule

Least Bittern

White-faced Ibis

GENERAL INDEX

Swallow-tailed Kite

humans and birds
    adapting to humans, 14–17, 83
    bird behavior observation, 15–17
    birds watching humans, 2, 15–17
    colonial birds/nesting, 44–45
    human etiquette/guidelines, 16–17
    nest boxes, 14–15
hybrids concept, 67–70, 73, 89

isotherms, 129

kleptoparasitism, 192–93, 194

lekking, 236–39
loafing, 124–25

maintenance behavior. *See* bathing; ectoparasites;
    molt cycle; preening; *specific aspects*
mate attraction. *See* courtship
migration
    altitudinal migration, 284
    connectivity among birds, 290–91
    differential migration, 283–84, 285
    evolution of, 295–96, 298
    examples, 273–77
    flyways, 284
    high-tech tracking methods, 277
    how birds find their way, 288–90
    leapfrog migration, 283, 284, 285
    local populations, 281–83
    long-/short-distance migration, 277–81, 282
    molt migration, 287–88
    observing/migration calendar, 270–73
    rarity/vagrancy, 297, 298–301
    reverse migration, 129, 300–301
    risks/advantages, 291–95
    roosting, 125
    timing, 284–87
molt cycle
    basic strategies, 113–15
    overview, 106–15, 116
molt migration, 287–88
morality and nature, 38

nanotags use, 277, 291
nest/nesting
    brood parasitism, 267–69
    caring for young, 242, 262–67
    construction, 250–52
    delayed plumage maturation, 261–62
    eggs, 35, 252, 254–58
    failure, 252–54
    functions, 243–44
    hatching, 257–58
    hatchling development types, 258
    humans observing, 243
    humans providing nests, 14–15, 44–45

monogamy/polygamy, 259–61
nest boxes, 14–15
nesting cooperation, 151–53, 154
sperm competition, 261
types, 245–50
See also courtship
nictitating membrane, 52, 53

origins of birds, 19–20

pair bonding. *See* courtship
peck order/hierarchies, 155, 157
physiology. *See* anatomy/physiology
play, 45–46
plumage. *See* feathers
populations concept, 74–75
postures/meanings, 4–5, 26, 42
predation by birds, 196, 197–99
predation of birds
    being cautious, 11
    flocking behavior, 82
    optimal foraging/theory, 199, 201
    overview, 196–99
predator harassment/mobbing, 199, 200–210
preening, 117–18
provisioning/protection. *See* nest/nesting

reverse sexual size dimorphism, 179
roosting, 125–27, 148

satellite tracking, 277
search images and feeding, 165, 170–72
senses/brain
    "bird brain" term, 49
    brain anatomy, 49–50
    hearing overview, 55–57
    mirrors, 48–49
    reaction time, 50–51
    smell, 57–60
    taste, 60–61, 62
    touch, 61–63
    vision/eyes (overview), 27, 50, 51–55
    See also auditory bird
sexual selection/display, 90, 233–41
sleeping, 13–14, 126–27
smell sense, 57–60
social behavior
    agonistic behavior, 157–62
    cooperation examples, 151–55, 156–57
    peck order/hierarchies, 155, 157
    social context overview, 146–51
    social network theory, 162–63
    See also specific aspects
social network theory, 162–63
species
    concept, 65–70
    hybrids concept, 67–70, 73, 89

OPPOSITE: Song Sparrow **BELOW:** Pied-billed Grebe

Purchase Peterson Field Guide titles wherever books are sold.
For more information on Peterson Field Guides, visit **www.petersonfieldguides.com.**

PETERSON FIELD GUIDES®

Roger Tory Peterson's innovative format uses accurate, detailed drawings to pinpoint key field marks for quick recognition of species and easy comparison of confusing look-alikes.

BIRDS

Birds of Northern Central America

Bird Sounds of Eastern North America

Bird Sounds of Western North America

Birds of North America

Birds of Eastern and Central North America

Western Birds

Feeder Birds of Eastern North America

Hawks of North America

Hummingbirds of North America

Warblers

Eastern Birds' Nests

PLANTS AND ECOLOGY

Eastern and Central Edible Wild Plants

Eastern and Central Medicinal Plants and Herbs

Western Medicinal Plants and Herbs

Eastern Forests

Eastern Trees

Western Trees

Eastern Trees and Shrubs

Ferns of Northeastern and Central North America

Mushrooms

Venomous Animals and Poisonous Plants

Wildflowers of Northeastern and North-Central North America

MAMMALS

Animal Tracks

Mammals

INSECTS

Insects

Eastern Butterflies

Moths of Northeastern North America

Moths of Southeastern North America

REPTILES AND AMPHIBIANS

Eastern Reptiles and Amphibians

Western Reptiles and Amphibians

FISHES

Freshwater Fishes

SPACE

Stars and Planets

GEOLOGY

Rocks and Minerals

PETERSON FIRST GUIDES®

The first books the beginning naturalist needs, whether young or old. Simplified versions of the full-size guides, they make it easy to get started in the field, and feature the most commonly seen natural life.

Astronomy

Birds

Butterflies and Moths

Caterpillars

Clouds and Weather

Fishes

Insects

Mammals

Reptiles and Amphibians

Rocks and Minerals

Seashores

Shells

Trees

Urban Wildlife

Wildflowers

PETERSON FIELD GUIDES FOR YOUNG NATURALISTS

This series is designed with young readers ages eight to twelve in mind, featuring the original artwork of the celebrated naturalist Roger Tory Peterson.

Backyard Birds

Birds of Prey

Songbirds

Butterflies

Caterpillars

PETERSON FIELD GUIDES® COLORING BOOKS®

Fun for kids ages eight to twelve, these color-your-own field guides include color stickers and are suitable for use with pencils or paint.

Birds

Butterflies

Reptiles and Amphibians

Wildflowers

Shells

Mammals

PETERSON REFERENCE GUIDES®

Reference Guides provide in-depth information on groups of birds and topics beyond identification.

Behavior of North American Mammals

Birding by Impression

Woodpeckers of North America

Sparrows of North America

PETERSON AUDIO GUIDES

Birding by Ear: Eastern/Central

Bird Songs: Eastern/Central

PETERSON FIELD GUIDE / *BIRD WATCHER'S DIGEST* BACKYARD BIRD GUIDES

Identifying and Feeding Birds

Bird Homes and Habitats

The Young Birder's Guide to Birds of North America

The New Birder's Guide to Birds of North America

E-books